SCHAUM'S OUTLINE OF

THEORY AND PROBLEMS

of

COLLEGE CHEMISTRY

FIFTH EDITION

•

BY

DANIEL SCHAUM, B.S.

EDITED AND REVISED BY

JEROME L. ROSENBERG, Ph.D.

Professor of Chemistry
University of Pittsburgh

•

SCHAUM'S OUTLINE SERIES

McGRAW-HILL BOOK COMPANY

New York, St. Louis, San Francisco, Toronto, Sydney

ISBN 07-053713-5

19 20 21 22 23 24 25 SH SH 7 6 5 4

Preface

Lord Kelvin said: "I often say that when you can measure what you are speaking about, and express it in numbers, you know something about it." This book aims to help the student to see clearly the principles of chemistry and of simple mathematics associated with the problem work of a course in college chemistry. It is intended also to be of service to advanced students who feel the need for a comprehensive review in the practical application of the fundamental theory.

The previous edition of this book has been very favorably received and adopted as a supplementary problems book by more than four hundred and fifty colleges. In this fifth edition many chapters have been revised and enlarged to keep pace with recent advances and with current trends in redefining the scope of the first chemistry course. A completely new chapter on the structure of matter has been added, reflecting the increasing emphasis in the newer textbooks on the relationship between structure and properties and between structure and reactivity. This chapter includes analytical problems relating to the chemical bond and to molecular and crystal geometry. The thermochemistry chapter was substantially rewritten in terms of enthalpies of reaction instead of the loosely defined heats of reaction. Practice in the use of thermochemical tables is included. The scope of the final chapter was extended to include, in addition to nuclear chemistry, problems dealing with the quantum nature of light.

The three chapters on ionic equilibrium and electrochemistry were considerably expanded to increase the usefulness of this book for courses in which analytical chemistry occupies a major segment of the first year chemistry course. Titration curves and elementary indicator theory are now treated. Many types of multiple equilibria have been introduced in this edition, including the cases of two weak acids, amphoterism, successive addition of ligands, simultaneous solution of two salts with a common ion, the solution of salts under conditions of limited acidity, the solution of salts by successive complexation, and the relationship between electrode potentials and ionic equilibria. The use of electrochemical tables in the determination of electrode reactions in electrolysis is discussed.

Some changes were made in the order of presentation. The discussion of the nucleonic makeup of atoms was brought into an early chapter so that the C^{12} reference for the atomic weight scale can be appreciated by the students. Some rearrangements were made to integrate empirical formula problems with composition problems within the same chapter, to present the material on oxidation-reduction in a more compact manner, and to shorten and simplify the treatment of equivalent weights. An explicit treatment of the balancing of ordinary equations was introduced. The chapter on standard solutions was made more general by including the use of molar and formal solutions in reactions of complex stoichiometry. The method of treating mass balance in chemical equilibrium problems was given more detailed expression. Numerical data of old problems were changed to keep up with the most recent accepted values for physical and chemical constants. As in previous editions, problems of previously used types were added to represent applied fields of chemistry of current interest.

Dimensional units are used throughout in order to stimulate their constant application by the student, and an explanation of the systems of units and of dimensions appears in the first chapter. Emphasis is placed on the use of the correct number of significant figures in line with scientific and engineering practice. The use of significant figures, exponents, and logarithms is explained in the appendix.

The author stresses the common-sense method of chemical reasoning, instead of mere substitution in a formula. The ideas and reasonings affecting the solution of problems are presented as directly as possible and in the most simple and familiar terms, using word equations and mathematics of the simplest sort.

This book includes a large number of representative problems with complete and detailed solutions, covering the subject matter of a first year course in college chemistry. Both the solved problems and the supplementary problems for student practice are arranged so as to present a natural development of each topic, and they include a wide range of applications in both pure and applied chemistry. A simplified outline of the principles necessary to understand and to solve the problems is given at the beginning of each chapter. This book is in no way a condensation of ordinary text material; it is intended to be a very comprehensive problem book.

DANIEL SCHAUM

February, 1966

CONTENTS

CONTENTS

Chapter 1

Measurements

INTRODUCTION

Most of the calculations in chemistry and physics are concerned with measurements of different kinds of *dimensions,* e.g. length, velocity, volume, mass, energy. Every measurement includes both a number and a unit. The *unit* simultaneously identifies the kind of dimension and the magnitude of the reference quantity used as a convenient basis for comparison. Many common units are used for the dimension of length, e.g. inch, yard, mile, centimeter, meter, kilometer. The *number* obviously indicates how many of the reference units are contained in the quantity being measured. Thus the statement that the length of a room is 20 feet means that the length of the room is 20 times the length of the foot, which in this case is the unit of length chosen for comparison.

SYSTEMS OF MEASUREMENT

Dimensional calculations are greatly simplified if the unit for each kind of measure is expressed in terms of the units of selected reference dimensions. The three selected reference dimensions of mechanics are *length, mass,* and *time.* As examples of relating other quantities to the reference dimensions, the unit of speed is defined as unit length per unit time; the unit of volume is the cube of the unit of length, etc. Unfortunately, there are several systems of units widely established in the English-speaking nations, and typical calculations very often are concerned with converting values from one system to another. We will reserve for later chapters the introduction of additional reference dimensions for describing electrical and thermal phenomena.

METRIC SYSTEM

The metric system is used by all scientists, and is the legal system in all nations except Great Britain and the United States. The version of the metric system most commonly used by chemists is called the *centimeter-gram-second* or CGS system after its three reference units: the *centimeter* (cm) of length, the *gram* (g) of mass, and the *second* (sec) of time. Other units in this system are derived from these three reference units. For example, the unit for volume in the *cgs* system is the cubic centimeter (cm^3 or cc), since volume = length × length × length = cm × cm × cm = cm^3; the unit for speed is the unit for length (distance) divided by the unit for time, or the cm per sec, since speed = $\dfrac{distance}{time} = \dfrac{cm}{sec}$; the unit for density is the unit for mass divided by the unit for volume, or the gram per cm^3 (g/cm^3 or g/cc), since density = $\dfrac{mass}{volume} = \dfrac{g}{cm^3}$.

Another version of the metric system, the *meter-kilogram-second* or MKS system is more commonly used in some other branches of science and is, in fact, the basis of the International System of units.

Compound units like g/cm^3 and cm/sec are often written in the form g cm^{-3} and cm sec^{-1}. For an explanation of negative exponents, see Appendix A.

Note that the term *per* in a word definition is equivalent to *divided by* in the mathematical notation.

Note. Symbols and abbreviations for unit terms do not require to be punctuated by a period mark except when the abbreviation resembles another word, e.g. *in.* for *inch.*

U. S. OR BRITISH SYSTEM

The U. S. or British system is used by most engineers and is the legal system in these two nations. The three reference units in this system are the *foot* of length, the *pound* of mass, and the *second* of time. Other units in this system are derived from these three reference units. For example, unit volume is the cubic foot (ft^3), unit speed is the foot per second (ft/sec), unit density is the pound per cubic foot (lb/ft^3).

LENGTH

The metric and British units of length are based on definite distances marked on bars that are preserved as standards. (The ultimate international standard is the natural wavelength of a sharply defined orange-red line in the spectrum of krypton.)

The metric standard of length is 100 centimeters or one *meter* (m). It is divided into a thousand equal parts called *millimeters* (mm). The millimeter in turn is divided into a thousand equal parts called *microns* (μ). *Micro* means one millionth, and a micron is one millionth of a meter. A thousandth of a micron is called a *millimicron* (mμ) or a *nanometer* (nm), the prefix *nano* meaning one billionth (10^{-9}). The *angstrom* (Å), equal to .0001 micron, is most convenient for measurements of atomic and molecular dimensions.

$$1 \text{ meter} = 100 \text{ centimeters} = 1000 \text{ millimeters}$$

1 kilometer	(km)	= 1000 m	1 micron	(μ)	= 10^{-4} cm
1 meter	(m)	= 100 cm	1 millimicron	(mμ)	= 10^{-7} cm
1 centimeter	(cm)	= 10 mm	1 nanometer	(nm)	= 10^{-7} cm = 1 mμ
			1 angstrom	(Å)	= 10^{-8} cm

All British units of length are easily converted to metric units by remembering just one interrelation, the definition of the inch:

$$1 \text{ inch} = 2.54 \text{ cm}$$

The following relationships are just a few of those that follow from the basic definitions.

CONVERSION FACTORS FOR LENGTH

1 inch (in.) = 2.54 cm	1 centimeter = 0.3937 in.
1 foot (ft) = 30.48 cm	1 meter = 3.281 ft
1 mile (mi) = 1.609 km	1 kilometer = 0.6214 mi

VOLUME

The natural unit of capacity is based on a standard unit of length, e.g. the cubic centimeter or the cubic meter. For a long time another unit existed, the *liter*, which was not simply related to the units of length. In 1964, an International Conference on Weights and Measures redefined the liter to equal exactly one cubic decimeter or 1000 cubic centimeters, or 28 parts per million less than its previous value. As a result of this change, the following relationships are identically true.

$$1 \text{ liter } (l) = 1000 \text{ milliliters (ml)} = 1000 \text{ cc}$$

$$1 \text{ milliliter (ml)} = 1 \text{ cc}$$

The British unit of volume is the *cubic foot* (cu ft or ft³), which is equal to 12 in. × 12 in. × 12 in. = 1728 cubic inches (cu in. or in³). An equally frequent engineering unit of liquid measure is the U.S. *gallon,* which is defined as 231 cubic inches. Either of these or their legally related measures can be converted to exact metric equivalents by remembering the linear factor of 2.54 cm per inch.

CONVERSION FACTORS FOR VOLUME

1 liter	= 1.057 liq. quarts	1 cubic foot	= 28.32 liters
1 liq. quart	= 0.9463 liter	1 cubic inch	= 16.39 cc or ml
1 gallon	= 3.785 liters	1 fluid ounce	= 29.57 cc or ml

MASS

Mass is a measure of quantity of matter. Although all measures and properties other than mass are apt to change in physical and especially chemical changes, the total mass of any system undergoing various changes remains constant. (The total mass of a system can change only if some mass is converted into energy, or energy into mass. In reactions involving atomic nuclei, like radioactivity and nuclear fission, the change in mass may be significant. In ordinary chemical reactions the change in mass is too small to be detected.)

Weight, which is the gravitational pull on any mass, is commonly used as a practical measure of mass because this pull or force per unit mass (the gravitational acceleration) does not vary a great deal from point to point on the earth's surface. To avoid terrestrial variations in weight which would show up in delicate spring scales, masses are compared with each other by matching weights in the lever-arm type of analytical balance.

The MKS unit of mass is the *kilogram* (kg), defined in terms of a standard kept at an international laboratory in France. The *gram,* is, of course, one thousandth of the kilogram. A smaller unit is the *milligram* (mg), which is one thousandth of a gram (0.001 g). A still smaller unit is the *gamma* (γ), or *microgram,* which is one millionth of a gram (10^{-6} g).

$$1 \text{ kilogram (kg)} = 1000 \text{ g} \qquad 1 \text{ gram (g)} = 1000 \text{ milligrams (mg)}$$

The British or U.S. unit of mass is the commercial or avoirdupois pound, otherwise defined as 7000 grains. There are 16 ounces in one pound (avoir).

CONVERSION FACTORS FOR MASS

1 pound (avoir)	= 453.6 g	1 kilogram	= 2.205 lb (avoir)
1 ounce (avoir)	= 28.35 g	1 gram	= 15.43 grains

The U.S. short ton is 2000 lb, the British long ton is 2240 lb, and the metric ton (1000 kg) is 2205 lb.

USE AND MISUSE OF UNITS

The unit terms (e.g. cm, kg, g/ml, ft/sec) must be regarded as a necessary part of the completely defined measure. It is as foolish to separate the number of a measure from its unit as it is to separate the laboratory reagent bottles from their labels. The unit is a necessary part of the quantity, as the x in $2x$, or the y^2 in $5y^2$.

In every mathematical operation the unit terms must be carried along with the numbers and must undergo the same mathematical operations as the numbers. Quantities cannot be added or subtracted directly unless they have the same units as well as the same dimensions. For example, it is obvious that we cannot add 5 hours (time) to 20 miles/hour (speed), since *time* and *speed* have different physical significance. If we are to add 2 lb (mass) and 4 kg (mass), we must first convert lb to kg or kg to lb. Any number of quantities, however, can be combined in multiplication or division, in which the units as well as the numbers obey the algebraic laws of squaring, cancellation, etc. Thus:

(1) $6 \text{ ml} + 2 \text{ ml} = 8 \text{ ml}$ $(\text{ml} + \text{ml} = \text{ml})$

(2) $5 \text{ cm} \times 2 \text{ cm}^2 = 10 \text{ cm}^3$ $(\text{cm} \times \text{cm}^2 = \text{cm}^3)$

(3) $3 \text{ ft}^3 \times 200 \text{ lb/ft}^3 = 600 \text{ lb}$ $\left(\text{ft}^3 \times \dfrac{\text{lb}}{\text{ft}^3} = \text{lb} \right)$

(4) $2 \text{ sec} \times 3 \text{ ft/sec}^2 = 6 \text{ ft/sec}$ $\left(\text{sec} \times \dfrac{\text{ft}}{\text{sec}^2} = \text{ft/sec} \right)$

(5) $\dfrac{15 \text{ g}}{3 \text{ g/ml}} = 5 \text{ ml}$ $\left(\dfrac{\text{g}}{\text{g/ml}} = \text{g} \times \dfrac{\text{ml}}{\text{g}} = \text{ml} \right)$

MAGNITUDES

An important characteristic of an answer to any practical problem is the correct order of magnitude. This is the proper location of the decimal point or, in the use of the exponential notation, the assignment of the proper power of 10. If this is wrong, though all else is very exact, the answer is worthless, often costly. Thus the first problem discipline is getting, by visual inspection, the correct order of magnitude. Only after this are slide rules and log tables justified for obtaining an accurate numerical solution.

Consider the multiplication: $122 \text{ g} \times 0.0518 = 6.32 \text{ g}$. Visual inspection shows that 0.0518 is a little more than 1/20, and 1/20 of 122 is a little more than 6. Hence the answer should be a little more than 6 g, which it is. If the answer were given incorrectly as 63.2 g or 0.632 g, visual inspection or mental checking of the result would indicate that the decimal point had been misplaced.

SIGNIFICANT FIGURES, EXPONENTS, LOGARITHMS, SLIDE RULE

The student is urged to review the sections in the Appendix on Significant Figures, Exponents, and Logarithms, and to purchase an inexpensive slide rule. (A slide rule adequate for most of the problems in this book, together with a book of instructions, can be purchased for a few dollars or less at most college bookstores.) This knowledge will enable him to give correct numerical answers and to reduce greatly the time required for numerical operations.

Solved Problems

1.1. The following examples illustrate the conversions among the various units of length, volume, or weight (mass).

1 mile $= 1.609 \text{ km} = 1609 \text{ m} = 1.609 \times 10^3 \text{ m} = 1.609 \times 10^5 \text{ cm} = 1.609 \times 10^6 \text{ mm}$

1 yard $= 0.9144 \text{ m} = 91.44 \text{ cm} = 914.4 \text{ mm} = 9.144 \times 10^5 \mu \text{ (microns)}$

1 inch $= 2.54 \text{ cm} = 0.0254 \text{ m} = 2.54 \times 10^{-2} \text{ m} = 25.4 \text{ mm} = 2.54 \times 10^4 \mu$
         $= 2.54 \times 10^7 \text{ nm}$

1 quart $= 0.9463 l = 946.3 \text{ ml (or cc)}$

1 cubic foot $= 0.02832 \text{ cubic meter (m}^3\text{)} = 28.32 l = 2.832 \times 10^4 \text{ ml (or cc)}$

1 pound (avoir) $= 0.4536 \text{ kg} = 453.6 \text{ g} = 4.536 \times 10^5 \text{ mg}$

1 metric ton $= 1000 \text{ kg} = 2205 \text{ lb (avoir)}$

1.2. Convert 5.00 inches to (*a*) centimeters, (*b*) millimeters, (*c*) meters.

    (*a*) 5.00 in. $= 5.00$ in. $\times 2.54$ cm/in. $= 12.7 \dfrac{\text{in. cm}}{\text{in.}} = 12.7$ cm

        The procedure can be understood easily in terms of the word definition of the conversion factor. 2.54 is the number of cm *per* in., that is, the number of cm in one in. Thus the number of cm in 5 in. is 5×2.54.

        More formally, the conversion factor may be considered to be a statement of equality between 2.54 cm and 1 in. Since 2.54 cm $= 1$ in., then

$$2.54 \text{ cm per in.} = 2.54 \text{ cm/in.} = \frac{2.54 \text{ cm}}{1 \text{ in.}} = \frac{2.54 \text{ cm}}{2.54 \text{ cm}} = 1$$

Thus the conversion factor, 2.54 cm/in., is mathematically equal to 1. Any quantity may be multiplied or divided by the conversion factor without changing the essential value of the quantity, since multiplication or division by one leads to an identity. In this problem the multiplication 5.00 in. $\times$ 2.54 cm/in. leads to the useful result of 12.7 cm, a statement of the same quantity of length in different units. The division of 5.00 in. by 2.54 cm/in., although mathematically allowed, does not lead to a useful result because the quotient, 1.97 in²/cm, is not expressed in a simple unit which has a physical interpretation. In general, any proper conversion factor is equal to 1 and may be used as a multiplier or divisor of any physical quantity. The choice of multiplication or division depends on the particular problem.

    (*b*) 12.7 cm $= 12.7$ cm $\times 10$ mm/cm $= 127 \dfrac{\text{cm mm}}{\text{cm}} = 127$ mm

    (*c*) 12.7 cm $= \dfrac{12.7 \text{ cm}}{100 \text{ cm/m}} = 0.127 \dfrac{\text{cm m}}{\text{cm}} = 0.127$ m

        Here it was appropriate to divide by the conversion factor. If by mistake we had multiplied 12.7 cm $\times 100$ cm/m, the answer would have been expressed in cm²/m, since $\text{cm} \times \dfrac{\text{cm}}{\text{m}} = \dfrac{\text{cm}^2}{\text{m}}$. We would immediately realize our error, seeing that the answer is not expressed in meters.

1.3. Convert (*a*) 14.0 cm and (*b*) 7.00 meters to inches.

    (*a*) 14.0 cm $= \dfrac{14.0 \text{ cm}}{2.54 \text{ cm/in.}} = 5.51$ in.   Or, 14.0 cm $\times 0.3937$ in./cm $= 5.51$ in.

    (*b*) 7.00 m $= \dfrac{700 \text{ cm}}{2.54 \text{ cm/in.}} = 276$ in.   Or, 7.00 m $\times 39.37$ in./m $= 276$ in.

1.4. How many square inches in one square meter?

$$1 \text{ m} = \frac{100 \text{ cm}}{2.54 \text{ cm/in.}} = 39.37 \text{ in.} \qquad (1 \text{ m})^2 = (39.37 \text{ in.})^2 = 1550 \text{ in}^2$$

1.5. (*a*) How many cubic centimeters in one cubic meter? (*b*) How many liters in one cubic meter?

    (*a*) $1 \text{ m}^3 = 1 \text{ m} \times 1 \text{ m} \times 1 \text{ m} = 100 \text{ cm} \times 100 \text{ cm} \times 100 \text{ cm} = 1{,}000{,}000 \text{ cm}^3$

       Or, $1 \text{ m}^3 = (100 \text{ cm})^3 = (10^2 \text{ cm})^3 = 10^6 \text{ cm}^3 = 1{,}000{,}000 \text{ cm}^3$

    (*b*) $1 \text{ m}^3 = \dfrac{1{,}000{,}000 \text{ cm}^3}{1000 \text{ cm}^3/l} = 1000 \, l$ Or, $1 \text{ m}^3 = \dfrac{10^6 \text{ cm}^3}{10^3 \text{ cm}^3/l} = 10^3 \, l = 1000 \, l$

1.6. Find the capacity in liters of a box 0.6 m long, 10 cm wide, and 50 mm deep.

    Convert meters and millimeters to centimeters.

$$\begin{aligned} \text{Volume} &= 0.6 \text{ m} \times 10 \text{ cm} \times 50 \text{ mm} \\ &= 60 \text{ cm} \times 10 \text{ cm} \times 5 \text{ cm} = 3000 \text{ cm}^3 = 3 \text{ liters} \end{aligned}$$

1.7. Determine the weight of 66 lb of sulfur in (a) kilograms and (b) grams. (c) Find the weight of 3.4 kg of copper in pounds.

(a) $66 \text{ lb} = \dfrac{66 \text{ lb}}{2.2 \text{ lb/kg}} = 30 \text{ kg}$ Or, $66 \text{ lb} = 66 \text{ lb} \times 0.454 \text{ kg/lb} = 30 \text{ kg}$

(b) $30 \text{ kg} = 30,000 \text{ g}$ Or, $66 \text{ lb} = 66 \text{ lb} \times 454 \text{ g/lb} = 30,000 \text{ g}$

(c) $3.4 \text{ kg} = 3.4 \text{ kg} \times 2.2 \text{ lb/kg} = 7.5 \text{ lb}$

1.8. A uniform steel bar is 16.0 inches long and weighs 6 lb 4 oz. Determine the weight of the bar in grams per centimeter of length.

$16.0 \text{ in.} = 16.0 \text{ in.} \times 2.54 \text{ cm/in.} = 40.6 \text{ cm}$

$6\frac{1}{4} \text{ lb} = 6\frac{1}{4} \text{ lb} \times 454 \text{ g/lb} = 2840 \text{ g}$ $\text{Grams/cm} = \dfrac{2840 \text{ g}}{40.6 \text{ cm}} = 69.9 \text{ g/cm}$

Or, $\dfrac{6\frac{1}{4} \text{ lb}}{16.0 \text{ in.}} = \dfrac{6\frac{1}{4} \text{ lb}}{16.0 \text{ in.}} \times \dfrac{454 \text{ g}}{1 \text{ lb}} \times \dfrac{1 \text{ in.}}{2.54 \text{ cm}} = 69.9 \text{ g/cm}$

It should be understood that $\dfrac{454 \text{ g}}{1 \text{ lb}} = \dfrac{1 \text{ lb}}{1 \text{ lb}} = 1$ and $\dfrac{1 \text{ in.}}{2.54 \text{ cm}} = \dfrac{1 \text{ in.}}{1 \text{ in.}} = 1.$

1.9. A pressure of one atmosphere is equal to 1033 grams per square centimeter. Express this pressure in pounds per square inch.

$$1033 \text{ g/cm}^2 = 1033 \frac{\text{g}}{\text{cm}^2} \times \frac{1 \text{ lb}}{453.6 \text{ g}} \times \frac{(2.54 \text{ cm})^2}{(1 \text{ in.})^2} = 14.69 \text{ lb/in}^2$$

It should be understood that $\dfrac{(2.54 \text{ cm})^2}{(1 \text{ in.})^2} = \dfrac{(1 \text{ in.})^2}{(1 \text{ in.})^2} = 1.$

In general, any proper conversion factor is equal to 1. Hence any conversion factor raised to any power is also equal to 1.

1.10. Fatty acids spread spontaneously on water to form a monomolecular film. A benzene solution containing 0.1 mm³ of stearic acid is dropped into a tray full of water. The acid is insoluble in water but spreads on the surface to form a continuous film area of 400 cm² after all of the benzene has evaporated. What is the average film thickness in angstroms?

$1 \text{ mm}^3 = (0.1 \text{ cm})^3 = (10^{-1} \text{ cm})^3 = 10^{-3} \text{ cm}^3$. Then $0.1 \text{ mm}^3 = 10^{-4} \text{ cm}^3$.

$\text{Film thickness in cm} = \dfrac{\text{volume}}{\text{area}} = \dfrac{10^{-4} \text{ cm}^3}{400 \text{ cm}^2} = \dfrac{10^{-4} \text{ cm}^3}{4 \times 10^2 \text{ cm}^2} = 2.5 \times 10^{-7} \text{ cm}$

$\text{Film thickness in angstroms} = 2.5 \times 10^{-7} \text{ cm} \times 10^8 \text{ Å/cm} = 25 \text{ Å}$

1.11. The apothecaries' *minim* is legally defined as follows: 60 minims per fluid dram, 8 fluid drams per fluid ounce, 16 fluid ounces per liquid pint, 8 liquid pints per gallon of 231 cubic inches. Convert the minim to cubic centimeters.

$$1 \text{ minim} = 1 \text{ minim} \times \frac{1 \text{ fl dr}}{60 \text{ minim}} \times \frac{1 \text{ fl oz}}{8 \text{ fl dr}} \times \frac{1 \text{ pt}}{16 \text{ fl oz}} \times \frac{1 \text{ gal}}{8 \text{ pt}} \times \frac{231 \text{ in}^3}{1 \text{ gal}} \times \frac{(2.54 \text{ cm})^3}{(1 \text{ in.})^3}$$

$$= 0.0616 \text{ cm}^3 \text{ (or 0.0616 ml)}$$

Note that all the units cancel out, except cm³ which appears in the answer.

1.12. New York City's 7.8 million people have a daily per capita consumption of 140 gallons of water. How many tons of sodium fluoride (45% fluorine by weight) would be required per year to give this water a tooth-strengthening dose of 1 part (by weight) fluorine per million parts water? One U.S. gallon of water at normal room temperature weighs 8.34 lb.

Weight of water, in tons, required per year

$$= (7.8 \times 10^6) \times 140 \times 365 \, \frac{\text{gal water}}{\text{yr}} \times \frac{8.34 \text{ lb water}}{1 \text{ gal water}} \times \frac{1 \text{ ton}}{2000 \text{ lb}}$$

$$= 1.66 \times 10^9 \, \frac{\text{ton water}}{\text{yr}}$$

Note that all units cancel out, except $\dfrac{\text{ton water}}{\text{yr}}$ which appears in the result.

Weight of sodium fluoride, in tons, required per year

$$= 1.66 \times 10^9 \, \frac{\text{ton water}}{\text{yr}} \times \frac{1 \text{ ton fluorine}}{10^6 \text{ ton water}} \times \frac{1 \text{ ton sodium fluoride}}{0.45 \text{ ton fluorine}}$$

$$= 3.7 \times 10^3 \, \frac{\text{ton sodium fluoride}}{\text{yr}}$$

Note that all units cancel out, except $\dfrac{\text{ton sodium fluoride}}{\text{yr}}$ which appears in the answer.

Supplementary Problems

1.13. (a) Express 3.69 meters in kilometers, in centimeters, and in millimeters.
(b) Express 36.24 millimeters in centimeters and in meters.
Ans. (a) 0.00369 km, 369 cm, 3690 mm; (b) 3.624 cm, 0.03624 m

1.14. Determine the number of (a) millimeters in 10 inches, (b) feet in 5 meters, (c) centimeters in 4 ft 3 in.   *Ans.* 254 mm, 16.4 ft, 130 cm

1.15. Convert the molar volume of 22.4 liters to milliliters, to cubic centimeters, to cubic meters, and to cubic feet.   *Ans.* 22,400 ml, 22,400 cm^3, 0.0224 m^3, 0.791 ft^3

1.16. Express the volume of 50 gallons of gasoline in (a) liters, (b) cubic meters.
Ans. 189 l, 0.189 m^3

1.17. Express the weight (mass) of 32 g of oxygen in milligrams, in kilograms, and in pounds (avoir).
Ans. 32,000 mg, 0.032 kg, 0.0705 lb

1.18. How many grams in 5.00 lb of copper sulfate? How many pounds in 4.00 kg of mercury? How many milligrams in 1 lb 2 oz of sugar?   *Ans.* 2270 g, 8.82 lb, 510,000 mg

1.19. Convert the weight of 500 lb of coal to (a) kilograms, (b) metric tons, (c) U.S. short tons, (d) British long tons.   *Ans.* 227 kg, 0.227 metric ton, 0.250 short ton, 0.223 long ton

1.20. The color of light depends on its wavelength. The longest visible rays, at the red end of the visible spectrum, are 0.000078 cm in length. Express this length in microns, in millimicrons, in nanometers, and in angstroms.   *Ans.* 0.78 μ, 780 mμ, 780 nm, 7800 Å

1.21. Determine the number of (a) cubic centimeters in a cubic inch, (b) cubic inches in a liter, (c) cubic feet in a cubic meter.   *Ans.* 16.4 cm^3, 61.0 in^3, 35.3 ft^3

1.22. The density of water is 1.000 gram per cubic centimeter (g/cm^3) at 4°C. Calculate the density of water in pounds per cubic foot at the same temperature.   *Ans.* 62.4 lb/ft^3

1.23. What is the average speed, in miles per hour, of a sprinter in doing the 100 meter dash in 10.1 seconds?   *Ans.* 22 mi/hr

1.24. In a crystal of platinum, centers of individual atoms are 2.8 Å apart along the direction of closest packing. How many atoms would lie on a one-inch length of a line in this direction?
Ans. 9×10^7 atoms

1.25. The grocer's familiar pound (avoirdupois) is defined from the ancient *grain*, as equal to 7000 grains. The druggist's "pound apothecary" is defined as follows:

$$20 \text{ grains/scruple}, \quad 3 \text{ scruples/dram}, \quad 8 \text{ drams/oz-apoth}, \quad 12 \text{ oz-apoth/lb-apoth}$$

How many grains are in (*a*) the apothecary ounce and in (*b*) the apothecary pound? (*c*) How does the avoirdupois pound compare with the apothecary pound? (*d*) How does the familiar ounce of 1/16 pound compare with the apothecary ounce?
Ans. (*a*) 480 grains, (*b*) 5760 grains, (*c*) 1 lb avoir = 1.215 lb apoth, (*d*) 1 oz avoir = 0.911 oz apoth

1.26. The price of platinum is quoted on the basis of the troy ounce, which is the same as the apothecary's ounce. At $95.00 per troy ounce for platinum, what would be the cost of the metal that goes into a 22 ounce (avoirdupois) flute? Use the answers to Problem 1.25.   *Ans.* $1900

1.27. The silica gel which is used to protect sealed overseas shipments from moisture seepage has a surface area of 6.0×10^6 cm^2 per gram. What is this surface area in square feet per gram?
Ans. 6.5×10^3 ft^2/g

1.28. The blue iridescence of butterfly wings is due to striations which are 0.15 micron apart, measured by the electron microscope. What is this distance in millionths of an inch? How does this spacing compare with the wavelength of blue light, about 4500 angstroms?
Ans. 6 millionths inch, 1/3 wavelength of blue light

1.29. The thickness of a soap bubble film at its thinnest (bimolecular) stage is about 60 angstroms. (*a*) What is this thickness in inches? (*b*) How does this thickness compare with the wavelength of yellow sodium light, which is 0.5890 micron?
Ans. (*a*) 2.4×10^{-7} in.,   (*b*) about 0.01 of wavelength of yellow light

1.30. An average man requires about 2.00 mg of riboflavin (vitamin B$_2$) per day. How many pounds of cheese would a man have to eat per day if this were his only source of riboflavin and if the cheese contained 5.5×10^{-6} gram of riboflavin per gram?   *Ans.* 0.80 lb/day

1.31. When a sample of healthy human blood is diluted to 200 times its initial volume and microscopically examined in a layer 0.10 mm thick, an average of 30 red corpuscles are found in each 100×100 micron square. (*a*) How many red cells are in a cubic millimeter of blood? (*b*) The red blood cells have an average life of one month, and the adult blood volume is about 5 liters. How many of these red cells are generated every second in the sternal bone marrow of the adult?
Ans. 6×10^6 cells/mm^3, 1×10^7 cells/sec

1.32. There is some reason to think that the length of the day, determined from the earth's period of rotation, is increasing uniformly by about 0.001 second every century. What is this variation in parts per billion?   *Ans.* 3×10^{-4} sec per 10^9 sec

1.33. The bromine content of average ocean water is 65 parts by weight per million. Assuming 100% recovery, how many gallons of ocean water must be processed to produce one pound of bromine. Assume that the density of sea water is 1.0 g/cc.   *Ans.* 1.8×10^3 gallons

1.34. An important physical quantity has the value 1.987 calories or 0.08206 liter atmospheres. What is the multiplying conversion factor to be used in converting liter-atmospheres to calories?
Ans. 24.22 cal/*l*-atm

1.35. A porous catalyst for chemical reactions had an internal surface area of 800 square meters per cc of bulk material. Fifty percent of the bulk volume consists of the pores (holes) while the other fifty percent of the volume is made up of the solid substance. Assume that the pores are all cylindrical tubules of uniform diameter d and length l and that the measured internal surface area is the total area of the curved surfaces of the tubules. What is the diameter of each pore? *Hint:* Find the number of tubules per bulk cc, n, in terms of l and d, by using the formula for the volume of a cylinder $V = \frac{1}{4}\pi d^2 l$. Then apply the surface formula to the cylindrical surfaces of n tubules ($S_{\text{cyl}} = \pi dl$).   *Ans.* 25 Å

Chapter 2

Density and Specific Gravity

DENSITY

Density denotes the mass or quantity of matter of a substance contained in one unit of its volume. A dense substance is one which has a large quantity of matter in a small volume.

$$\text{Density} \ = \ \text{mass per unit volume} \ = \ \frac{\text{mass of body}}{\text{volume of body}}$$

The density of solids and liquids is expressed usually in grams per cubic centimeter (g/cc or g/cm^3), in grams per milliliter (g/ml), or in pounds per cubic foot (lb/ft^3). The density of gases is expressed in grams per liter (g/l) or in grams per cubic centimeter (g/cc).

$$\text{Density of water at } 4°C \ = \ 1.000 \text{ g/cm}^3 \ = \ 1.000 \text{ g/ml}$$
$$= \ 62.4 \text{ lb/ft}^3$$

SPECIFIC GRAVITY

Specific gravity of a body is that number which denotes the ratio of the mass (or weight) of a body and the mass (or weight) of an equal volume of a substance taken as a standard. Solids and liquids are referred to water as standard, while gases are often referred to air as standard. Then

$$\text{Specific gravity} \ = \ \frac{\text{mass of solid or liquid}}{\text{mass of an equal volume of water at } 4°C}$$

The water standard is sometimes taken at 4°C, the temperature at which water has its maximum density. Otherwise a specified temperature for the water standard may be indicated. Since the density of water does not vary by more than half a per cent over the temperature range 0°C to 30°C, the rounded value of 1.00 g/cc may be used as the standard for specific gravity computations where the experimental substance is weighed and compared to water at ordinary laboratory temperatures. If greater accuracy is desired, substances are usually described by densities rather than specific gravities.

If a piece of aluminum weighs 2.70 times as much as an equal volume of water, its specific gravity is 2.70 in any system of measures. If this fact applies to the temperature range from 0°C to 30°C, where the density of water is 1.00 g/cc (or 1.00 g/ml), the density of aluminum is 2.70 g/cc (or 2.70 g/ml). Within the precision discussed above, the density and specific gravity of any substance are numerically the same when the density is expressed in grams per cc (or in grams per ml).

In the British system, the specific gravity of aluminum is also 2.70. But the density of aluminum in British units = 2.70 × density of water = 2.70 × 62.4 lb/ft^3 = 168 lb/ft^3.

The specific gravity of a substance is the same in any system of units, since it expresses the quotient of the mass of the substance divided by the mass of an equal volume of water. It is expressed by a pure number without units.

9

Solved Problems

2.1. Calculate the density and specific gravity of a body that weighs 420 g and has a volume of 52 cc.

$$\text{Density} \;=\; \frac{\text{mass}}{\text{volume}} \;=\; \frac{420\text{ g}}{52\text{ cc}} \;=\; 8.1\text{ g/cc}$$

Since the density and specific gravity are numerically the same when the density is expressed in g/cc (or in g/ml), the specific gravity of the body is 8.1.

2.2. What volume will 300 g of mercury occupy? Density of mercury is 13.6 g/ml.

$$\text{Volume} \;=\; \frac{\text{mass}}{\text{density}} \;=\; \frac{300\text{ g}}{13.6\text{ g/ml}} \;=\; 22.1\text{ ml}$$

2.3. The specific gravity of cast iron is 7.20. Calculate its density (a) in grams per cubic centimeter and (b) in pounds per cubic foot.

(a) Cast iron is 7.20 times as dense as water. The density of water in the metric system is 1.00 g/cc. Hence the density of cast iron is 7.20 g/cc.

(b) The density of water in the British system is 62.4 lb/ft³. Hence the density of cast iron = 7.20 × 62.4 lb/ft³ = 449 lb/ft³.

2.4. A casting of an alloy in the form of a disc weighed 50.0 grams. The disc was 0.250 inches thick and had a circular cross-section of diameter 1.380 inches. What is the density of the alloy?

$$\text{Volume of the cylinder} \;=\; \frac{\pi d^2 h}{4} \;=\; \frac{\pi \times 1.380^2 \times 0.250}{4}\text{ in}^3 \times \frac{16.39\text{ cc}}{1\text{ in}^3} \;=\; 6.13\text{ cc}$$

$$\text{Density of the alloy} \;=\; \frac{\text{mass}}{\text{volume}} \;=\; \frac{50.0\text{ g}}{6.13\text{ cc}} \;=\; 8.15\text{ g/cc}$$

2.5. The density of zinc is 455 lb/ft³. Find (a) the specific gravity of zinc and (b) the weight of 9.00 cc of zinc.

(a) Specific gravity of zinc $\;=\; \dfrac{\text{mass of 1 ft}^3\text{ of zinc}}{\text{mass of 1 ft}^3\text{ of water}} \;=\; \dfrac{455\text{ lb}}{62.4\text{ lb}} \;=\; 7.29$

(b) 1 cc of zinc weighs 7.29 g. Then 9.00 cc weighs 9.00 × 7.29 g = 65.6 g.

2.6. A specific gravity bottle weighs 220 g when empty, 380 g when filled with water, and 351 g when filled with kerosene. Find the specific gravity of kerosene and the capacity of the bottle.

$$\text{Sp gr kerosene} \;=\; \frac{\text{mass of kerosene}}{\text{mass of equal volume of water}} \;=\; \frac{(351-220)\text{g}}{(380-220)\text{g}} \;=\; 0.819$$

$$\text{Capacity (volume) of bottle} \;=\; \frac{\text{mass of water}}{\text{density of water}} \;=\; \frac{(380-220)\text{g}}{1.00\text{ g/cc}} \;=\; 160\text{ cc}$$

2.7. Find the volume in gallons of 400 lb of cottonseed oil, specific gravity 0.926. One gallon of water weighs 8.34 lb.

$$\text{Density of oil} \;=\; 0.926 \times \text{density of water} \;=\; 0.926 \times 8.34\text{ lb/gal}$$

$$\text{Volume} \;=\; \frac{\text{mass}}{\text{density}} \;=\; \frac{400\text{ lb}}{0.926 \times 8.34\text{ lb/gal}} \;=\; 51.8\text{ gallons}$$

2.8. Battery acid has a specific gravity of 1.285 and contains 38.0% by weight H_2SO_4. How many grams of pure H_2SO_4 are contained in a liter of battery acid?

 1 ml of acid weighs 1.285 g. Then 1 liter of acid weighs 1285 g.

 Since 38.0% by weight of the acid is pure H_2SO_4, the number of grams of H_2SO_4 in 1 liter of battery acid = 0.380×1285 g = 488 g.

 Percentage indicates the number of parts per hundred; 38 g of pure H_2SO_4 is contained in 100 g of the acid. The percentage number divided by 100 is the fractional composition; then $0.38 \times$ weight of acid = weight of H_2SO_4 content.

 Formally, the solution can be written as follows:

$$\text{Weight of } H_2SO_4 \;=\; 1285 \text{ g acid} \times \frac{38 \text{ g } H_2SO_4}{100 \text{ g acid}} \;=\; 488 \text{ g } H_2SO_4$$

 Here the conversion factor $\dfrac{38 \text{ g } H_2SO_4}{100 \text{ g acid}}$ is taken to be equal to 1. Although the condition, 38 g H_2SO_4 = 100 g acid, is not a universal truth in the same sense that 1 inch always equals 2.54 cm, the condition is a rigid one of association of 38 g H_2SO_4 with every 100 g acid for this particular acid. Mathematically, these two quantities may be considered to be equal for this problem, since one of the quantities implies the other. Liberal use will be made in subsequent chapters of unit conversion factors that are to be used only for particular cases, in addition to the unit factors that are universally valid.

2.9. (a) Calculate the weight of pure HNO_3 per ml of the concentrated acid which assays 69.8% by weight HNO_3 and has a specific gravity of 1.42. (b) Calculate the weight of pure HNO_3 in 60.0 ml of concentrated acid. (c) What volume of the concentrated acid contains 63.0 g of pure HNO_3?

 (a) 1 ml of acid weighs 1.42 g. Since 69.8% of the total weight of the acid is pure HNO_3, then the number of grams of HNO_3 in 1 ml of acid = 0.698×1.42 g = 0.991 g.

 (b) Weight of HNO_3 in 60.0 ml of acid = 60.0 ml $\times 0.991$ g/ml = 59.5 g HNO_3.

 (c) 63.0 g HNO_3 is contained in $\dfrac{63.0 \text{ g}}{0.991 \text{ g/ml}}$ = 63.6 ml acid.

Supplementary Problems

2.10. Find the density and specific gravity of ethyl alcohol if 80.0 cc weighs 63.3 g.
  Ans. 0.791 g/cc, 0.791

2.11. Find the volume of 40 kg of carbon tetrachloride (sp gr 1.60).   *Ans.* 25 l

2.12. Determine the weight of 20.0 cubic feet of aluminum (sp gr 2.70).   *Ans.* 3370 lb

2.13. The heaviest solid (osmium) and the lightest room temperature liquid (butane) have specific gravities of 22.5 and 0.6 respectively. Calculate the density of osmium in lb/ft³ and of butane in kg/l.
  Ans. 1400 lb/ft³, 0.6 kg/l

2.14. Air weighs about 8 lb per 100 cubic feet. What is its density in grams per cubic foot and in grams per liter?   *Ans.* 36 g/ft³, 1.3 g/l

2.15. What is the density of a steel ball which has a diameter of 0.750 cm and weighs 1.765 g? Volume of a sphere of radius r is $(4/3)\pi r^3$.   *Ans.* 7.99 g/cc

2.16. A wooden block, $10'' \times 6.0'' \times 2.0''$, weighs 3 lb 10 oz. What is the density of the wood? *Ans.* 0.84 g/cc

2.17. An alloy was machined into a flat disc, 3.15 cm in diameter and 0.45 cm thick, with a hole 0.75 cm in diameter drilled through the center. The disc weighed 20.2 g. What is the density of the alloy? *Ans.* 6.1 g/cc

2.18. A drum holds 200 lb of water or 132 lb of gasoline. (*a*) What is the specific gravity of gasoline? (*b*) What is the density of gasoline in g/cm³ and in lb/ft³? (*c*) What is the capacity of the drum in gallons? One gallon of water weighs 8.34 lb.   *Ans.* 0.66, 0.66 g/cm³, 41.2 lb/ft³, 24.0 gal

2.19. A glass vessel weighed 20.2376 g when empty and 20.3102 g when filled with water at 4°C to an etched mark. The same vessel was then dried and filled with a solution to the same mark at 4°. The vessel was now found to weigh 20.3300 g. What is the density of the solution?   *Ans.* 1.273 g/cc

2.20. The lightness of a General Electric plastic foam is illustrated by the fact that a piece $13 \times 9.5 \times 2.5$ inches weighs only 350 grams. Du Pont cellulose sponges, $7 \times 12 \times 2.5$ cm, weigh 12 grams. *Foamglas* is light as balsa, only 10 lb/ft³, two-thirds as heavy as cork. Calculate the specific gravity of these synthetics and of cork.   *Ans.* 0.069, 0.057, 0.16, 0.24

2.21. A sample of concentrated sulfuric acid is 95.7% H_2SO_4 by weight and its density is 1.84 g/cc. (*a*) How many grams of pure H_2SO_4 are contained in 1000 ml of the acid? (*b*) How many ml of acid contain 100 g of pure H_2SO_4?   *Ans.* 1760 g, 56.8 ml

2.22. Analysis shows that 20.0 ml of concentrated hydrochloric acid of density 1.18 g/cc contains 8.36 g HCl. (*a*) Find the weight of HCl per ml of acid solution. (*b*) Find the percent by weight of HCl in the concentrated acid.   *Ans.* 0.418 g HCl/ml acid, 35.4% HCl

2.23. An electrolytic tin-plating process gives a coating 30 millionths of an inch thick. How many square meters can be coated with one kilogram of tin, density 7.3 g/cc?   *Ans.* 180 m²

2.24. A piece of gold leaf (sp gr 19.3) weighing 1.93 mg can be beaten further into a transparent film covering an area of 14.5 cm². (*a*) What is the volume of 1.93 mg of gold? (*b*) What is the thickness of the transparent film in angstroms?   *Ans.* 1.00×10^{-4} cm³, 690 Å.

2.25. A piece of capillary tubing was calibrated in the following manner. A clean sample of the tubing weighed 3.247 g. A thread of mercury, drawn into the tube, occupied a length of 2.375 cm, as observed under a microscope. The weight of the tube with the mercury was 3.489 g. The density of mercury is 13.60 g/cc. Assuming that the capillary bore is a uniform cylinder, find the diameter of the bore.   *Ans.* 0.98 mm

Chapter 3

Temperature Measurement

TEMPERATURE

Temperature may be defined as that property of a body which determines the flow of heat. Two bodies are at the same temperature if there is no transfer of heat when they are placed together. Temperature is a fundamental concept, an intrinsic dimension which cannot be defined in terms of mass, length and time dimensions.

CELSIUS AND FAHRENHEIT SCALES

Although the absolute standard measure of temperature is a gas thermometer, a liquid thermometer of the type commonly used in laboratories may be employed to illustrate the thermometric scales. Two fixed points are chosen for standardizing a thermometer, usually the freezing point and boiling point of water under one atmosphere pressure. The thermometer liquid (mercury, for example) is allowed to come to the temperature of a fixed point, and the height of the liquid in the thermometer which corresponds to the fixed point is noted.

On the Celsius scale (°C) the freezing point of water is defined as 0° and the boiling point as 100° at one atmosphere pressure. The distance between the two fixed point levels on the thermometer is divided into 100 equal parts, and each division corresponds to 1°C. Equally spaced divisions may be extended above and below the fixed points. (Although Celsius is the official name of this scale accepted by international scientific agreement, it is frequently called the centigrade scale, especially in the United States. The term Celsius and centigrade will be used interchangeably in this book.) Such a thermometer is adequate for most applications. For precise measurements, more elaborate calibration procedures are required.

On the Fahrenheit scale (°F) the freezing point of water is defined as 32° and the boiling point as 212°. The distance between the two fixed point levels on the thermometer is divided into 180 equal parts, and each division corresponds to 1°F. Equally spaced divisions may be extended above and below the fixed points.

INTERCONVERSION OF CELSIUS AND FAHRENHEIT

Between the freezing and boiling points of water there are 100 Celsius or 180 Fahrenheit intervals. Then 100 Celsius intervals = 180 Fahrenheit intervals. Hence

$$1 \text{ Celsius (centigrade) interval} = \frac{180}{100} = \frac{9}{5} \text{ Fahrenheit intervals}$$

$$1 \text{ Fahrenheit interval} = \frac{100}{180} = \frac{5}{9} \text{ Celsius interval}$$

But the freezing point of water is 0° on the Celsius scale and 32° on the Fahrenheit scale; or, 0°C = 32°F. Then

$$\text{Temperature Celsius} = 5/9 \times (\text{temperature Fahrenheit} - 32)$$

$$\text{Temperature Fahrenheit} = (9/5 \times \text{temperature Celsius}) + 32$$

ABSOLUTE KELVIN SCALE

All gases held at constant volume show a uniform increase in pressure with increasing temperature over a wide range of experimental conditions. Experiments show that the pressure of a gas increases by 1/273 of its pressure at 0°C for each degree (°C) the temperature is raised above 0°C. Similarly, when the temperature of the gas is lowered the pressure decreases by 1/273 of its pressure at 0°C for each degree (°C) the temperature is lowered. It follows that a gas continuing to show such behavior would no longer exert any pressure when the temperature becomes 273° below 0°C. This temperature, −273°C, (at which gas molecules would cease to move, according to the kinetic theory) is called the *absolute zero of temperature.* In practice all gases, on cooling, liquefy or solidify before −273°C is reached. Absolute zero is defined in terms of the hypothetical gas that would follow the same laws at low temperatures that real gases obey at higher temperatures. (Although the exact value for the absolute zero is accepted as −273.15°C, −273°C will be sufficiently accurate for the problems in this book.)

The absolute zero of temperature, −273°C, is taken as the zero point on the absolute Kelvin scale (°K). The Kelvin and Celsius scales differ only in the choice of the zero point.

$$\text{Kelvin temperature} \ = \ 273 \ + \ \text{Celsius temperature}$$

RANKINE OR FAHRENHEIT ABSOLUTE SCALE

In the *Rankine* (°R) scale the value of the degree interval is the same as the Fahrenheit degree interval, but the zero point corresponds to 0°K or −273.15°C. This scale is commonly used in engineering.

Since $-273.15°C = \frac{9}{5}(-273.15) + 32 = -459.7°F$, $0°R = -459.7°F$. With rounding off,

$$\text{Rankine temperature} \ = \ 460 \ + \ \text{Fahrenheit temperature}$$

Thus, $0°F = 460°R$, $40°F = 500°R$, $-40°F = 420°R$.

Because the Rankine and Kelvin scales have the same zero, they are simply related to each other.

$$\text{Rankine temperature} \ = \ \frac{9}{5} \times \text{Kelvin temperature}$$

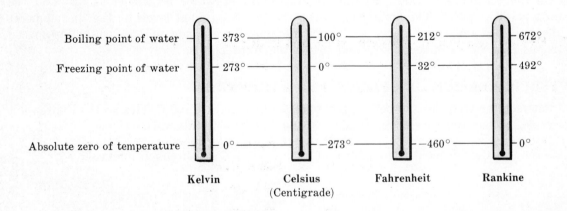

	Kelvin	Celsius (Centigrade)	Fahrenheit	Rankine
Boiling point of water	373°	100°	212°	672°
Freezing point of water	273°	0°	32°	492°
Absolute zero of temperature	0°	−273°	−460°	0°

Solved Problems

3.1. Ethyl alcohol boils at 78.5°C and freezes at −117°C at one atmosphere pressure. Convert these temperatures to the Fahrenheit scale.

$$\text{Fahrenheit} = 9/5 \times \text{Celsius} + 32$$

$$\text{Boiling point, °F} = 9/5 \times 78.5 + 32 = 141 + 32 = 173°F$$

$$\text{Freezing point, °F} = 9/5 \times (-117) + 32 = -211 + 32 = -179°F$$

3.2. Mercury boils at 675°F and solidifies at −38.0°F at one atmosphere pressure. Express these temperatures in centigrade units.

$$\text{Centigrade} = 5/9 \times (\text{Fahrenheit} - 32)$$

$$\text{Boiling point, °C} = 5/9 \times (675 - 32) = 5/9 \times 643 = 357°C$$

$$\text{Solidification point, °C} = 5/9 \times (-38.0 - 32) = 5/9 \times (-70.0) = -38.9°C$$

3.3. Change 40°C and −5°C to the Kelvin scale.

$$\text{Kelvin temperature} = \text{Celsius temperature} + 273$$

$$40°C = x°K = 40 + 273 = 313°K$$

$$-5°C = x°K = -5 + 273 = 268°K$$

3.4. Convert 220°K and 498°K to the centigrade scale.

$$\text{Centigrade temperature} = \text{Kelvin temperature} - 273$$

$$220°K = x°C = 220 - 273 = -53°C$$

$$498°K = x°C = 498 - 273 = 225°C$$

3.5. Express −22°F in degrees centigrade and in degrees Kelvin.

$$\text{Centigrade} = 5/9 \times (°F - 32) = 5/9 \times (-22 - 32) = 5/9 \times (-54) = -30°C$$

$$\text{Kelvin} = 273 + \text{centigrade} = 273 - 30 = 243°K$$

3.6. Convert: (a) 440°F and (b) −75°F to the Rankine scale and to the Kelvin scale.

$$\text{Rankine temperature} = \text{Fahrenheit temperature} + 460$$

$$(a) \qquad °R = 440 + 460 = 900°R$$

$$(b) \qquad °R = -75 + 460 = 385°R$$

$$\text{Kelvin temperature} = \tfrac{5}{9} \times \text{Rankine temperature}$$

$$(a) \qquad °K = \tfrac{5}{9} \times 900 = 500°K$$

$$(b) \qquad °K = \tfrac{5}{9} \times 385 = 214°K$$

Supplementary Problems

3.7. (a) Convert 68°F to C; 5°F to C; 176°F to C.   *Ans.* (a) 20°C, −15°C, 80°C

  (b) Convert 30°C to F; 5°C to F; −20°C to F.   (b) 86°F, 41°F, −4°F

3.8. Convert the following temperatures: −195.5°C to F; −430°F to C; 1705°C to F.
Ans. −319.9°F, −256.7°C, 3101°F

3.9. The temperature of dry ice (sublimation temperature at normal pressure) is −109°F. Is this hotter or colder than the temperature of boiling ethane (a component of bottled gas) which is −88°C?
Ans. hotter

3.10. Gabriel Fahrenheit in 1714 suggested for the zero point on his scale the lowest temperature then obtainable from a mixture of salts and ice, and for his 100°-point he suggested the highest known normal animal temperature. Express these "extremes" in degrees centigrade.
Ans. −17.8°C, 37.8°C

3.11. Convert 300°K, 760°K, 180°K, to degrees centigrade.   *Ans.* 27°C, 487°C, −93°C

3.12. Express 8°K, 273°K, in degrees Fahrenheit.   *Ans.* −445°F, 32°F

3.13. Convert 14°F to degrees Celsius and degrees Kelvin.   *Ans.* −10°C, 263°K

3.14. At what temperature have the Celsius and Fahrenheit readings the same numerical value?
Ans. −40°

3.15. Convert 105°R, 340°R, 520°R, to degrees Fahrenheit and to degrees Kelvin.
Ans. −355°F, −120°F, 60°F; 58°K, 189°K, 289°K

3.16. Convert 27°K, 452°K, to degrees Rankine.   *Ans.* 49°R, 814°R

3.17. A water-stabilized electric arc was reported to reach a temperature of 25,600°F. On the absolute scale, what is the ratio of this temperature to that of an oxy-acetylene flame at 3500°C?   *Ans.* 3.84

3.18. Construct a temperature scale on which the freezing and boiling points of water are 100° and 400°, respectively, and the degree interval is a constant multiple of the Celsius degree interval. What will the absolute zero be on this scale and what will be the melting point of sulfur, which is 444.6°C?
Ans. −719°, 1433.8°

Chapter 4

Atomic Weights, Molecular Weights, and Moles

ATOMS

In the form of the atomic theory proposed by John Dalton in 1805, all atoms of a given element were thought to be identical. Chemists in the following decades set themselves the task of finding the relative masses of the atoms of the different elements by precise quantitative chemical analysis. Over a hundred years after Dalton's proposal was made, investigations with radioactive substances showed that not all atoms of a given element are identical.

Modern atomic weight tables must recognize this fact because one of the properties which distinguishes these different kinds of atoms for a given element is the atomic mass. Each element can exist in several *isotopic* forms, and all atoms of the same isotope are identical.

NUCLEI

Every atom has a positively charged nucleus which contains over 99.9% of the total mass of the atom. Although the structure of the nucleus is not thoroughly understood, every nucleus may be described as being made up of two different kinds of particles. These two kinds of particles, called *nucleons*, are the *proton* and the *neutron*. Of these two, only the proton is charged and the sign of its charge is positive. The total charge of any nucleus is equal to the number of protons times the charge of one proton. The magnitude of the protonic charge may indeed be considered *the* fundamental unit of charge for atomic and nuclear phenomena, since no smaller charge than this has ever been discovered in nature. The nuclear charge expressed in this unit is then equal to the number of protons in the nucleus.

The atoms of all isotopes of any one element have the same number of protons. This number is called the atomic number, Z, and is a characteristic of the element. The nuclei of the different isotopes differ, however, in the number of neutrons and therefore in the total number of nucleons per nucleus. The total number of nucleons, A, is called the *mass number*, and the number of neutrons is therefore $(A - Z)$. Atoms of the different isotopic forms of an element are called nuclides and are distinguished by using the mass number as a superscript to the elementary symbol on either the left or right. Thus ^{15}N or N^{15} refers to the nitrogen of mass number 15.

RELATIVE ATOMIC WEIGHTS

The masses of the individual atoms are very small. Even the heaviest atom has a mass less than 5×10^{-22} g. It has been convenient, therefore, to define a special unit in which the masses of the atoms can be expressed without having to use exponents. This unit is called the *unified atomic mass unit* and is designated by the symbol u. It is defined as exactly 1/12 the mass of a C^{12} atom. Thus the mass of the C^{12} atom is exactly $12u$. The masses of other isotopic atoms can be expressed in this same unit. For example, the mass of the Na^{23} atom is $22.9898\,u$. Compilations of masses of all the atoms of specified isotopic form are available. These are listed as nuclidic masses, and the symbol u is usually omitted. A portion of such a table is given in Chapter 20.

Most chemical reactions do not discriminate among the various isotopes. For example, the percentages of iron atoms which are Fe^{54}, Fe^{56}, Fe^{57}, and Fe^{58} are 5.82, 91.66, 2.19, and 0.33 respectively, in all iron ores, meteorites, and iron compounds prepared synthetically. For chemical purposes it is of interest to know the *average mass* of an iron atom in this natural isotopic mixture. These average masses are also tabulated in terms of the unit u and are given a different name to distinguish them from the nuclidic masses. The average masses are called the *relative atomic weights*, the *chemical atomic weights*, or simply the *atomic weights*. It should be emphasized that they are the averaged masses for the atoms of the elements as they exist in nature. These are the values listed inside the cover of this book and form the basis for practically all chemical weight calculations. The atomic weights are determined either by carrying out precise chemical analyses on a chemical compound of known formula in which all the atomic weights are known but one, or by making a physical determination of the masses of all the stable nuclides of an element and averaging them according to their relative proportions in natural ores of the element.

MOLE

The average chemical experiment involves the reaction of enormous numbers of molecules. It has been convenient to define a new term, the *mole*, as a collection of a large fixed number of fundamental chemical units, comparable to the quantity that might be used in a real experiment. A mole of atoms of any element has been defined as that amount of substance containing the same number of atoms as exactly 12 g of pure C^{12}. This number is called Avogadro's number, N_A; its value, as determined experimentally, is 6.023×10^{23}. The value of the unified atomic mass unit, u, may now be related to the gram scale, since 1 atom of C^{12} has a mass of $12\,u$. Then

$$\text{mass of 1 mole of } C^{12} \text{ atoms} \;=\; N_A \times \text{mass of 1 } C^{12} \text{ atom} \;=\; N_A \times 12\,u$$

$$=\; 12 \text{ g}$$

or
$$1\,u \;=\; \frac{1}{N_A}\text{g} \;=\; \frac{1}{6.023 \times 10^{23}}\text{g} \;=\; 1.660 \times 10^{-24} \text{ g}$$

Let us now consider a mole of atoms of some other element of atomic weight W. The average mass of an atom of this element is Wu, or $\dfrac{W}{N_A}$ g. Then the mass of a mole (N_A) of such atoms is $N_A \times \left(\dfrac{W}{N_A}\right)$ grams, or simply W grams. In other words, the mass in grams of a mole of atoms of any element is equal to the atomic weight. A mole of atoms is often called a *gram-atom*, and is abbreviated *g-at*, and the atomic weight may be given the units *grams per gram-atom*.

SYMBOLS, FORMULAS, FORMULA WEIGHTS

The symbol of an element is used to designate that element, as distinct from other elements. In addition, the symbol is often used to designate a particular amount of that element, one gram-atom. Thus Au sometimes means simply gold. In other contexts it refers to 1 gram-atom, or 197.0 grams of gold.

A formula indicates the relative number of atoms of each element in a substance. Fe_3O_4 refers to a compound in which 3 atoms of iron are present for every 4 atoms of oxygen. If the subscripts are the smallest possible set of integral numbers that express the relative numbers of atoms, the formula is called an empirical formula. Fe_3O_4 is such an example. Fe_3O_4 is a descriptive notation for this particular oxide of iron. Fe_3O_4 is also used in some contexts to mean a particular amount of the compound, the amount that contains 3 gram-atoms of iron and 4 gram-atoms of oxygen. This amount of compound contains N_A simplest formula units (where the formula unit contains 3 iron atoms and 4 oxygen atoms) and is, therefore, called a *mole* of Fe_3O_4. It is also called one *gram-formula weight* (*gfw*). It is composed of (3×55.85) or 167.55 grams of iron and (4×16.00) or 64.00 grams of oxygen combined to make $(167.55 + 64.00)$ or 231.55 grams of compound. 231.55 is called the formula weight of the compound. In general, the formula weight is equal to the sum of numbers, each number being the product of the atomic weight of one of the elements appearing in the compound multiplied by the number of atoms of that element indicated in the formula. The formula weight may be given the units *grams per gram-formula weight*.

MOLECULAR FORMULAS, MOLECULAR WEIGHTS, MOLES

A formula that expresses not only the relative number of atoms of each element but also the actual number of atoms of each element in one molecule of the compound is called a molecular formula. The formula weight is then called the molecular weight.

For example: The molecular formula for benzene is C_6H_6. This indicates that the benzene molecule is composed of 6 atoms of carbon and 6 atoms of hydrogen, and the molecular weight of benzene is $6 \times 12.011 + 6 \times 1.008 = 78.114$.

The empirical or simplest formula for benzene is CH; it indicates only that benzene is composed of the elements carbon and hydrogen in the ratio of 1 atom of C to 1 atom of H. The empirical formula does not represent the actual number of atoms of C and of H in one molecule of benzene, and thus it does not represent the molecular weight of benzene.

To determine the molecular formula of a compound it is necessary to know the molecular weight of the compound. The molecular formula is a whole number multiple $(1, 2, 3,$ etc.$)$ of the empirical formula.

The formula C_6H_6 may be used to mean 78.114 grams of benzene. This amount of compound contains $6 N_A$ atoms (6 g-at) of carbon and $6 N_A$ atoms (6 g-at) of hydrogen. It contains N_A molecules (C_6H_6 units) and is, therefore, called a *mole* of benzene. Avogadro's number may thus be given the units *molecules per mole* as well as *atoms per gram-atom*.

In short, the mass in grams of one mole of any chemical entity is equal to the formula weight of that entity: atom, molecule, ion, or simplest collection of atoms in the fixed proportions of a chemical compound. A mole always contains N_A of the individual chemical units.

When the unit of mass is not the gram, other units may be defined such as the *pound-mole, kilogram-mole, ton-mole,* etc. Thus a *mole* of oxygen is 32 grams of oxygen (O_2), a *pound-mole* of oxygen is 32 pounds of oxygen, a *ton-mole* of oxygen is 32 tons of oxygen, etc. The number of chemical particles is of course different for each kind of mole defined this way. In the rest of this book the *mole* will refer to the *gram-mole* unless a statement is made to the contrary.

Solved Problems

ATOMIC WEIGHTS

4.1. It has been found by mass spectrometric analysis that the relative abundances of the various isotopic atoms of silicon in nature are: 92.21% Si^{28}, 4.70% Si^{29}, and 3.09% Si^{30}. The nuclidic masses of these three species are 27.977, 28.976, and 29.974, respectively. Calculate the atomic weight of silicon from these data.

The atomic weight is the average mass of the three nuclides, each weighted according to its own relative abundance.

$$W = \frac{(92.21 \times 27.977) + (4.70 \times 28.976) + (3.09 \times 29.974)}{100}$$

$$= 28.086$$

Atomic weight problems are among the few types encountered in this book for which the slide rule or four-place logarithm table may seem inadequate. Actually, the computations in this particular case can be performed without resort to a five-place logarithm table by noting that only the first of the three products above need be known to five significant figures. This term can be evaluated as follows:

$$\frac{92.21 \times 27.977}{100} = .9221 \times 27.977 = (1 - .0779) \times 27.977$$

$$= 27.977 - .0779 \times 27.977$$

$$= 27.977 - 2.179 = 25.798$$

The other terms need be evaluated to only 4 and 3 significant figures only, and are 1.362 and .926 respectively.

4.2. Naturally occurring carbon contains two isotopes, C^{12} and C^{13}, the nuclidic masses of which are 12.00000 and 13.0034. What are the percentage abundances of the two isotopes in a sample of carbon whose atomic weight is 12.01112?

Let x = % abundance of C^{13}; then $(100 - x)$ is % of C^{12}.

$$W = 12.01112 = \frac{12.00000(100 - x) + 13.0034x}{100}$$

$$= 12.0000 + \frac{(13.0034 - 12.0000)x}{100} = 12.0000 + .010034x$$

Thus

$$x = \frac{12.01112 - 12.00000}{.010034} = \frac{.01112}{.010034} = 1.109\% \ C^{13}$$

and

$$100 - x = 98.891\% \ C^{12}$$

4.3. Before 1961, a physical atomic weight scale was used whose basis was an assignment of the value 16.00000 to O^{16}. On our present scale, O^{16} has a nuclidic mass of 15.9949. What would have been the physical atomic weight of C^{12} on the old scale?

The ratio of the masses of any two nuclides must be independent of the establishment of the reference point.

$$\left(\frac{W \text{ of } C^{12}}{W \text{ of } O^{16}}\right)_{\text{old scale}} = \left(\frac{W \text{ of } C^{12}}{W \text{ of } O^{16}}\right)_{\text{new scale}} = \frac{12.0000}{15.9949}$$

or

$$(W \text{ of } C^{12})_{\text{old scale}} = 12.0000 \times \frac{16.0000}{15.9949} = 12.0000 \times \left(\frac{15.9949 + .0051}{15.9949}\right)$$

$$= 12.0000 \times \left(1 + \frac{.0051}{15.9949}\right) = 12.0000 + .0038$$

$$= 12.0038$$

4.4. A 1.5276 g sample of $CdCl_2$ was converted to metallic cadmium and non-cadmium-containing products by an electrolytic process. The weight of the metallic cadmium was 0.9367 g. If the atomic weight of chlorine is taken as 35.453, what must be the atomic weight of Cd from this experiment?

Throughout this book we will refer to mass in terms of the chemist's unit, the mole. We will use the symbol n_X to refer to a mole (or gram-atom or gram-formula weight) of X. In this case, we calculate first the number of g-at of Cl in the weighed sample.

$$\text{Weight of } CdCl_2 \quad = \quad 1.5276 \text{ g}$$
$$\text{Weight of Cd in } CdCl_2 \quad = \quad 0.9367 \text{ g}$$
$$\overline{\text{Weight of Cl in } CdCl_2 \quad = \quad 0.5909 \text{ g}}$$

$$n_{Cl} \quad = \quad \frac{0.5909 \text{ g}}{35.453 \text{ g/g-at}} \quad = \quad 0.016667 \text{ g-at}$$

From the formula $CdCl_2$ we see that the number of g-at of Cd is exactly half the number of g-at of Cl.

$$n_{Cd} \;=\; \tfrac{1}{2}n_{Cl} \;=\; \tfrac{1}{2}(0.016667) \;=\; 0.008333 \text{ g-at}$$

The atomic weight is the weight per g-at.

$$W_{Cd} \quad = \quad \frac{0.9367 \text{ g}}{0.008333 \text{ g-at}} \quad = \quad 112.41 \text{ g/g-at}$$

4.5. In a chemical determination of the atomic weight of vanadium, 2.8934 g of pure $VOCl_3$ was allowed to undergo a set of reactions as a result of which all the chlorine contained in this compound was converted to AgCl. The weight of the resulting AgCl was 7.1801 g. Assuming the atomic weights of Ag and Cl are 107.870 and 35.453, what is the experimental value for the atomic weight of vanadium?

This problem is similar to the preceding one, except that n_{Cl} must be obtained by way of n_{AgCl}. The three Cl atoms of $VOCl_3$ are converted to 3 formula units of AgCl, the formula weight of which is 143.323 (the sum of 107.870 and 35.453).

$$n_{AgCl} \quad = \quad \frac{7.1801 \text{ g}}{143.323 \text{ g/gfw}} \quad = \quad 0.050097 \text{ gfw}$$

From the formula, $n_{Cl} \;=\; n_{AgCl} \;=\; 0.050097 \text{ g-at Cl}$

Also, from the formula of $VOCl_3$,

$$n_V \;=\; \tfrac{1}{3}n_{Cl} \;=\; \tfrac{1}{3}(0.050097) \;=\; 0.016699 \text{ g-at V}$$

To find the amount of vanadium in the weighed sample of $VOCl_3$, we must subtract the weights of chlorine and oxygen contained. If we designate the mass of any substance or chemical constituent X by m_X, then $m_X = n_X \times FW_X$, where FW_X is the formula weight of X.

$$m_{Cl} \;=\; n_{Cl} \times W_{Cl} \;=\; 0.050097 \text{ g-at} \times 35.453 \text{ g/g-at} \;=\; 1.7761 \text{ g Cl}$$

Noting from the formula $VOCl_3$ that $n_O = n_V$,

$$m_O \;=\; n_O \times W_O \;=\; 0.016699 \text{ g-at} \times 15.999 \text{ g/g-at} \;=\; 0.2672 \text{ g O}$$

By difference, $m_V \;=\; m_{VOCl_3} - m_O - m_{Cl}$

$$= \;(2.8934 - 0.2672 - 1.7761) \text{ g} \;=\; 0.8501 \text{ g}$$

Then $W_V \;=\; \dfrac{m_V}{n_V} \;=\; \dfrac{0.8501 \text{ g}}{0.016699 \text{ g/g-at}} \;=\; 50.91 \text{ g/g-at}$

FORMULA WEIGHTS AND MOLES

4.6. Determine (a) the molecular weight of sulfuric acid, H_2SO_4, and (b) the formula weight of potassium chloroiridate, K_2IrCl_6.

The molecular weight of a compound is the sum of weights of all the atoms in the molecular formula of the compound.

$$
\begin{array}{llll}
(a) & 2\,\text{H} & = & 2 \times 1.008 & = & 2.016 \\
    & 1\,\text{S} & = & 1 \times 32.064 & = & 32.064 \\
    & 4\,\text{O} & = & 4 \times 15.9994 & = & 63.998 \\
    & \text{Molecular weight} & & & = & 98.078
\end{array}
\qquad
\begin{array}{llll}
(b) & 2\,\text{K} & = & 2 \times 39.1 & = & 78.2 \\
    & 1\,\text{Ir} & = & 1 \times 192.2 & = & 192.2 \\
    & 6\,\text{Cl} & = & 6 \times 35.45 & = & 212.7 \\
    & \text{Formula weight} & & & = & 483.1
\end{array}
$$

Note that the atomic weights are not all known to the same number of significant figures or to the same number of decimal places in u units (unified atomic mass units). In general, the rules for significant figures discussed in Appendix B apply. In (b), there is no point in writing the atomic weight of K as 39.102, since the value for Ir is known to only 0.1 u. Note also that in order to express 6 times the atomic weight of Cl to 0.1 u it was necessary to use the atomic weight to 0.01 u. Similarly, an extra figure was used in the atomic weight for oxygen to give the maximum significance to the last digit in the sum column.

4.7. (a) How many grams of H_2S are contained in 0.400 mole of H_2S?

(b) How many gram-atoms of H and of S are contained in 0.400 mole of H_2S?

(c) How many grams of H and of S are contained in 0.400 mole of H_2S?

(d) How many molecules of H_2S are contained in 0.400 mole H_2S?

(e) How many atoms of H and of S are contained in 0.400 mole H_2S?

Atomic weight of H = 1.01, of S = 32.06. Molecular weight of H_2S = $2 \times 1.01 + 32.06 = 34.08$.

Note that it was not necessary to express the molecular weight to 0.001 u even though the atomic weights are known to this significance. Since the limiting factor in this problem is n_{H_2S}, known to one part in 400, the value 34.08 (expressed to one part in over 3000) for the molecular weight is more than adequate for this problem.

(a) Number of grams of compound = number of moles $\times$ weight of 1 mole

      "   "   "   " H_2S = 0.400 mole $\times$ 34.08 g/mole = 13.63 g H_2S

(b) One mole of H_2S contains 2 gram-atoms of H and 1 gram-atom of S.

       Then 0.400 mole of H_2S contains $2 \times 0.400 = 0.800$ gram-atom H and 0.400 gram-atom S.

(c) Number of grams of element = number of gram-atoms $\times$ weight of 1 gram-atom

      "   "   "   "   H = 0.800 g-atom $\times$ 1.008 g/g-atom = 0.806 g H

      "   "   "   "   S = 0.400 g-atom $\times$ 32.06 g/g-atom = 12.82 g S

(d) Number of molecules = number of moles $\times$ number of molecules in 1 mole

                       = 0.400 mole $\times 6.02 \times 10^{23}$ molecules/mole = 2.41×10^{23} molecules

(e) Number of atoms of element = number of gram-atoms $\times$ number of atoms per gram-atom

      "   "   "   "   H = 0.800 g-at $\times 6.02 \times 10^{23}$ atoms/g-at = 4.82×10^{23} atoms H

      "   "   "   "   S = 0.400 g-at $\times 6.02 \times 10^{23}$ atoms/g-at = 2.41×10^{23} atoms S

4.8. How many gram-atoms are contained in (a) 10.02 g calcium, (b) 92.91 g phosphorus? (c) How many moles of phosphorus are contained in 92.91 g phosphorus if the formula of the molecule is P_4? (d) How many atoms are contained in 92.91 g phosphorus? (e) How many molecules are contained in 92.91 g phosphorus?

Atomic weight of Ca = 40.08, of P = 30.974.

Hence 1 g-at Ca = 40.08 g Ca, 1 g-at P = 30.974 g P.

(a) Gram-atoms of Ca = $\dfrac{\text{grams of Ca}}{\text{atomic weight of Ca}}$ = $\dfrac{10.02\ \text{g}}{40.08\ \text{g/g-atom}}$ = 0.250 g-at Ca

(b) Gram-atoms of P = $\dfrac{\text{grams of P}}{\text{atomic weight of P}}$ = $\dfrac{92.91\ \text{g}}{30.974\ \text{g/g-atom}}$ = 3.000 g-at P

(c) Molecular weight of P_4 = 4×30.974 = 123.90

      Moles of P_4 = $\dfrac{\text{grams of P}}{\text{molecular weight of } P_4}$ = $\dfrac{92.91\ \text{g}}{123.90\ \text{g/mole}}$ = 0.7500 mole P_4

(d) Atoms of P = 3.000 g-atom $\times 6.023 \times 10^{23}$ atoms/g-atom = 1.807×10^{24} atoms P

(e) Molecules of P_4 = 0.7500 moles $\times 6.023 \times 10^{23}$ molecules/mole = 4.517×10^{23} molecules P_4

4.9. How many moles are represented by (a) 9.54 g SO_2, (b) 85.16 g NH_3?

> Atomic weight of S = 32.06, of O = 16.00, of N = 14.007, of H = 1.008.
>
> Molecular weight of SO_2 = 32.06 + 2 × 16.00 = 64.06.
>
> Molecular weight of NH_3 = 14.007 + 3 × 1.008 = 17.031.
>
> Hence, 1 mole of SO_2 = 64.06 g SO_2, and 1 mole of NH_3 = 17.031 g NH_3.

(a) Moles of SO_2 = $\dfrac{\text{grams of } SO_2}{\text{molecular weight of } SO_2}$ = $\dfrac{9.54 \text{ g}}{64.06 \text{ g/mole}}$ = 0.1489 moles SO_2

(b) Moles of NH_3 = $\dfrac{\text{grams of } NH_3}{\text{molecular weight of } NH_3}$ = $\dfrac{85.16 \text{ g}}{17.031 \text{ g/mole}}$ = 5.000 moles NH_3

Supplementary Problems

ATOMIC WEIGHTS

4.10. Naturally occurring argon consists of three isotopes, the atoms of which occur in the following abundances: 0.337% Ar^{36}, 0.063% Ar^{38}, and 99.600% Ar^{40}. The nuclidic masses of these isotopes are 35.968, 37.963, and 39.962 respectively. Calculate the atomic weight of argon from these data. *Ans.* 39.947

4.11. Naturally occurring boron consists of 80.20% B^{11} (nuclidic mass = 11.009) and 19.80% of another isotope. To account for the atomic weight, 10.811, what must be the nuclidic mass of the other stable isotope?   *Ans.* 10.01

4.12. The nuclidic masses of Cl^{35} and Cl^{37} are 34.9689 and 36.9659 respectively. These are the only naturally occurring chlorine isotopes. What percentage distributions account for the atomic weight, 35.4527?   *Ans.* 24.23% Cl^{37}

4.13. The nuclidic masses of N^{14} and N^{15} are 14.0031 and 15.0001 respectively. To account for an atomic weight of 14.0067, what must be the ratio of N^{15} to N^{14} atoms in natural nitrogen?   *Ans.* 0.0036

4.14. At one time there was a chemical atomic weight scale based on the assignment of the value 16.0000 to naturally occurring oxygen. What would have been the atomic weight, on such a table, of silver on the basis of current information? The atomic weights of oxygen and silver on the present table are 15.9994 and 107.870.   *Ans.* 107.874

4.15. The nuclidic mass of Sr^{90} had been determined on the old physical scale (O^{16} = 16.0000) as 89.936. Recompute this to the present atomic weight scale, on which O^{16} is 15.9949.   *Ans.* 89.907

4.16. In a chemical atomic weight determination, the tin content of 3.7692 g of $SnCl_4$ was found to be 1.7170 g. If the atomic weight of chlorine is taken as 35.453, what is the determined value for the atomic weight of tin from this experiment?   *Ans.* 118.64

4.17. A 12.5843 g sample of $ZrBr_4$ was dissolved and, after several chemical steps, all of the combined bromine was precipitated as AgBr. The silver content of the AgBr was found to be 13.2160 g. Assume the atomic weights of silver and bromine to be 107.870 and 79.909. What value was obtained for the atomic weight of Zr from this experiment?   *Ans.* 91.21

4.18. The atomic weight of sulfur was determined by decomposing 6.2984 g of Na_2CO_3 with sulfuric acid and weighing the resultant Na_2SO_4 formed. The weight was found to be 8.4380 g. Assuming known values for the atomic weights of C, O, and Na as 12.011, 15.9994, and 22.990 respectively, what value is computed for the atomic weight of sulfur?   *Ans.* 32.019

FORMULA WEIGHTS AND MOLES

4.19. Determine the molecular weights (or formula weights) to 0.01 u for the following: NaOH, HNO_3, F_2, S_8, $Ca_3(PO_4)_2$, $Fe_4[Fe(CN)_6]_3$.   *Ans.* 40.00, 63.02, 38.00, 256.51, 310.19, 859.28

4.20. How many grams of each of the constituent elements are contained in one mole or one gram-formula weight of the following compounds: (a) CH_4, (b) Fe_2O_3, (c) Ca_3P_2? How many atoms of each element are contained in the same amount of compound?

Ans. (a) 12.01 g C, 4.032 g H; 6.02×10^{23} atoms C, 2.41×10^{24} atoms H

      (b) 111.69 g Fe, 48.00 g O; 1.204×10^{24} atoms Fe, 1.81×10^{24} atoms O

      (c) 120.24 g Ca, 61.95 g P; 1.81×10^{24} atoms Ca, 1.204×10^{24} atoms P

4.21. Calculate the number of grams in a mole (or gram-formula weight) of each of the following common substances: (a) calcite $CaCO_3$, (b) quartz SiO_2, (c) cane sugar $C_{12}H_{22}O_{11}$, (d) gypsum $CaSO_4 \cdot 2 H_2O$, (e) white lead $Pb(OH)_2 \cdot 2 PbCO_3$.   *Ans.* 100.09 g, 60.09 g, 342.3 g, 172.2 g, 775.7 g

4.22. How many pounds are in a pound-mole (or pound-formula weight) of each of the following ores: (a) galena PbS, (b) smithsonite $ZnCO_3$, (c) malachite $CuCO_3 \cdot Cu(OH)_2$?

Ans. 239 lb, 125 lb, 221 lb

4.23. What is the average weight of (a) one hydrogen atom, (b) one oxygen atom, (c) one uranium atom?

Ans. (a) 1.67×10^{-24} g, (b) 2.66×10^{-23} g, (c) 3.95×10^{-22} g

4.24. What is the weight of one molecule of (a) CH_3OH, (b) $C_{60}H_{122}$, (c) $C_{1200}H_{2000}O_{1000}$?

Ans. (a) 5.32×10^{-23} g, (b) 1.40×10^{-21} g, (c) 5.38×10^{-20} g

4.25. How many gram-atoms of the element are contained in (a) 32.7 g Zn, (b) 7.09 g Cl, (c) 95.4 g Cu, (d) 4.31 g Fe, (e) 0.378 g S?   *Ans.* 0.500, 0.200, 1.50, 0.0772, 0.0118 g-atom

4.26. How many moles are represented by: (a) 24.5 g H_2SO_4, (b) 4.00 g O_2?

Ans. 0.250 mole, 0.125 mole

4.27. (a) How many gram-atoms of Ba and of Cl are contained in 107.0 g of $Ba(ClO_3)_2 \cdot H_2O$?

(b) How many molecules of water of hydration are in this same amount?

Ans. (a) 0.332 g-at Ba, 0.664 g-at Cl, (b) 2.00×10^{23} molecules H_2O

4.28. How many gram-atoms of Fe and of S are contained in (a) 1 g-f wt of FeS_2 (pyrite), (b) 1 kg of FeS_2? (c) How many grams of S are contained in exactly 1 kg of FeS_2?

Ans. (a) 1 g-at Fe, 2 g-at S, (b) 8.33 g-at Fe, 16.7 g-at S, (c) 535 g S

Chapter 5

Formulas and Composition Calculations

INTRODUCTION

It will be assumed in this and all succeeding chapters that the atomic weights are known. Values from the inside front cover of this book should be taken except where special attention is given to samples of matter not containing the normal isotopic distribution. If the atomic weights are known, a chemical formula automatically implies full information as to the composition. *Vice versa*, full information about the weight composition of a compound implies the empirical formula. In this chapter, we deal with problems which relate the formula to the composition and relate one system of expressing composition to another.

EMPIRICAL FORMULAS

An empirical formula expresses the relative numbers of atoms of the different elements in a compound with the smallest possible set of integers. These integers may be found by converting analytical weight composition data to the n values, or number of gram-atoms, of each element in some fixed weight of the compounds. Consider a compound which analyzes 17.09% magnesium, 37.93% aluminum, and 44.98% oxygen. (Unless stated to the contrary, percentage is a *weight* percentage, i.e. number of grams of the element per 100 g of the compound.) The following tabular form is a convenient systematic scheme for handling the data.

(1) E (element)	(2) m_E (mass of E per fixed amount of compound, in this case 100 g)	(3) W_E (Atomic weight of E)	(4) n_E (number of gram-atoms of E) $= \dfrac{m_E}{W_E}$	(5) $\dfrac{n_E}{\text{smallest } n_E}$ $= \dfrac{n_E}{.703}$
Mg	17.09 g	24.31 g/g-at	.703 g-at	1.00
Al	37.93 g	26.98 g/g-at	1.406 g-at	2.00
O	44.98 g	16.00 g/g-at	2.812 g-at	4.00

The numbers in column (4) give the numbers of g-at of the component elements in the fixed weight of compound, 100 g, chosen as the basis. Any set of numbers obtained by multiplying or dividing each of the numbers in column (4) by the same factor will be in the same ratio to each other as the numbers in (4). The set in column (5) is such a set, obtained by dividing each of the n_E values in (4) by the *smallest* entry in (4), .703. The numbers in column (5) show that the relative numbers of gram-atoms, and therefore of atoms, of Mg, Al, and O in this compound are $1:2:4$. Therefore the empirical formula is $MgAl_2O_4$.

COMPOSITION FROM FORMULA

The existence of a formula for a compound implies that fixed relationships exist between the weights of any two elements in the compound or between the weight of any element and the weight of the compound as a whole. These relationships can best be seen by writing the formula in vertical form, as illustrated with the compound Al_2O_3.

(1)	(2)	(3)	(4)	(5)
	n per gfw of compound	W_E (atomic weight of element)	m_E per gfw of compound $= n_E \times W_E$	m_E per g compound
Al_2	2 g-at	27.0 g/g-at	54.0 g	$\dfrac{54.0 \text{ g Al}}{102.0 \text{ g Al}_2\text{O}_3} = .529 \text{ g Al/g Al}_2\text{O}_3$
O_3	3 g-at	16.0 g/g-at	48.0 g	$\dfrac{48.0 \text{ g O}}{102.0 \text{ g Al}_2\text{O}_3} = .471 \text{ g O/g Al}_2\text{O}_3$
Al_2O_3	1 gfw		FW = 102.0	Check: 1.000

The sum of the entries in column (4) for the elements equals the formula weight (FW) of the compound. The entries in column (5) represent the *fractional* content of the various elements in the compound. These numbers are really dimensionless (g/g) and are the same in any unit of mass. Thus, 1 gram (pound, ton, etc.) of Al_2O_3 contains 0.529 gram (pound, ton, etc.) of Al and 0.471 gram (pound, ton, etc.) of O. It is obvious that the sum of the constituent fractions of any compound must equal 1.000.

The *percentage* of aluminum in Al_2O_3 is the number of parts by weight of Al in 100 parts by weight of Al_2O_3. It follows that the *percentage* is expressed by a number 100 times as great as the fraction. Thus, the percentages of aluminum and oxygen are 52.9% and 47.1%, respectively. The sum of the constituent percentages of any compound must equal 100.0%.

Sometimes the composition of a substance with respect to a particular element is expressed in terms of a simple compound that can be prepared from that element. For example, the aluminum content of a glass may be expressed as 1.3% Al_2O_3. This means that if all the aluminum contained in 100 g of the glass were converted to Al_2O_3, the weight of the Al_2O_3 would be 1.3 g. This convention is not meant to imply that the aluminum in the glass is in the chemical form Al_2O_3. The use of the oxide notation may have arisen for analytical convenience if the aluminum had been determined by separation, conversion to Al_2O_3, and weighing of the oxide. In many cases oxide notations are the result of historical errors in the assignment of chemical structures to complex substances. Whatever the origin, it is a straightforward procedure to convert the data in such a form to direct elementary composition, or *vice versa*, by the use of such a *quantitative factor* as found in column (5). The ratio $\dfrac{54.0 \text{ g Al}}{102.0 \text{ g Al}_2\text{O}_3}$ and its reciprocal, $\dfrac{102.0 \text{ g Al}_2\text{O}_3}{54.0 \text{ g Al}}$, are called quantitative factors, and may be used as unit factors in numerical problems. Although the numerator and denominator of one of these factors are not equal to each other in the strict sense that one inch always equals 2.54 cm, in the *particular* compound Al_2O_3, 54.0 g Al implies 102.0 g Al_2O_3, and *vice versa*. The two quantities are said to be chemically *equivalent* in this compound, and the quantitative factors may be used as unit conversion factors for all problems involving Al_2O_3.

NON-STOICHIOMETRIC FACTORS

A similar use of conversion factors limited to particular cases is common even when the relative proportions are not fixed by a chemical formula. Consider a silver alloy used for jewelry containing 86% silver. Factors based on this composition, such as $\dfrac{.86 \text{ g Ag}}{1 \text{ g alloy}}$ or $\dfrac{100 \text{ g alloy}}{86 \text{ g Ag}}$, may be used as unit factors for all problems involving alloys of this particular composition.

Solved Problems

CALCULATION OF FORMULAS

5.1. Derive the empirical formula of a hydrocarbon that on analysis gave the following percentage composition: C = 85.63%, H = 14.37%.

The tabular solution, as applied to 100 g of compound, is as follows:

E	m_E	W_E	$n_E = \dfrac{m_E}{W_E}$	$\dfrac{n_E}{7.129 \text{ g-at}}$
C	85.63 g	12.011 g/g-at	7.129 g-at	1.000
H	14.37 g	1.008 g/g-at	14.26 g-at	2.000

where E = element, m_E = mass of element per 100 g of compound, W_E = atomic weight of element, n_E = number of g-at of element per 100 g of compound.

The procedure of dividing each n_E by n_C is equivalent to finding the number of atoms of each element for every one atom of carbon. The ratio of H to C atoms is 2:1. Hence, the empirical formula is CH_2. The molecular formula might be CH_2, C_2H_4, C_3H_6, C_4H_8, etc., since each of these formulas implies the same percentage composition as CH_2.

5.2. A compound on analysis gave the following percentage composition: K = 26.57%, Cr = 35.36%, O = 38.07%. Derive the empirical formula of the compound.

The standard tabular solution, as applied to 100 g of compound, follows:

(1) E	(2) m_E	(3) W_E	(4) $n_E = \dfrac{m_E}{W_E}$	(5) $\dfrac{n_E}{0.6800 \text{ g-at}}$	(6) $\dfrac{n_E}{0.6800 \text{ g-at}} \times 2$
K	26.57 g	39.10 g/g-at	0.6800 g-at	1.000	2
Cr	35.36 g	52.00 g/g-at	0.6800 g-at	1.000	2
O	38.07 g	16.00 g/g-at	2.379 g-at	3.499	7

In distinction to the previous examples, the numbers in column (5) are not all integers. The ratio of the numbers of atoms of two elements in a compound must be the ratio of small whole numbers, in order to satisfy one of the postulates of Dalton's atomic theory. Allowing for experimental and calculational uncertainty, we say that the entry for oxygen in column (5), 3.499, is within the allowed error of being 3.500 or 7/2, indeed the ratio of small whole numbers. By rounding off in this way and multiplying each of the entries in column (5) by 2, we arrive at the smallest set of integers that correctly represent the relative numbers of atoms in the compound, as tabulated in column (6). The formula is thus $K_2Cr_2O_7$.

5.3. A 15.00 gram sample of an unstable hydrated salt, $Na_2SO_4 \cdot x\,H_2O$, was found to contain 7.05 grams of water. Determine the empirical formula of the salt.

15.00 g of the hydrated salt contains 7.05 g H_2O and (15.00 − 7.05) = 7.95 g Na_2SO_4.

Hydrates are compounds containing water molecules loosely bound to the other components. H_2O may usually be removed intact by heating such compounds and may then be replaced by wetting. The Na_2SO_4 and H_2O groups may thus be considered as the units of which the compound is made. Formula weights and moles are used in place of atomic weights and gram-atoms in the previous problems. Another slight difference is that the analytical data are expressed on the basis of 15.00 grams, rather than 100 grams of compound. Since any fixed weight of compound is equally satisfactory, there is no need to convert to percentages. The tabulation for 15.00 grams of compound follows.

X	m_X	FW_X	$n_X = \dfrac{m_X}{FW_X}$	$\dfrac{n_X}{0.0559 \text{ mole}}$
Na_2SO_4	7.95 g	142.1 g/mole	0.0559 mole	1.00
H_2O	7.05 g	18.02 g/mole	0.391 mole	6.99

The mole ratio of H_2O to Na_2SO_4 is within the allowed error of being 7 to 1, and the empirical formula is $Na_2SO_4 \cdot 7 H_2O$.

5.4. A 2.500 gram sample of uranium was heated in the air. The resulting oxide weighed 2.949 g. Determine the empirical formula of the oxide.

2.949 g of the oxide contains 2.500 g U and $(2.949 - 2.500) = .449$ g O.

A calculation on the basis of 2.949 g of the oxide shows 2.672 g-at oxygen per g-at uranium. The smallest multiplying integer that will give whole numbers is 3.

$$\frac{n_O}{n_U} = \frac{2.672 \text{ g-at O}}{1.000 \text{ g-at U}} = \frac{3 \times 2.672 \text{ g-at O}}{3 \times 1.000 \text{ g-at U}} = \frac{8.02 \text{ g-at O}}{3.00 \text{ g-at U}}$$

The empirical formula is U_3O_8.

Emphasis must be placed on the importance of carrying out the computations to as many significant figures as the analytical precision requires. If the numbers in the ratio $1:2.67$ had been multiplied by 2 to give $2:5.34$ and these numbers had been rounded off to $2:5$, the wrong formula would have been obtained. This would not have been justified because it would have assumed an error of 34 parts in 500 in the analysis of oxygen. The weight of oxygen, 0.449 g, indicates a possible error of only a few parts in 500. When the multiplying factor 3 was used, the rounding off was from 8.02 to 8.00, the assumption being made that the analysis of oxygen may have been in error by 2 parts in 800; this amount of error is more reasonable.

5.5. A 1.367 g sample of an organic compound was combusted in a stream of air to yield 3.002 g CO_2 and 1.640 g H_2O. If the original compound contained only C, H and O, what is its empirical formula?

It is necessary to use quantitative factors for CO_2 and H_2O to find how much C and H are present in the combustion products and, thus, in the original sample.

$$m_C = \frac{1 \text{ g-at C}}{1 \text{ mole } CO_2} \times 3.002 \text{ g } CO_2 = \frac{12.01 \text{ g C}}{44.01 \text{ g } CO_2} \times 3.002 \text{ g } CO_2 = .819 \text{ g C}$$

$$m_H = \frac{2 \text{ g-at H}}{1 \text{ mole } H_2O} \times 1.640 \text{ g } H_2O = \frac{2 \times 1.008 \text{ g H}}{18.02 \text{ g } H_2O} \times 1.640 \text{ g } H_2O = .1835 \text{ g H}$$

The amount of oxygen in the original sample cannot be obtained from the weight of combustion products, since the CO_2 and H_2O contain oxygen that came partly from the combined oxygen in the compound and partly from the air stream used in the combustion process. The oxygen content of the sample can be obtained, however, by difference.

$$m_O = m_{compd.} - m_C - m_H = 1.367 - .819 - .184 = .364 \text{ g}$$

The problem can now be solved by the usual procedures. The numbers of g-at of the elements in 1.367 g compound are found to be: C, 0.0682; H, 0.1820; O, 0.0228. These numbers are in the ratio $3:8:1$, and the empirical formula is C_3H_8O.

COMPOSITION PROBLEMS

5.6. A strip of electrolytically pure copper weighing 3.178 g is strongly heated in a stream of oxygen until it is all converted to the black oxide. The resultant black powder weighs 3.978 g. What is the percentage composition of this oxide?

Total weight of black oxide $= 3.978$ g

Weight of copper in oxide $= 3.178$ g

Weight of oxygen in oxide $= 0.800$ g

$$\text{Fraction of copper} \ = \ \frac{\text{weight of copper in oxide}}{\text{total weight of oxide}} \ = \ \frac{3.178 \text{ g}}{3.978 \text{ g}} \ = \ 0.799 \ = \ 79.9\%$$

$$\text{Fraction of oxygen} \ = \ \frac{\text{weight of oxygen in oxide}}{\text{total weight of oxide}} \ = \ \frac{0.800 \text{ g}}{3.978 \text{ g}} \ = \ 0.201 \ = \ 20.1\%$$

$$\text{Check:} \qquad 100.0\%$$

This means that 1 part by weight of oxide contains 0.799 part by weight of copper and 0.201 part by weight of oxygen; or 100 parts by weight of oxide contain 79.9 parts by weight of copper and 20.1 parts by weight of oxygen.

Note that the percentages are obtained by multiplying 0.799 and 0.201 by 100.

If a slide rule is used, the calculations may not be so precise, but the sum of the percentages for the check should be between 99.8 and 100.2%.

5.7. (a) Determine the percentage of iron in each of the following compounds: $FeCO_3$, Fe_2O_3, Fe_3O_4.

(b) How many grams of iron could be obtained from 2000 grams of Fe_2O_3?

(a) Formula weight of $FeCO_3$ = 115.86, of Fe_2O_3 = 159.70, of Fe_3O_4 = 231.55.

$$\text{Fraction of Fe in } FeCO_3 \ = \ \frac{1 \text{ atomic weight Fe}}{\text{formula weight } FeCO_3} \ = \ \frac{55.85}{115.86} \ = \ 0.4820 \ = \ 48.20\%$$

$$\text{Fraction of Fe in } Fe_2O_3 \ = \ \frac{2 \text{ atomic weights Fe}}{\text{formula weight } Fe_2O_3} \ = \ \frac{2 \times 55.85}{159.70} \ = \ 0.6994 \ = \ 69.94\%$$

$$\text{Fraction of Fe in } Fe_3O_4 \ = \ \frac{3 \text{ atomic weights Fe}}{\text{formula weight } Fe_3O_4} \ = \ \frac{3 \times 55.85}{231.55} \ = \ 0.7237 \ = \ 72.37\%$$

(b) Weight of Fe in 2000 grams Fe_2O_3 = 0.6994 × 2000 grams = 1399 grams Fe

Quantitative Factor Method.

Weight of Fe in 2000 grams Fe_2O_3

$$= \ 2000 \text{ g } Fe_2O_3 \times \frac{2 \text{ g-at Fe}}{1 \text{ gfw } Fe_2O_3} \ = \ 2000 \ \cancel{\text{g } Fe_2O_3} \times \frac{2 \times 55.85 \text{ g Fe}}{159.70 \ \cancel{\text{g } Fe_2O_3}} \ = \ 1399 \text{ g Fe}$$

Note that the Fe_2O_3's cancel out, leaving the answer in terms of g Fe.

Gram-Formula Weight Method.

A slight modification of this procedure achieves the same result in a more detailed fashion by invoking the explicit computation of n, the number of g-at or gfw of each substance.

$$n_{Fe_2O_3} \ = \ \frac{2000 \text{ g } Fe_2O_3}{159.70 \text{ g } Fe_2O_3/\text{gfw } Fe_2O_3} \ = \ 12.524 \text{ gfw } Fe_2O_3$$

By examination of the formula, we can write

$$n_{Fe} \ = \ 2 \, n_{Fe_2O_3} \ = \ 2 \times 12.524 \ = \ 25.05 \text{ g-at Fe}$$

Then the required weight of Fe = 25.05 g-at Fe × 55.85 g Fe/g-at Fe = 1399 g Fe.

5.8. Given the formula K_2CO_3, determine the percentage composition of potassium carbonate.

One formula weight of K_2CO_3 contains

2 atomic weights of K	=	2 × 39.102 =	78.204 parts by weight of K
1 atomic weight of C	=	1 × 12.011 =	12.011 parts by weight of C
3 atomic weights of O	=	3 × 15.999 =	47.997 parts by weight of O
Formula weight of K_2CO_3		=	138.212 parts by weight

$$\text{Fraction of K in K}_2\text{CO}_3 = \frac{2 \text{ atomic weights K}}{\text{formula weight K}_2\text{CO}_3} = \frac{2 \times 39.102}{138.21} = 0.5658 = 56.58\%$$

$$\text{Fraction of C in K}_2\text{CO}_3 = \frac{1 \text{ atomic weight C}}{\text{formula weight K}_2\text{CO}_3} = \frac{12.011}{138.21} = 0.0869 = 8.69\%$$

$$\text{Fraction of O in K}_2\text{CO}_3 = \frac{3 \text{ atomic weights O}}{\text{formula weight K}_2\text{CO}_3} = \frac{3 \times 15.999}{138.21} = 0.3473 = 34.73\%$$

Check: 100.00%

5.9. (a) Calculate the percentage of CaO in $CaCO_3$.

(b) How many pounds of CaO can be obtained from one ton of limestone that is 97.0% $CaCO_3$?

(a) Formula weight of $CaCO_3$ = 100.1, of CaO = 56.1.

The quantitative factor can be written by considering the conservation of Ca atoms. 1 gfw CaO contains the same amount of Ca (1 g-at) as 1 gfw $CaCO_3$.

$$\text{Fraction of CaO in CaCO}_3 = \frac{\text{formula weight CaO}}{\text{formula weight CaCO}_3} = \frac{56.1}{100.1} = 0.560 = 56.0\%$$

(b) Weight of $CaCO_3$ in 1 ton limestone $= 0.970 \times 2000 \text{ lb} = 1940 \text{ lb CaCO}_3$

Weight of CaO $=$ fraction of CaO in $CaCO_3 \times$ weight of $CaCO_3$
$= 0.560 \times 1940 \text{ lb} = 1090 \text{ lb CaO in 1 ton limestone}$

5.10. How much 58.0% sulfuric acid solution is needed to provide 150 g of H_2SO_4?

$$\frac{100 \text{ g solution}}{58.0 \text{ g H}_2\text{SO}_4} = \frac{x \text{ g solution}}{150 \text{ g H}_2\text{SO}_4}$$

or $\qquad x \text{ g solution} = 150 \text{ g H}_2\text{SO}_4 \times \dfrac{100 \text{ g solution}}{58.0 \text{ g H}_2\text{SO}_4} = 259 \text{ g solution}$

Note that "g H_2SO_4" cancels out, leaving the answer in terms of "g solution". Note also that the last equation above, derived from the preceding one by a rearrangement of terms, could have been written directly by a factor-label procedure.

5.11. How much calcium is in the amount of $Ca(NO_3)_2$ that contains 20.0 g of nitrogen?

It is not necessary to find the weight of $Ca(NO_3)_2$ containing 20.0 g of nitrogen. A relationship between two component elements of a compound may be found directly from the formula.

$$\text{Weight of Ca} = 20.0 \text{ g N} \times \frac{1 \text{ g-at Ca}}{2 \text{ g-at N}} = 20.0 \text{ g N} \times \frac{40.1 \text{ g Ca}}{2 \times 14.0 \text{ g N}} = 28.6 \text{ g Ca}$$

5.12. (a) How much H_2SO_4 could be produced from 500 kg of sulfur?

(b) How many kilograms of Glauber's salt, $Na_2SO_4 \cdot 10\,H_2O$, could be obtained from 1.000 kg H_2SO_4?

Atomic weight of S = 32.06. Formula weight of H_2SO_4 = 98.08, of $Na_2SO_4 \cdot 10\,H_2O$ = 322.2.

(a) The formula H_2SO_4 indicates that 1 gram-atom S (32.06 g S) will give 1 mole H_2SO_4 (98.08 g H_2SO_4). Then

32.06 kg S will give 98.08 kg H_2SO_4

1 kg S will give $\dfrac{98.08}{32.06}$ kg H_2SO_4

and $\qquad$ 500 kg S will give $500 \times \dfrac{98.08}{32.06}$ kg $H_2SO_4 = 1530$ kg H_2SO_4

(b) One mole of H_2SO_4 (98.08 g H_2SO_4) will give 1 gfw $Na_2SO_4 \cdot 10\,H_2O$ (322.2 g $Na_2SO_4 \cdot 10\,H_2O$), since each substance contains one sulfate (SO_4) group. Then

98.08 kg H_2SO_4 will give 322.2 kg $Na_2SO_4 \cdot 10\,H_2O$

and $\qquad$ 1 kg H_2SO_4 will give $\dfrac{322.2}{98.08} = 3.285$ kg $Na_2SO_4 \cdot 10\,H_2O$

Another Method.

(a) Number of kg of H_2SO_4 produced from 500 kg of S

$$= \ 500 \text{ kg } S \times \frac{\text{formula weight } H_2SO_4}{\text{atomic weight } S} \ = \ 500 \text{ kg} \times \frac{98.08}{32.06} \ = \ 1530 \text{ kg } H_2SO_4$$

Note that the S's cancel out, leaving the answer in terms of kg H_2SO_4.

(b) Number of kg of $Na_2SO_4 \cdot 10\,H_2O$ produced from 1 kg of H_2SO_4

$$= \ 1 \text{ kg } H_2SO_4 \times \frac{\text{formula wt } Na_2SO_4 \cdot 10\,H_2O}{\text{formula wt } H_2SO_4} \ = \ 1 \text{ kg} \times \frac{322.2}{98.08}$$

$$= \ 3.285 \text{ kg } Na_2SO_4 \cdot 10\,H_2O$$

The H_2SO_4's cancel out, leaving the answer in terms of kg $Na_2SO_4 \cdot 10\,H_2O$.

5.13. How many tons of $Ca_3(PO_4)_2$ must be treated with carbon and sand in an electric furnace to make one ton of phosphorus?

Atomic weight of P = 30.97;   formula weight of $Ca_3(PO_4)_2$ = 310.2.

The formula $Ca_3(PO_4)_2$ indicates that 2 g-at of P (2×30.97 g = 61.94 g P) is contained in 1 gfw of $Ca_3(PO_4)_2$ (310.2 g $Ca_3(PO_4)_2$).   Then

61.94 tons P is contained in 310.2 tons $Ca_3(PO_4)_2$

and   1 ton P is contained in $\dfrac{310.2}{61.94}$ = 5.01 tons $Ca_3(PO_4)_2$

5.14. A 5.82 gram silver coin is dissolved in nitric acid. When sodium chloride is added to the solution all the silver is precipitated as AgCl. The AgCl precipitate weighs 7.20 grams. Determine the percentage of silver in the coin.

$$\text{Fraction of Ag in AgCl} \ = \ \frac{\text{atomic weight Ag}}{\text{formula weight AgCl}} \ = \ \frac{107.9}{143.3} = 0.753.$$

Weight of Ag in 7.20 g AgCl = 0.753×7.20 g = 5.42 g Ag.

Hence the 5.82 g coin contains 5.42 g Ag.

$$\text{Fraction of Ag in coin} \ = \ \frac{\text{weight of Ag in coin}}{\text{total weight of coin}} \ = \ \frac{5.42 \text{ g}}{5.82 \text{ g}} \ = \ 0.931 \ = \ 93.1\% \text{ Ag}$$

5.15. A sample of impure sulfide ore contains 42.34% Zn. Find the percentage of pure ZnS in the sample.

Atomic weight of Zn = 65.37;   formula weight of ZnS = 97.43.

The formula ZnS shows that 1 formula weight ZnS contains 1 atomic weight Zn.

$$\text{Fraction of Zn in ZnS} \ = \ \frac{\text{atomic weight Zn}}{\text{formula weight ZnS}} \ = \ \frac{65.37}{97.43} \ = \ 0.6709 \ = \ 67.09\%$$

If the sample were 100% ZnS, it would contain 67.09% Zn. But since the sample contains only 42.34% Zn, it is $\dfrac{42.34}{67.09} \times 100\%$ = 63.11% pure ZnS.

5.16. A sample of potato starch was ground in a ball mill to give a starch-like molecule of lower molecular weight. The product analyzed 0.086% phosphorus. If each molecule is assumed to contain one atom of phosphorus, what is the average molecular weight of the material?

One gram-atom of phosphorus (31.0 g P) is contained in 1 mole of the material.

Since 0.086 g P is contained in 100 g of material, then

$$31.0 \text{ g P is contained in } \ \frac{31.0}{0.086} \times 100 \text{ g} \ = \ 3.6 \times 10^4 \text{ g of material}$$

Hence the average molecular weight of the material is 3.6×10^4.

Another Method.

The problem can be stated as an equation, "How many grams equals 1 mole?" Successive multiplications and divisions by proper conversion factors will transform 1 mole into the corresponding number of grams.

$$x \text{ g starch} = 1 \text{ mole starch} \times \frac{1 \text{ g-at P}}{1 \text{ mole starch}} \times \frac{31.0 \text{ g P}}{1 \text{ g-at P}} \times \frac{1 \text{ g starch}}{0.00086 \text{ g P}}$$

$$= 3.6 \times 10^4 \text{ g starch}$$

Note that "mole starch", "g-at P" and "g P" cancel out, leaving the answer in terms of "g starch".

5.17. A granulated sample of aircraft alloy (Al, Mg, Cu) weighing 8.72 g was first treated with alkali to dissolve the aluminum, then with very dilute HCl to dissolve the magnesium, leaving a residue of copper. The residue after alkali-boiling weighed 2.10 g, and the acid-insoluble residue from this weighed 0.69 g. What is the composition of the alloy?

$$\text{Weight of Al} = 8.72 \text{ g} - 2.10 \text{ g} = 6.62 \text{ g}$$
$$\text{Weight of Mg} = 2.10 \text{ g} - 0.69 \text{ g} = 1.41 \text{ g}$$
$$\text{Weight of Cu} = 0.69 \text{ g}$$

$$\text{Fraction of Al} = \frac{\text{weight of Al}}{\text{weight of sample}} = \frac{6.62 \text{ g}}{8.72 \text{ g}} = 0.759 = 75.9\%$$

$$\text{Fraction of Mg} = \frac{\text{weight of Mg}}{\text{weight of sample}} = \frac{1.41 \text{ g}}{8.72 \text{ g}} = 0.162 = 16.2\%$$

$$\text{Fraction of Cu} = \frac{\text{weight of Cu}}{\text{weight of sample}} = \frac{0.69 \text{ g}}{8.72 \text{ g}} = 0.079 = 7.9\%$$

$$\text{Check:} \quad 100.0\%$$

5.18. A Pennsylvania bituminous coal is analyzed as follows: Exactly 2.500 g is weighed into a fused silica crucible. After drying for one hour at 110°C the moisture-free residue weighs 2.415 g. The crucible next is covered with a vented lid and strongly heated until no volatile matter remains. The residual coke button weighs 1.528 g. The crucible is then heated without the cover until all specks of carbon have disappeared, and the final ash weighs 0.245 g. What is the Proximate Analysis of this coal, that is – percents of moisture, volatile combustible matter (VCM), fixed carbon (FC), and ash?

$$\text{Moisture} = 2.500 \text{ g} - 2.415 \text{ g} = 0.085 \text{ g Moisture}$$
$$\text{VCM} = 2.415 \text{ g} - 1.528 \text{ g} = 0.887 \text{ g VCM}$$
$$\text{FC} = 1.528 \text{ g} - 0.245 \text{ g} = 1.283 \text{ g FC}$$
$$\text{Ash} = 0.245 \text{ g Ash}$$
$$\text{Total} = 2.500 \text{ g Coal}$$

$$\text{Fraction of moisture} = \frac{\text{weight of moisture}}{\text{weight of coal}} = \frac{0.085 \text{ g}}{2.500 \text{ g}} = 0.034 = 3.4\%$$

$$\text{Fraction of VCM} = \frac{\text{weight of VCM}}{\text{weight of coal}} = \frac{0.887 \text{ g}}{2.500 \text{ g}} = 0.355 = 35.5\%$$

Similarly, the other percentages are calculated to be: 51.3% FC, 9.8% Ash.

5.19. On the "dry basis" a sample of coal analyzes as follows: volatile combustible matter, 21.06%; fixed carbon, 71.80%; ash, 7.14%. If the moisture present in the coal is 2.49%, what is the analysis on the "wet basis"?

On the wet basis, % water = 2.49%. Then % of other components = 100% − 2.49% = 97.5% = 0.975 of percentages in dry sample.

On wet basis: % volatile matter $= 0.975 \times 21.06\% = 20.5\%$

% carbon $= 0.975 \times 71.80\% = 70.0\%$

% ash $= 0.975 \times 7.14\% = 7.0\%$

% water $= 2.5\%$

Check: $\overline{100.0\%}$

5.20. When the Bayer process is used for recovering aluminum from siliceous ores, some aluminum is always lost because of the formation of an unworkable "mud" having the following average formula: $3\,Na_2O \cdot 3\,Al_2O_3 \cdot 5\,SiO_2 \cdot 5\,H_2O$. Since aluminum and sodium ions are always in excess in the solution from which this precipitate is formed, the precipitation of the silicon in the "mud" is complete. A certain ore contained 13% (by weight) kaolin $(Al_2O_3 \cdot 2\,SiO_2 \cdot 2\,H_2O)$ and 87% gibbsite $(Al_2O_3 \cdot 3\,H_2O)$. (a) What percent of the total aluminum in this ore is recoverable in the Bayer process? (b) How much aluminum is recoverable from 100 g of ore?

Formula weight of kaolin, $Al_2O_3 \cdot 2\,SiO_2 \cdot 2\,H_2O = 258$; of gibbsite, $Al_2O_3 \cdot 3\,H_2O = 156$.

(a) 100 g ore $= $ 13 g kaolin $+$ 87 g gibbsite.

Weight of Al in 13 g kaolin $=$ 13 g kaolin $\times \dfrac{2 \text{ gfw Al}}{1 \text{ gfw kaolin}} = 13 \times \dfrac{54.0}{258}$ g Al $= 2.7$ g Al

Weight of Al in 87 g gibbs. $=$ 87 g gibbs. $\times \dfrac{2 \text{ gfw Al}}{1 \text{ gfw gibbsite}} = 87 \times \dfrac{54.0}{156}$ g Al $= 30.1$ g Al

Total weight of Al in 100 g ore $= 2.7$ g $+ 30.1$ g $= 32.8$ g.

Kaolin has equal numbers of Al and Si atoms, and 13 g kaolin contains 2.7 g Al. The mud takes 6 Al atoms for 5 Si atoms. Hence the precipitation of all the Si from 13 g kaolin involves the loss of $6/5 \times 2.7$ g $= 3.2$ g Al.

Fraction of Al recoverable $= \dfrac{\text{recoverable Al}}{\text{total Al}} = \dfrac{(32.8 - 3.2)\text{g}}{32.8 \text{ g}} = 0.90 = 90\%$

(b) Weight of Al recoverable from 100 g ore $= 32.8$ g $- 3.2$ g $= 30$ g.

5.21. A clay was partially dried and then contained 50% silica and 7% water. The original clay contained 12% water. What is the percentage of silica in the original sample?

There is 50 g of silica in $(100 - 7)$g $= 93$ g of dry clay.

Also, there is $(100 - 12)$g $= 88$ g of dry constituents in 100 g of original clay.

Fraction of silica in original clay

$=$ fraction of dry constituents in original clay $\times$ fraction of silica in dry clay

$= \dfrac{88 \text{ g dry clay}}{100 \text{ g original}} \times \dfrac{50 \text{ g silica}}{93 \text{ g dry clay}} = 0.47 \dfrac{\text{silica}}{\text{original}} = 47\%$ silica in original.

Note that the term "dry clay" cancels out, leaving as an answer 0.47 part silica per part original clay, or 47% silica in original clay.

5.22. Two unblended manganese ores contain 40% and 25% of manganese, respectively. How many pounds of each ore must be mixed to give 100 lb of blended ore containing 35% of manganese?

Let $x =$ pounds of 40% ore required; then $100 - x =$ pounds of 25% ore required.

Mn from 40% ore $+$ Mn from 25% ore $=$ total Mn in 100 lb of mixture

$0.40x$ lb Mn $+ 0.25(100 - x)$lb Mn $= 0.35 \times 100$ lb Mn

Solving, $x = 67$ lb of 40% ore. Then $100 - x = 33$ lb of 25% ore.

5.23. A nugget of gold and quartz weighs 100 g. Specific gravity of gold = 19.3, of quartz = 2.65, of nugget = 6.4. Determine the weight of gold in the nugget.

Let x = grams of gold in nugget; then $(100 \text{ g} - x)$ = grams of quartz in nugget.

Volume of nugget = volume of gold in nugget + volume of quartz in nugget

$$\frac{100 \text{ g}}{6.4 \text{ g/cm}^3} = \frac{x}{19.3 \text{ g/cm}^3} + \frac{100 \text{ g} - x}{2.65 \text{ g/cm}^3} \quad \text{from which} \quad x = 68 \text{ g gold}$$

Supplementary Problems

CALCULATION OF FORMULAS

5.24. Derive the empirical formulas of the substances having the following percentage composition: (a) Fe = 63.53%, S = 36.47%; (b) Fe = 46.55%, S = 53.45%; (c) Fe = 53.73%, S = 46.27%.
Ans. (a) FeS, (b) FeS$_2$, (c) Fe$_2$S$_3$

5.25. A compound contains 21.6% sodium, 33.3% chlorine, 45.1% oxygen. Derive its empirical formula. Take Na = 23.0, Cl = 35.5, O = 16.0.   *Ans.* NaClO$_3$

5.26. When 1.010 grams of zinc vapor is burned in air, 1.257 grams of the oxide is produced. What is the empirical formula of the oxide?   *Ans.* ZnO

5.27. A compound has the following percentage composition: Na = 19.3%, S = 26.9%, O = 53.8%. Its molecular weight is 238. Derive its molecular formula.   *Ans.* Na$_2$S$_2$O$_8$

5.28. Determine the simplest formula of a compound that has the following composition: Cr = 26.52%, S = 24.52%, O = 48.96%.   *Ans.* Cr$_2$S$_3$O$_{12}$ or Cr$_2$(SO$_4$)$_3$

5.29. A 3.245 g sample of a titanium chloride was reduced with sodium to metallic titanium. After the resultant sodium chloride was washed out, the residual titanium metal was dried and weighed 0.819 g. What is the empirical formula of titanium chloride?   *Ans.* TiCl$_4$

5.30. A compound contains 63.1% carbon, 11.92% hydrogen, and 24.97% fluorine. Derive its empirical formula.   *Ans.* C$_4$H$_9$F

5.31. An organic compound was found on analysis to consist of 47.37% carbon and 10.59% hydrogen. The balance was presumed to be oxygen. What is the empirical formula of the compound?
Ans. C$_3$H$_8$O$_2$

5.32. Derive the empirical formulas of the minerals that have the following composition:

(a) ZnSO$_4$ = 56.14%,   H$_2$O = 43.86%

(b) MgO = 27.16%,   SiO$_2$ = 60.70%,   H$_2$O = 12.14%

(c) Na = 12.10%,   Al = 14.19%,   Si = 22.14%,   O = 42.09%,   H$_2$O = 9.48%.

Ans. (a) ZnSO$_4 \cdot$ 7 H$_2$O,   (b) 2 MgO $\cdot$ 3 SiO$_2 \cdot$ 2 H$_2$O,   (c) Na$_2$Al$_2$Si$_3$O$_{10} \cdot$ 2 H$_2$O

5.33. A borane (a compound containing only boron and hydrogen) analyzed 88.45% boron. What is its empirical formula?   *Ans.* B$_5$H$_7$

5.34. An experimental catalyst used in the polymerization of butadiene has the following composition: 23.3% Co, 25.3% Mo, and 51.4% Cl. What is its empirical formula?   *Ans.* Co$_3$Mo$_2$Cl$_{11}$

5.35. A 1.500 g sample of a compound containing only C, H and O was burned completely. The only combustion products were 1.738 g CO$_2$ and 0.711 g H$_2$O. What is the empirical formula of the compound?   *Ans.* C$_2$H$_4$O$_3$

5.36. Elementary analysis showed that an organic compound contained C, H, N and O as its only elementary constituents. A 1.279 g sample was burned completely, as a result of which 1.60 g of CO$_2$ and 0.77 g of H$_2$O were obtained. A separately weighed 1.625 g sample contained 0.216 g nitrogen. What is the empirical formula of the compound?   *Ans.* C$_3$H$_7$O$_3$N

COMPOSITION

5.37. A fusible alloy is made by melting together 10.6 lb bismuth, 6.4 lb lead, and 3.0 lb tin. (*a*) What is the percentage composition of the alloy? (*b*) How much of each metal is required to make 70.0 g of alloy? (*c*) What weight of alloy can be made from 4.2 lb of tin?
Ans. (*a*) 53% Bi, 32% Pb, 15% Sn   (*b*) 37.1 g Bi, 22.4 g Pb, 10.5 g Sn   (*c*) 28 lb alloy

5.38. Calculate the percentage of copper in each of the following minerals: cuprite Cu_2O, copper pyrites $CuFeS_2$, malachite $CuCO_3 \cdot Cu(OH)_2$. How many tons of cuprite will give 500 tons of copper?
Ans. 88.82%, 34.62%, 57.47%, 563 tons

5.39. What is the nitrogen content (fertilizer rating) of NH_4NO_3? Of $(NH_4)_2SO_4$?
Ans. 35.0% N, 21.2% N

5.40. Determine the percentage composition of (*a*) silver chromate Ag_2CrO_4, (*b*) calcium pyrophosphate $Ca_2P_2O_7$.   *Ans.* (*a*) 65.02% Ag, 15.68% Cr, 19.30% O;   (*b*) 31.54% Ca, 24.38% P, 44.08% O

5.41. Determine the percentage composition of (*a*) UO_2F_2, (*b*) $C_3Cl_2F_6$.
Ans. (*a*) 77.28% U, 10.38% O, 12.34% F;   (*b*) 16.31% C, 32.09% Cl, 51.60% F

5.42. Find the percentage arsenic in a polymer having the empirical formula C_2H_8AsB.   *Ans.* 63.6% As

5.43. The specifications for a transistor material called for one boron atom in 10^{10} silicon atoms. What would be the boron content of one pound of such material?   *Ans.* 6×10^{-10} oz B

5.44. The purest form of carbon is prepared by decomposing pure sugar $C_{12}H_{22}O_{11}$ (driving off the contained H_2O). What is the maximum number of grams of carbon that could be obtained from one pound of sugar?   *Ans.* 191 g C

5.45. What weight of silver is present in 3.45 g Ag_2S?   *Ans.* 3.00 g Ag

5.46. Determine the weight of sulfur required to make 1000 lb H_2SO_4.   *Ans.* 327 lb S

5.47. What weight of CuO will be required to furnish 200 kg copper?   *Ans.* 250 kg CuO

5.48. How many pounds of metallic sodium and liquid chlorine can be obtained from one ton of salt? How many pounds of NaOH and how many pounds of hydrogen chloride?
Ans. 787 lb Na, 1213 lb Cl, 1370 lb NaOH, 1248 lb HCl

5.49. Compute the amount of zinc in a ton of ore containing 60.0% zincite ZnO.   *Ans.* 964 lb Zn

5.50. How much phosphorus is contained in 5.00 g of the compound $CaCO_3 \cdot 3\,Ca_3(PO_4)_2$? How much P_2O_5?   *Ans.* 0.902 g P, 2.07 g P_2O_5

5.51. A 10.00 gram sample of a crude ore contains 2.80 grams of HgS. What is the percentage of mercury in the ore?   *Ans.* 24.1% Hg

5.52. A procedure for analyzing the oxalic acid content of a solution involves the formation of the insoluble complex, $Mo_4O_3(C_2O_4)_3 \cdot 12\,H_2O$. (*a*) How many grams of this complex would form per gram of oxalic acid, $H_2C_2O_4$, if one gram-formula weight of the complex results from the reaction with three moles of oxalic acid? (*b*) How many grams of molybdenum are contained in the complex formed by reaction with one gram of oxalic acid?   *Ans.* (*a*) 3.4 g complex, (*b*) 1.42 g Mo

5.53. The arsenic content of an agricultural insecticide was reported as 28% As_2O_5. What is the % arsenic in this preparation?   *Ans.* 18% As

5.54. Express the potassium content of a fertilizer in % K_2O if its elementary potassium content is 4.5%.
Ans. 5.4% K_2O

5.55. A typical analysis of a pyrex glass showed 12.9% B_2O_3, 2.2% Al_2O_3, 3.8% Na_2O, 0.4% K_2O, and the balance SiO_2. Assume that the oxide percentages add up to 100%. What is the ratio of silicon to boron atoms in the glass?   *Ans.* 3.6

5.56. A piece of plumber's solder weighing 3.00 g was dissolved in dilute nitric acid, then treated with dilute H_2SO_4. This precipitated the lead as $PbSO_4$, which after washing and drying weighed 2.93 g. The solution was then neutralized to precipitate stannic acid, which was decomposed by heating, yielding 1.27 g SnO_2. What is the analysis of the solder as % Pb and % Sn?
Ans. 66.7% Pb, 33.3% Sn

5.57. A sample of impure cuprite Cu_2O contains 66.6% copper. What is the percentage of pure Cu_2O in the sample?   *Ans.* 75.0% Cu_2O

5.58. A cold cream sample weighing 8.41 g lost 5.83 g of moisture on heating to 110°C. The residue on extracting with water and drying lost 1.27 g of water-soluble glycerol. The balance was oil. Calculate the composition of this cream.   *Ans.* 69.3% moisture, 15.1% glycerol, 15.6% oil

5.59. A household cement gave the following analytical data: A 28.5 gram sample, on dilution with acetone, yielded a residue of 4.6 g of aluminum powder. The filtrate, on evaporation of the acetone and solvent, yielded 3.2 g of plasticized nitrocellulose which contained 0.8 g of benzene-soluble plasticizer. Determine the composition of this cement.
Ans. 16.2% Al, 72.6% solvent, 2.8% plasticizer, 8.4% nitrocellulose

5.60. A coal contains 2.4% water. After drying, the moisture-free residue contains 71.0% carbon. Determine the percentage of carbon on the "wet basis."   *Ans.* 69.3% C

5.61. A clay contains 45% silica and 10% water. What is the percentage of silica in the clay on a dry (water-free) basis?   *Ans.* 50% silica

5.62. A liter flask is filled with two liquids (A and B) of specific gravity 1.4 together. The specific gravity of liquid A is 0.8 and of liquid B 1.8. What volume of each enters the mixture? Assume no change of volume on mixing.   *Ans.* 400 ml of A, 600 ml of B

5.63. There are available 10 tons of a coal containing 2.5% sulfur, and also supplies of coal containing 0.80% and 1.10% sulfur respectively. How many tons of each of the latter should be mixed with the original 10 tons to give 20 tons containing 1.7% sulfur?   *Ans.* 6.7 tons of 0.80%, 3.3 tons of 1.10%

5.64. A sample of polystyrene prepared by heating styrene with tribromobenzoyl peroxide in the absence of air has the formula $Br_3C_6H_3(C_8H_8)_n$. The number n varies with the conditions of preparation. One sample of polystyrene prepared in this manner was found to contain 10.46% bromine. What is the value of n?   *Ans.* 19

5.65. One of the earliest methods for determining the molecular weight of proteins was based on chemical analysis. Hemoglobin was found to contain 0.335% iron. (*a*) If the hemoglobin molecule contains one atom of iron, what is its molecular weight? (*b*) If it contains 4 atoms of iron, what is its molecular weight?   *Ans.* (*a*) 1.67×10^4, (*b*) 6.7×10^4

5.66. A taconite ore consisted of 35.0% Fe_3O_4 and the balance siliceous impurities. How many tons of the ore must be processed in order to recover a ton of metallic iron (*a*) if there is 100% recovery, (*b*) if there is only 75% recovery?   *Ans.* 3.94 tons, 5.26 tons

5.67. A typical formulation for a cationic asphalt emulsion calls for 0.5% tallow amine emulsifier, and 70% asphalt, the rest consisting of water and water-soluble ingredients. How much asphalt can be emulsified per pound of the emulsifier?   *Ans.* 140 lb

5.68. Uranium hexafluoride, UF_6, is used in the gaseous diffusion process for separating uranium isotopes. How many pounds of elementary uranium can be converted to UF_6 per pound of combined fluorine?
Ans. 2.09 lb

Chapter 6

Calculations from Chemical Equations

INTRODUCTION

Calculations based on chemical equations are among the most important calculations in general chemistry, because of the large amount of descriptive and quantitative knowledge which is condensed into these equations. Knowledge about a chemical change is represented by an equation of formulas, just as each formula represents the composition of a substance in terms of the constituent atoms.

The balanced chemical equation is an algebraic equation with all of the reactants on the left side and all of the products on the right side; hence the equation sign usually is replaced by an arrow showing the rightward course of the reaction. If the reverse reaction also takes place, the double-arrow of equilibrium equations is used. Variable but important conditions such as temperature, pressure, catalysts, etc., may be noted above or below the equation arrow.

MOLECULAR RELATIONS FROM EQUATIONS

The *relative numbers of reacting and resulting molecules* are indicated by the coefficients of the formulas representing these molecules. For example, the combustion of ammonia in oxygen is described by the following balanced chemical equation

$$\underset{\text{4 molecules}}{4\,NH_3} \;+\; \underset{\text{3 molecules}}{3\,O_2} \;\longrightarrow\; \underset{\text{2 molecules}}{2\,N_2} \;+\; \underset{\text{6 molecules}}{6\,H_2O}$$

in which the algebraic coefficients 4, 3, 2 and 6 indicate that 4 molecules of NH_3 react with 3 molecules of O_2 to form 2 molecules of N_2 and 6 molecules of H_2O. This reaction is not reversible, hence only the single arrow is appropriate.

The balanced equation does not necessarily mean that if 4 molecules of NH_3 are mixed with 3 molecules of O_2 the reaction as indicated will go to completion. Some reactions between chemical substances occur almost instantaneously upon mixing, some occur completely after sufficient time has elapsed, and some reactions go to only a partial extent even after an infinite time. The common interpretation of the balanced equation for all categories is as follows. If a large number of NH_3 and O_2 molecules are mixed, a certain number of N_2 and H_2O molecules will be formed. At a given instant it is not necessary that either the NH_3 or O_2 is all consumed, but whatever reaction did occur took place in the molecular ratio prescribed by the equation. In other words, for every 4 molecules of NH_3 consumed by the reaction, 3 molecules of O_2 were also consumed, and the products were 2 molecules of N_2 and 6 of H_2O. This statement is true regardless of the actual extent of the reaction, whether 100% or a small fraction of a percent.

The atoms in *seven* indicated molecules ($4\,NH_3$, $3\,O_2$) rearrange to form *eight* molecules ($2\,N_2$, $6\,H_2O$); there is no algebraic rule governing these numbers of molecules, but the number of atoms on each side of the equation must balance for each element, since the equation obeys the laws of conservation of matter and of non-transmutability of the elements. Thus the equation is balanced and checked by counting the atoms of each kind ($4\,N$, $12\,H$, $6\,O$), not the molecules.

The number of atoms of any element occurring in a given substance is the product of the coefficient of the formula of the substance times the subscript of that element in the formula. Thus, $4\,NH_3$ represents 12 atoms of H because there are 3 atoms of H in each of 4 molecules of NH_3. Similarly, $6\,H_2O$ represents $12\,(6 \times 2)$ atoms of H. In some more complex formulas, several subscripts must be multiplied together before multiplying by the coefficient of the entire formula. Thus $3\,(NH_4)_2SO_4$ would represent 24 atoms of H because each of the 3 formula units of $(NH_4)_2SO_4$ contains $2\,(NH_4)$ radicals, each of which in turn contains 4 H atoms.

WEIGHT RELATIONS FROM EQUATIONS

All atoms and molecules have definite weights. These actual weights of atoms and molecules are proportional to their atomic and molecular weights. Hence the balanced equation also shows the *relative weights* of the reactants and resultants. For example, the chemical equation   [Molecular weight of $NH_3 = 17$, $O_2 = 32$, $N_2 = 28$, $H_2O = 18$]

$$\underline{4\,NH_3} \quad + \quad \underline{3\,O_2} \quad \longrightarrow \quad \underline{2\,N_2} \quad + \quad \underline{6\,H_2O}$$

4 moles	3 moles	2 moles	6 moles
$= 4 \times 17$ grams	$= 3 \times 32$ grams	$= 2 \times 28$ grams	$= 6 \times 18$ grams
4×17 parts	3×32 parts	2×28 parts	6×18 parts

shows that 4 moles of NH_3 (4×17 g NH_3) react with 3 moles of O_2 (3×32 g O_2) to form 2 moles of N_2 (2×28 g N_2) and 6 moles of H_2O (6×18 g H_2O).

Or, 4×17 parts by weight of NH_3 react with 3×32 parts by weight of O_2 to form 2×28 parts by weight of N_2 and 6×18 parts by weight of H_2O, where *any* unit of weight may be used. For example, 4×17 lb of NH_3 reacts with 3×32 lb of O_2 to form 2×28 lb of N_2 and 6×18 lb of H_2O.

In all cases, the sum of the reactant weights ($68 + 96$) must equal the sum of the resultant weights ($56 + 108$). (Law of Conservation of Mass)

These weight relations are particularly useful for several reasons:

(1) Weight relations are as exacting as the law of conservation of mass.

(2) Weight relations do not require any knowledge about the variable conditions; for example, whether the H_2O is water or steam.

(3) Weight relations do not require any knowledge of the true molecular formulas. In the above example, the weights or the numbers of atoms would be unchanged if the oxygen were assumed to be ozone, $2\,O_3$, instead of $3\,O_2$. In either case, the equation would be balanced with six oxygen atoms on each side. Similarly, if the water molecules were polymerized, weight relations would be the same whether the equation contained $6\,H_2O$, $3\,H_4O_2$, or $2\,H_6O_3$. This principle is very important in cases where the true molecular formulas are not known. Weight relations are valid for the many equations involving molecules that may dissociate (S_8, P_4, H_6F_6, N_2O_4, I_2, etc.) or those that associate to form complex polymers, such as the many industrially important derivatives of formaldehyde, starch, cellulose, nylon, synthetic rubbers, silicones, etc., regardless of whether empirical or molecular formulas are used.

Solved Problems

6.1. Balance the following skeleton equations:

    (a) $FeS_2 + O_2 \longrightarrow Fe_2O_3 + SO_2$,   (b) $C_7H_6O_2 + O_2 \longrightarrow CO_2 + H_2O$.

    (a) There are no fixed rules for balancing equations. Often a trial and error procedure must be used. It is commonly helpful to start with the most complex formula. Fe_2O_3 has two different elements and a greater total number of atoms than any of the other substances, so we might start with it. We note that oxygen atoms occur in pairs in the molecules O_2 and SO_2 but not in the formula unit of Fe_2O_3. If we write the equation with symbols representing the *integral* coefficients,

$$w\ FeS_2 + x\ O_2 \longrightarrow y\ Fe_2O_3 + z\ SO_2$$

then the total number of oxygen atoms on the left, $2x$, is even for any integral value of x. The total number on the right, $(3y + 2z)$ can be even or odd, depending on whether y is even or odd. We conclude from the required equality of $2x$ with $(3y + 2z)$ that y must be even. We can now try the smallest even number, 2, and proceed from there.

$$w\ FeS_2 + x\ O_2 \longrightarrow 2\ Fe_2O_3 + z\ SO_2$$

To balance iron atoms, w must equal 4.

$$4\ FeS_2 + x\ O_2 \longrightarrow 2\ Fe_2O_3 + z\ SO_2$$

To balance sulfur, z must equal 8.

$$4\ FeS_2 + x\ O_2 \longrightarrow 2\ Fe_2O_3 + 8\ SO_2$$

Finally, to balance oxygen, $2x = (6 + 16)$, or $x = 11$.

$$4\ FeS_2 + 11\ O_2 \longrightarrow 2\ Fe_2O_3 + 8\ SO_2$$

Note that the coefficient of the simplest substance, elementary oxygen in this case, was evaluated last. This is the usual consequence of beginning the balancing procedure with the most complex substance.

    (b) The most complex substance in this equation is $C_7H_6O_2$. We may assume 1 molecule of this substance and immediately write the coefficients for CO_2 and H_2O that will lead to balance of C and H respectively. The balance

$$C_7H_6O_2 + x\ O_2 \longrightarrow 7\ CO_2 + 3\ H_2O$$

of oxygen atoms is saved for last because an adjustment of x would not interfere with the balance of any other element. An arithmetic balance now demands that x equal 15/2, in an equation

$$C_7H_6O_2 + \frac{15}{2}\ O_2 \longrightarrow 7\ CO_2 + 3\ H_2O$$

that is arithmetically balanced but violates the rule of integral coefficients. The correct ratio is preserved and the fractions eliminated by multiplying each coefficient by 2.

$$2\ C_7H_6O_2 + 15\ O_2 \longrightarrow 14\ CO_2 + 6\ H_2O$$

This is the correct form.

6.2. Caustic soda, NaOH, is often prepared commercially by the reaction of Na_2CO_3 with slaked lime, $Ca(OH)_2$. How many grams of NaOH can be obtained by treating one kilogram (1000 g) of Na_2CO_3 with $Ca(OH)_2$?

First write the balanced equation for the reaction.

$$\underset{\substack{1\ gfw \\ =\ 106.0\ g}}{Na_2CO_3} + Ca(OH)_2 \longrightarrow \underset{\substack{2\ gfw \\ =\ 2 \times 40.0\ =\ 80.0\ g}}{2\ NaOH} + CaCO_3$$

Formula weight of $Na_2CO_3 = 106.0$, of NaOH $= 40.0$.

The equation states that 1 formula weight of Na_2CO_3 reacts with 1 formula weight of $Ca(OH)_2$ to give 2 formula weights of NaOH and 1 formula weight of $CaCO_3$. The problem concerns the weight relations between NaOH and Na_2CO_3. The other two substances, $Ca(OH)_2$ and $CaCO_3$, need not be considered in the subsequent calculations once the complete balanced equation has been obtained. Their weights might be desired in another problem involving this same reaction but will not be computed here.

First Method.

The equation indicates that 1 gfw Na_2CO_3 (106.0 g) gives 2 gfw NaOH (80.0 g). Thus

$$106.0 \text{ g } Na_2CO_3 \text{ gives } 80.0 \text{ g NaOH}$$

$$1 \text{ g } Na_2CO_3 \text{ gives } \frac{80.0}{106.0} \text{ g NaOH}$$

and $$1000 \text{ g } Na_2CO_3 \text{ gives } 1000 \times \frac{80.0}{106.0} \text{ g } = 755 \text{ g NaOH}$$

Gram-Formula Weight Method.

As in the preceding chapter, the symbol n with a subscript will be used to refer to the number of gfw, g-at, or moles of a substance. Consider 1000 g Na_2CO_3.

$$n_{Na_2CO_3} = \frac{1000 \text{ g}}{106.0 \text{ g/gfw}} = 9.43 \text{ gfw } Na_2CO_3$$

From the coefficients in the balanced equation, $n_{NaOH} = 2n_{Na_2CO_3} = 2 \times 9.43 = 18.86$ gfw NaOH.

Weight of NaOH = 18.86 gfw NaOH × 40.0 g NaOH/gfw NaOH = 754 g NaOH.

Note that the answers by the first two methods, 755 and 754, agree within the precision of the least precise factor used in the computation, 80.0 (to 1 part in 800).

Proportion Method.

Let x = number of grams of NaOH obtained from 1000 g Na_2CO_3.

The problem now reads: 106.0 g Na_2CO_3 (1 gfw Na_2CO_3) gives 80.0 g NaOH (2 gfw NaOH), hence 1000 g Na_2CO_3 will give x g NaOH. Then, by proportion,

$$\frac{106.0 \text{ g } Na_2CO_3}{80.0 \text{ g NaOH}} = \frac{1000 \text{ g } Na_2CO_3}{x}$$

from which $$x = 1000 \text{ g } Na_2CO_3 \times \frac{80.0 \text{ g NaOH}}{106.0 \text{ g } Na_2CO_3} = 755 \text{ g NaOH}$$

Note. It should be evident that 1000 *pounds* Na_2CO_3 will give 755 *pounds* NaOH, and 1000 *tons* Na_2CO_3 will give 755 *tons* NaOH.

Factor-Label Method.

As in the previous method, x = weight of NaOH obtained. The question is written as an equation, in which x is equated to the 1000 g Na_2CO_3, and the right side of the equation is manipulated by successive conversion factors until it has the desired units of g NaOH.

$$x = 1000 \text{ g } Na_2CO_3 \times \frac{1 \text{ gfw } Na_2CO_3}{106.0 \text{ g } Na_2CO_3} \times \frac{2 \text{ gfw NaOH}}{1 \text{ gfw } Na_2CO_3} \times \frac{40.0 \text{ g NaOH}}{1 \text{ gfw NaOH}} = 755 \text{ g NaOH}$$

6.3. Calculate the weight of lime (CaO) that can be prepared by heating 200 lb of limestone that is 95% pure $CaCO_3$.

Weight of pure $CaCO_3$ in 200 lb limestone = 0.95×200 lb = 190 lb $CaCO_3$.

Formula weight of $CaCO_3$ = 100, of CaO = 56.1. Then 1 lb-fw of $CaCO_3$ = 100 lb, of CaO = 56.1 lb.

The balanced equation for the reaction is

$$CaCO_3 \longrightarrow CaO + CO_2$$

1 lb-fw = 100 lb   1 lb-fw = 56.1 lb

First Method.

The equation shows that 1 lb-fw $CaCO_3$ (100 lb) gives 1 lb-fw CaO (56.1 lb). Thus

$$100 \text{ lb } CaCO_3 \text{ gives } 56.1 \text{ lb CaO}$$

$$1 \text{ lb } CaCO_3 \text{ gives } \frac{56.1}{100} \text{ lb CaO}$$

and $$190 \text{ lb } CaCO_3 \text{ gives } 190 \times \frac{56.1}{100} \text{ lb } = 107 \text{ lb CaO}$$

Pound-Formula Weight Method.

$$n_{\text{CaCO}_3} = \frac{190 \text{ lb CaCO}_3}{100 \text{ lb CaCO}_3/\text{lb-fw CaCO}_3} = 1.90 \text{ lb-fw CaCO}_3$$

$$n_{\text{CaO}} = n_{\text{CaCO}_3} = 1.90 \text{ lb-fw CaO}$$

Weight of CaO = 1.90 lb-fw CaO $\times$ 56.1 lb CaO/lb-fw CaO = 107 lb CaO

Factor-Label Method.

Weight of CaO = 190 lb CaCO$_3$ $\times \dfrac{1 \text{ lb-fw CaCO}_3}{100 \text{ lb CaCO}_3} \times \dfrac{1 \text{ lb-fw CaO}}{1 \text{ lb-fw CaCO}_3} \times \dfrac{56.1 \text{ lb CaO}}{1 \text{ lb-fw CaO}}$

= 107 lb CaO

6.4. The equation for the preparation of phosphorus in an electric furnace is

$$2 \text{ Ca}_3(\text{PO}_4)_2 + 6 \text{ SiO}_2 + 10 \text{ C} \longrightarrow 6 \text{ CaSiO}_3 + 10 \text{ CO} + \text{P}_4$$

Determine:

(a) The number of moles of phosphorus formed for each gfw of Ca$_3$(PO$_4$)$_2$ used.
(b) The number of grams of phosphorus formed for each gfw of Ca$_3$(PO$_4$)$_2$ used.
(c) The number of grams of phosphorus formed for each gram of Ca$_3$(PO$_4$)$_2$ used.
(d) The number of pounds of phosphorus formed for each pound of Ca$_3$(PO$_4$)$_2$ used.
(e) The number of tons of phosphorus formed for each ton of Ca$_3$(PO$_4$)$_2$ used.
(f) The number of gfw each of SiO$_2$ and C required for each gfw of Ca$_3$(PO$_4$)$_2$ used.

(a) From the equation, 1 mole P$_4$ is obtained for each 2 gfw Ca$_3$(PO$_4$)$_2$ used, or, $\frac{1}{2}$ mole per gfw.

(b) Molecular weight of P$_4$ = 124. Then $\frac{1}{2}$ mole P$_4$ = $\frac{1}{2} \times 124$ = 62 g of P$_4$.

(c) One gfw of Ca$_3$(PO$_4$)$_2$ (310 g) yields $\frac{1}{2}$ mole P$_4$ (62 g). Then 1 g Ca$_3$(PO$_4$)$_2$ gives 62/310 = 0.20 g P$_4$.

(d) 0.20 pound   (e) 0.20 ton

(f) From the equation, 1 gfw Ca$_3$(PO$_4$)$_2$ requires 3 gfw SiO$_2$ and 5 g-at C.

6.5. Most of the commercial hydrochloric acid is prepared by heating NaCl with concentrated H$_2$SO$_4$. How many pounds of sulfuric acid containing 90.0% H$_2$SO$_4$ by weight are needed for the production of 1000 lb of concentrated hydrochloric acid containing 42.0% HCl by weight?

(1) Weight of pure HCl in 1000 lb of 42.0% acid = 0.420 $\times$ 1000 lb = 420 lb.

(2) Determine the number of pounds of H$_2$SO$_4$ required to produce 420 lb HCl.

Molecular weight of H$_2$SO$_4$ = 98.1, of HCl = 36.46.

$$2 \text{ NaCl} \quad + \quad \underline{\text{H}_2\text{SO}_4} \quad \longrightarrow \quad \text{Na}_2\text{SO}_4 \quad + \quad \underline{2 \text{ HCl}}$$

$$\begin{array}{ccc} & 1 \text{ lb-mole} & \qquad\qquad 2 \text{ lb-moles} \\ & = 98.1 \text{ lb} & \qquad\qquad = 2 \times 36.46 = 72.92 \text{ lb} \end{array}$$

From the equation, 2 lb-moles HCl (72.92 lb) requires 1 lb-mole H$_2$SO$_4$ (98.1 lb). Thus

72.92 lb HCl requires 98.1 lb H$_2$SO$_4$

1 lb HCl requires $\dfrac{98.1}{72.92}$ lb H$_2$SO$_4$

and   420 lb HCl requires $420 \times \dfrac{98.1}{72.92}$ lb = 565 lb H$_2$SO$_4$

(3) Finally determine the number of pounds of sulfuric acid solution containing 90.0% H$_2$SO$_4$ that can be made from 565 lb of pure H$_2$SO$_4$.

0.900 lb of pure H$_2$SO$_4$ makes 1 lb of 90.0% solution; then 565 lb of pure H$_2$SO$_4$ will make

$$\frac{565 \text{ lb H}_2\text{SO}_4}{0.900 \dfrac{\text{lb H}_2\text{SO}_4}{\text{lb solution}}} = 628 \text{ lb solution}$$

6.6. The green-dyed 100-octane aviation gasoline used 4.00 ml of tetraethyl lead $(C_2H_5)_4Pb$, specific gravity 1.66, per gallon of product. This compound is made as follows.

$$4\,C_2H_5Cl + 4\,NaPb \longrightarrow (C_2H_5)_4Pb + 4\,NaCl + 3\,Pb$$

How many grams of ethyl chloride, C_2H_5Cl, are needed to make enough tetraethyl lead for one gallon of gasoline?

Molecular weight of $C_2H_5Cl = 64.5$, of $(C_2H_5)_4Pb = 323$.

Weight of 4.00 ml $(C_2H_5)_4Pb = 4.00$ ml $\times$ 1.66 g/ml $= 6.64$ g $(C_2H_5)_4Pb$ per gallon.

The equation shows that 1 mole $(C_2H_5)_4Pb$ requires 4 moles C_2H_5Cl.

Number of moles of $(C_2H_5)_4Pb$ in 6.64 g $= \dfrac{6.64\ g}{323\ g/mole} = 0.0206$ mole.

Hence 4×0.0206 mole $= 0.0824$ mole of C_2H_5Cl is needed.

Weight of C_2H_5Cl in 0.0824 mole $= 0.0824$ mole $\times$ 64.5 g/mole $= 5.31$ g C_2H_5Cl.

6.7. How many kilograms of pure H_2SO_4 could be obtained from one kilogram of pure iron pyrites (FeS_2) according to the following reactions?

$$4\,FeS_2 + 11\,O_2 \longrightarrow 2\,Fe_2O_3 + 8\,SO_2$$
$$2\,SO_2 + O_2 \longrightarrow 2\,SO_3$$
$$SO_3 + H_2O \longrightarrow H_2SO_4$$

Formula weight of $FeS_2 = 120$, of $H_2SO_4 = 98$.

First it should be noted that there is no by-product loss or other permanent loss of sulfur, so that it is not even necessary to balance the equations nor to use them further. Each incoming atom of sulfur produces an outgoing molecule of H_2SO_4. (One formula unit of FeS_2 contains 2 atoms of S, and one molecule of H_2SO_4 contains 1 atom of S.) Hence

$$FeS_2 \longrightarrow 2\ H_2SO_4$$
$$1\ gfw \longrightarrow 2\ moles\ H_2SO_4$$
$$n_{FeS_2} = \frac{1000\ g}{120\ g/gfw} = 8.33\ gfw\ FeS_2$$
$$n_{H_2SO_4} = 2\ n_{FeS_2} = 2 \times 8.33 = 16.66\ moles\ H_2SO_4$$

Weight of $H_2SO_4 = 16.66$ moles $\times$ 98 g/mole $= 1630$ g $= 1.63$ kg H_2SO_4.

Another Method.

One gfw FeS_2 (120 g FeS_2) yields 2 moles H_2SO_4 ($2 \times 98 = 196$ g H_2SO_4). Thus

$$120\ kg\ FeS_2\ yields\ 196\ kg\ H_2SO_4$$

and
$$1\ kg\ FeS_2\ yields\ \frac{196}{120}\ kg = 1.63\ kg\ H_2SO_4$$

6.8. A mixture containing 100 g H_2 and 100 g O_2 is sparked so that water is formed according to the reaction

$$2\,H_2 + O_2 \longrightarrow 2\,H_2O$$

How much water is formed?

Molecular weights: $H_2 = 2.02$, $O_2 = 32.0$, $H_2O = 18.0$.

The special feature of this problem is that the starting quantities of *two* reactants are specified. It is necessary first to determine which, if any, substance is in excess. The mole method is the simplest method for this type of problem.

Number of moles of $H_2 = \dfrac{100\ g}{2.02\ g/mole} = 49.5$ moles, of $O_2 = \dfrac{100\ g}{32.0\ g/mole} = 3.13$ moles.

If all the hydrogen were to be used, $\frac{1}{2}(49.5) = 24.8$ moles O_2 would be required. Obviously the hydrogen cannot all be used. Since O_2 is present in limiting amount, the calculations must be based on the weight of O_2. Counting only those moles which participate in the reaction,

$$n_{H_2O} = 2n_{O_2} = 2 \times 3.13 = 6.26 \text{ moles } H_2O$$

$$\text{Weight of } H_2O = 6.26 \text{ moles} \times 18.0 \text{ g/mole} = 113 \text{ g } H_2O$$

The amount of H_2 consumed $= 6.26$ moles $\times 2.02$ g/mole $= 13$ g.   The reaction mixture will contain, in addition to the 113 g H_2O, 87 g of unreacted H_2.

6.9. In one process for water-proofing, a fabric is exposed to $(CH_3)_2SiCl_2$ vapor. The vapor reacts with hydroxyl groups on the surface of the fabric or with traces of water to form the water-proofing film, $[(CH_3)_2SiO]_n$, by the reaction

$$n\,(CH_3)_2SiCl_2 + 2n\,OH^- \longrightarrow 2n\,Cl^- + n\,H_2O + [(CH_3)_2SiO]_n$$

where n stands for a large integer. The water-proofing film is deposited on the fabric layer upon layer. Each layer is 6 angstroms thick (the thickness of the $[(CH_3)_2SiO]$ group). How much $(CH_3)_2SiCl_2$ is needed to water-proof one side of a piece of fabric that is 1 meter by 2 meters with a film 300 layers thick? The density of the film is 1.0 g/cm^3.

$$
\begin{aligned}
\text{Mass of film} &= \text{volume of film} \times \text{density of film} \\
&= \text{area of film} \times \text{thickness of film} \times \text{density of film} \\
&= (100 \text{ cm} \times 200 \text{ cm})(300 \times 6 \text{ angstroms})(1.0 \text{ g/cm}^3) \\
&= (10^2 \text{ cm} \times 2 \times 10^2 \text{ cm})(3 \times 10^2 \times 6 \times 10^{-8} \text{ cm})(1.0 \text{ g/cm}^3) = 0.36 \text{ g}
\end{aligned}
$$

Formula weight of $(CH_3)_2SiO = 74$, of $(CH_3)_2SiCl_2 = 129$. This is a case where we purposely avoid using the molecular weight and use the formula weight instead. The equation indicates that 1 gfw (74 g) of $(CH_3)_2SiO$ in the film requires 1 mole (129 g) of $(CH_3)_2SiCl_2$. Then 0.36 g of $(CH_3)_2SiO$ requires $0.36 \text{ g} \times 129/74 = 0.63$ g of $(CH_3)_2SiCl_2$.

6.10. What is the % free SO_3 in an oleum (considered as a solution of SO_3 in H_2SO_4) that is labeled "109% H_2SO_4"? Such a designation refers to the total weight of pure H_2SO_4 that would be present after dilution of 100 g of the oleum, when all free SO_3 would combine with water to form H_2SO_4.

9 g H_2O will combine with all the free SO_3 in 100 g of the oleum to give a total of 109 g H_2SO_4.

The equation $H_2O + SO_3 \longrightarrow H_2SO_4$ indicates that 1 mole H_2O (18 g) combines with 1 mole SO_3 (80 g). Then 9 g H_2O combines with $\frac{9}{18} \times 80$ g $= 40$ g SO_3.

Thus 100 g of the oleum contains 40 g SO_3, or the % free SO_3 in the oleum is 40%.

6.11. $KClO_4$ may be made by the following series of reactions:

$$
\begin{aligned}
Cl_2 + 2\,KOH &\longrightarrow KCl + KClO + H_2O \\
3\,KClO &\longrightarrow 2\,KCl + KClO_3 \\
4\,KClO_3 &\longrightarrow 3\,KClO_4 + KCl
\end{aligned}
$$

How much Cl_2 is needed to prepare 100 g $KClO_4$ by the above sequence?

The mole method or the factor-label method are the simplest routes to the solution of this problem.

Mole Method.

$$
\begin{aligned}
n_{KClO} &= n_{Cl_2} \\
n_{KClO_3} &= \tfrac{1}{3}n_{KClO} = \tfrac{1}{3}n_{Cl_2} \\
n_{KClO_4} &= \tfrac{3}{4}n_{KClO_3} = \tfrac{3}{4}(\tfrac{1}{3}n_{Cl_2}) = \tfrac{1}{4}n_{Cl_2} \\
n_{KClO_4} &= \frac{100 \text{ g } KClO_4}{139 \text{ g } KClO_4/\text{gfw } KClO_4} = 0.720 \text{ gfw } KClO_4 \\
n_{Cl_2} &= 4 \times 0.720 = 2.88 \text{ moles } Cl_2
\end{aligned}
$$

Weight of Cl_2 = 2.88 moles Cl_2 × 71.0 g Cl_2/mole Cl_2 = 204 g Cl_2

Factor-Label Method.

$$x \text{ g } Cl_2 = \frac{100 \text{ g } KClO_4}{139 \text{ g } KClO_4/\text{gfw } KClO_4} \times \frac{4 \text{ gfw } KClO_3}{3 \text{ gfw } KClO_4}$$

$$\times \frac{3 \text{ gfw } KClO}{1 \text{ gfw } KClO_3} \times \frac{1 \text{ mole } Cl_2}{1 \text{ gfw } KClO} \times \frac{71.0 \text{ g } Cl_2}{1 \text{ mole } Cl_2} = 204 \text{ g } Cl_2$$

Supplementary Problems

BALANCING EQUATIONS

Balance the following equations.

6.12. $BCl_3 + P_4 + H_2 \longrightarrow BP + HCl$

6.13. $C_2H_2Cl_4 + Ca(OH)_2 \longrightarrow C_2HCl_3 + CaCl_2 + H_2O$

6.14. $(NH_4)_2Cr_2O_7 \longrightarrow N_2 + Cr_2O_3 + H_2O$

6.15. $Zn_3Sb_2 + H_2O \longrightarrow Zn(OH)_2 + SbH_3$

6.16. $HClO_4 + P_4O_{10} \longrightarrow H_3PO_4 + Cl_2O_7$

6.17. $C_6H_5Cl + SiCl_4 + Na \longrightarrow (C_6H_5)_4Si + NaCl$

6.18. $Sb_2S_3 + HCl \longrightarrow H_3SbCl_6 + H_2S$

6.19. $IBr + NH_3 \longrightarrow NI_3 + NH_4Br$

6.20. $KrF_2 + H_2O \longrightarrow Kr + O_2 + HF$

6.21. $Na_2CO_3 + C + N_2 \longrightarrow NaCN + CO$

6.22. $K_4Fe(CN)_6 + H_2SO_4 + H_2O \longrightarrow K_2SO_4 + FeSO_4 + (NH_4)_2SO_4 + CO$

6.23. $Fe(CO)_5 + NaOH \longrightarrow Na_2Fe(CO)_4 + Na_2CO_3 + H_2O$

6.24. $H_3PO_4 + (NH_4)_2MoO_4 + HNO_3 \longrightarrow (NH_4)_3PO_4 \cdot 12\, MoO_3 + NH_4NO_3 + H_2O$

WEIGHT-WEIGHT PROBLEMS

6.25. Consider the combustion of amyl alcohol, $C_5H_{11}OH$.

$$2\, C_5H_{11}OH + 15\, O_2 \longrightarrow 10\, CO_2 + 12\, H_2O$$

(a) How many moles of O_2 are needed for the combustion of 1 mole of amyl alcohol?

(b) How many moles of H_2O are formed for each mole of O_2 consumed?

(c) How many grams of CO_2 are produced for each mole of amyl alcohol burned?

(d) How many grams of CO_2 are produced for each gram of amyl alcohol burned?

(e) How many tons of CO_2 are produced for each ton of amyl alcohol burned?

Ans. (a) 7.5 moles O_2, (b) 0.80 moles H_2O, (c) 220 g CO_2, (d) 2.49 g CO_2, (e) 2.49 tons CO_2

6.26. A portable hydrogen generator utilizes the reaction $CaH_2 + 2\, H_2O \longrightarrow Ca(OH)_2 + 2\, H_2$. How many grams of H_2 can be produced by a 50 g cartridge of CaH_2?   *Ans.* 4.8 g H_2

6.27. Iodine is made by the reaction $2\, NaIO_3 + 5\, NaHSO_3 \longrightarrow 3\, NaHSO_4 + 2\, Na_2SO_4 + H_2O + I_2$. To produce each pound of iodine, how much $NaIO_3$ and how much $NaHSO_3$ must be used? *Ans.* 1.56 lb $NaIO_3$, 2.05 lb $NaHSO_3$

6.28. How much $KClO_3$ must be heated to obtain 3.50 g of oxygen?   *Ans.* 8.94 g $KClO_3$

6.29. What weight of ferric oxide will be produced by the complete oxidation of 100 g of iron?
$4\, Fe + 3\, O_2 \longrightarrow 2\, Fe_2O_3$ *Ans.* 143 g

6.30. (*a*) How many pounds of ZnO will be formed when 1 pound of zinc blende, ZnS, is strongly heated in air? $2\,ZnS + 3\,O_2 \longrightarrow 2\,ZnO + 2\,SO_2$

 (*b*) How many tons of ZnO will be formed from 1 ton ZnS?

 (*c*) How many kilograms of ZnO will be formed from 1 kilogram ZnS?

 Ans. 0.835 lb, 0.835 ton, 0.835 kg

6.31. In a rocket motor fueled with butane, C_4H_{10}, how many kilograms of liquid oxygen should be provided with each kilogram of butane to provide for complete combustion?

$2\,C_4H_{10} + 13\,O_2 \longrightarrow 8\,CO_2 + 10\,H_2O$ *Ans.* 3.58 kg

6.32. Chloropicrin, CCl_3NO_2, can be made cheaply for use as an insecticide by a process which utilizes the reaction

$$CH_3NO_2 + 3\,Cl_2 \longrightarrow CCl_3NO_2 + 3\,HCl$$

How much nitromethane, CH_3NO_2, is needed to form 500 g of chloropicrin?   *Ans.* 186 g

6.33. Ethyl alcohol (C_2H_5OH) is made by the fermentation of glucose ($C_6H_{12}O_6$) as indicated by the equation

$$C_6H_{12}O_6 \longrightarrow 2\,C_2H_5OH + 2\,CO_2$$

How many pounds of alcohol can be made from 2000 lb of glucose?   *Ans.* 1020 lb

6.34. How many pounds of 83.4% pure salt cake (Na_2SO_4) could be produced from 250 lb of 94.5% pure salt?

$2\,NaCl + H_2SO_4 \longrightarrow Na_2SO_4 + 2\,HCl$ *Ans.* 344 lb

6.35. How many kilograms of H_2SO_4 can be prepared from 1 kg of cuprite, Cu_2S, if each atom of S in Cu_2S is converted into 1 molecule of H_2SO_4?   *Ans.* 0.616 kg

6.36. (*a*) How much bismuth nitrate, $Bi(NO_3)_3 \cdot 5\,H_2O$, would be formed from a solution of 10.4 grams of bismuth in nitric acid? $Bi + 4\,HNO_3 + 3\,H_2O \longrightarrow Bi(NO_3)_3 \cdot 5\,H_2O + NO$

 (*b*) What weight of 30.0% nitric acid (containing 30.0% HNO_3 by weight) is required to react with this weight of bismuth?

 Ans. (*a*) 24.1 g, (*b*) 41.8 g

6.37. One of the reactions used in the petroleum industry for improving the octane rating of fuels is

$$C_7H_{14} \longrightarrow C_7H_8 + 3\,H_2$$

The two hydrocarbons appearing in this equation are both liquids; the hydrogen formed is a gas. What is the percentage reduction in liquid weight accompanying the completion of the above reaction?   *Ans.* 6.2%

6.38. In the Mond process for purifying nickel, the volatile nickel carbonyl, $Ni(CO)_4$, is produced by the following reaction.

$$Ni + 4\,CO \longrightarrow Ni(CO)_4$$

How much CO is used up in volatilizing each pound of nickel?   *Ans.* 1.91 lb CO

6.39. When copper is heated with an excess of sulfur, Cu_2S is formed. How many grams of Cu_2S could be produced if 100 g of copper is heated with 50 g of sulfur?   *Ans.* 125 g Cu_2S

6.40. The reduction of Cr_2O_3 by Al proceeds quantitatively on ignition of a suitable fuse.

$$2\,Al + Cr_2O_3 \longrightarrow Al_2O_3 + 2\,Cr$$

(*a*) How much metallic chromium can be made by bringing to reaction temperature a mixture of 5.0 lb Al with 20.0 lb Cr_2O_3? (*b*) Which reactant remains at the completion of the reaction and how much?   *Ans.* (*a*) 9.6 lb Cr, (*b*) 5.9 lb Cr_2O_3

6.41. A mixture of 1 ton of CS_2 and 2 tons of Cl_2 is passed through a hot reaction tube where the following reaction takes place: $CS_2 + 3\,Cl_2 \longrightarrow CCl_4 + S_2Cl_2$

(*a*) How much CCl_4 can be made by complete reaction of the limiting starting material?

(*b*) Which starting material is in excess, and how much of it remains unreacted?

 Ans. (*a*) 1.45 tons CCl_4, (*b*) 0.28 tons CS_2

6.42. A gram (dry weight) of green algae was able to absorb 4.7×10^{-3} moles CO_2 per hour by photosynthesis. If the fixed carbon atoms were all stored after photosynthesis as starch, $(C_6H_{10}O_5)_n$, how long would it take for the algae to double their own weight? Neglect the increase in photosynthetic rate due to the increasing amount of living matter.   *Ans.* 7.9 hr

6.43. Carbon disulfide, CS_2, can be made from by-product SO_2. The overall reaction is

$$5\,C\ +\ 2\,SO_2\ \longrightarrow\ CS_2\ +\ 4\,CO$$

How much CS_2 can be produced from 450 lb of waste SO_2 with excess coke, if the SO_2 conversion is 82%?   *Ans.* 219 lb

6.44. A 50.0 gram sample of impure zinc reacts with 129 ml of hydrochloric acid which has specific gravity 1.18 and contains 35.0% HCl by weight. What is the percent of metallic zinc in the sample? Assume that the impurity is inert to HCl.   *Ans.* 96% Zn

6.45. The chemical formula of the chelating agent, Versene, is $C_2H_4N_2(C_2H_2O_2Na)_4$. If each mole of this compound could bind one gram-atom of Ca^{++}, what would be the rating of pure Versene expressed as mg $CaCO_3$ bound per gram of chelating agent? Here the Ca^{++} is expressed in terms of the amount of $CaCO_3$ it could form.   *Ans.* 264 mg $CaCO_3$ per g

6.46. In a typical electric furnace product of crude CaC_2 made by the following reaction

$$CaO\ +\ 3\,C\ \longrightarrow\ CaC_2\ +\ CO$$

the product is 85% CaC_2 and 15% unreacted CaO. (a) How much CaO is to be added to the furnace charge for each 50 tons of CaC_2 produced? (b) How much CaO is to be added to the furnace charge for each 50 tons of crude product?   *Ans.* (a) 53 tons CaO, (b) 45 tons CaO

6.47. The plastics industry uses large amounts of phthalic anhydride, $C_8H_4O_3$, made by the controlled oxidation of naphthalene.

$$2\,C_{10}H_8\ +\ 9\,O_2\ \longrightarrow\ 2\,C_8H_4O_3\ +\ 4\,CO_2\ +\ 4\,H_2O$$

Since some of the naphthalene is oxidized to other products, only 70% of the maximum yield predicted by the above equation is actually obtained. How much phthalic anhydride would be produced in practice by the oxidation of 100 lb of $C_{10}H_8$?   *Ans.* 81 lb

6.48. The empirical formula of a commercial ion exchange resin is $C_8H_7SO_3Na$. The resin can be used to soften water according to the reaction

$$Ca^{++}\ +\ 2\,C_8H_7SO_3Na\ \longrightarrow\ (C_8H_7SO_3)_2Ca\ +\ 2\,Na^+$$

What would be the maximum uptake of Ca^{++} by the resin, expressed in g-at per gram of resin? *Ans.* 0.0024 g-at Ca^{++}/g

6.49. The insecticide DDT is made by the following reaction.

$$CCl_3CHO\ (chloral)\ +\ 2\,C_6H_5Cl\ (chlorobenzene)\ \longrightarrow\ (ClC_6H_4)_2CHCCl_3\ (DDT)\ +\ H_2O$$

If 100 lb of chloral were reacted with 100 lb of chlorobenzene, how much DDT would be formed? Assume the reaction goes to completion without side reactions or losses.   *Ans.* 157 lb DDT

6.50. Commercial sodium "hydrosulfite" is 90% pure $Na_2S_2O_4$. How much of the commercial product could be made by using 100 tons of zinc with a sufficient supply of the other reactants? The reactions are

$$Zn\ +\ 2\,SO_2\ \longrightarrow\ ZnS_2O_4$$
$$ZnS_2O_4\ +\ Na_2CO_3\ \longrightarrow\ ZnCO_3\ +\ Na_2S_2O_4$$

Ans. 296 tons

6.51. Fluorocarbon polymers can be made by fluorinating polyethylene according to the reaction

$$(CH_2)_n\ +\ 4n\,CoF_3\ \longrightarrow\ (CF_2)_n\ +\ 2n\,HF\ +\ 4n\,CoF_2$$

where n is a large integer. The CoF_3 can be regenerated by the reaction

$$2\,CoF_2\ +\ F_2\ \longrightarrow\ 2\,CoF_3$$

(a) If the HF formed in the first reaction cannot be reused, how many grams of fluorine are consumed per gram of fluorocarbon produced, $(CF_2)_n$? (b) If the HF can be recovered and electrolyzed to hydrogen and fluorine, and if this fluorine is used for regenerating CoF_3, what is the net consumption of fluorine per gram of fluorocarbon?   *Ans.* (a) 1.52 g, (b) 0.76 g

6.52. The Scholler wood-sugar process yields fermenting sugars ($C_6H_{12}O_6$) equal to 35% of the dry weight of the wood sawdust used. Calculate from the following proven reactions how many pounds of butadiene (C_4H_6) rubber can be obtained from 2000 lb of dry sawdust.

$$C_6H_{12}O_6 \longrightarrow 2\ C_2H_5OH + 2\ CO_2$$

$$2\ C_2H_5OH + O_2 \longrightarrow 2\ C_2H_4O + 2\ H_2O$$

$$2\ C_2H_4O + H_2 \longrightarrow C_4H_6 + 2\ H_2O$$

Ans. 210 lb

6.53. One method of preparing hydrogen peroxide, H_2O_2, utilizes the following series of reactions.

$$2\ NH_4HSO_4 \longrightarrow H_2 + (NH_4)_2S_2O_8$$

$$(NH_4)_2S_2O_8 + 2\ H_2O \longrightarrow 2\ NH_4HSO_4 + H_2O_2$$

Assuming 100% efficiency of each stage, what is the consumption of NH_4HSO_4 and of H_2O per 100 g of 85% (by weight) H_2O_2 solution?   *Ans.* 0.0 g NH_4HSO_4, 105 g H_2O

6.54. One type of silicone rubber is prepared by polymerizing $(CH_3)_2SiCl_2$. This important intermediate, in turn, is made according to the equation

$$SiCl_4 + 2\ CH_3MgCl \longrightarrow (CH_3)_2SiCl_2 + 2\ MgCl_2$$

The by-product $MgCl_2$ may be decomposed electrolytically into magnesium and chlorine, both of which are needed for the synthesis of CH_3MgCl. If the magnesium is to be recovered completely for recycling, how much magnesium must be produced electrolytically per pound of $(CH_3)_2SiCl_2$?
Ans. 0.38 lb

6.55. A fluorine disposal plant was constructed to carry out the reactions

$$F_2 + 2\ NaOH \longrightarrow \tfrac{1}{2}\ O_2 + 2\ NaF + H_2O$$

$$2\ NaF + CaO + H_2O \longrightarrow CaF_2 + 2\ NaOH$$

As the plant operated, enough lime (CaO) was added from time to time to just bring about complete precipitation of the fluoride as CaF_2. Over a period of operation, 2000 lb of fluorine was fed into the plant and 10,000 lb of lime was required. What was the percent utilization of the lime?
Ans. 29.5%

6.56. Consider the following two equations used in the preparation of $KMnO_4$.

$$MnO_2 + 4\ KOH + O_2 \longrightarrow K_2MnO_4 + 2\ H_2O$$

$$3\ K_2MnO_4 + 4\ CO_2 + 2\ H_2O \longrightarrow 2\ KMnO_4 + 4\ KHCO_3 + MnO_2$$

How much oxygen gas must be consumed in order to make 100 g of $KMnO_4$?   *Ans.* 30.4 g

6.57. Hydrazoic acid, HN_3, may be made by the following sequence of reactions:

$$N_2 + 3\ H_2 \longrightarrow 2\ NH_3$$

$$4\ NH_3 + Cl_2 \longrightarrow N_2H_4 + 2\ NH_4Cl$$

$$4\ NH_3 + 5\ O_2 \longrightarrow 4\ NO + 6\ H_2O$$

$$2\ NO + O_2 \longrightarrow 2\ NO_2$$

$$2\ NO_2 + 2\ KOH \longrightarrow KNO_2 + KNO_3 + H_2O$$

$$2\ KNO_2 + H_2SO_4 \longrightarrow K_2SO_4 + 2\ HNO_2$$

$$N_2H_4 + HNO_2 \longrightarrow HN_3 + 2\ H_2O$$

If there is no recovery of NH_4Cl or of KNO_3, how much hydrogen and how much chlorine must be used in the preparation of each 100 g of HN_3?   *Ans.* 42 g H_2, 165 g Cl_2

Chapter 7

Measurement of Gases

GAS VOLUMES

Gases, on account of their greater compressibility and thermal expansivity as compared with liquids and solids, occupy volumes that depend very sensitively on the external variables, pressure and temperature. Special attention must therefore be given to factors influencing the volume of gases.

PRESSURE

Pressure is defined as the force acting on a unit area of surface.

$$\text{Pressure} = \frac{\text{force acting perpendicular to an area}}{\text{area over which the force is distributed}}$$

$$\text{Pressure (in g/cm}^2) = \frac{\text{force (in grams)}}{\text{area (in square centimeters)}}$$

$$\text{Pressure (in lb/in}^2) = \frac{\text{force (in pounds)}}{\text{area (in square inches)}}$$

The pressure due to a column of fluid is

$$\text{Pressure} = \text{height} \times \text{density of fluid}$$

$$\text{Pressure (in g/cm}^2) = \text{height (in cm)} \times \text{density (in g/cm}^3)$$

$$\text{Pressure (in lb/in}^2) = \text{height (in in.)} \times \text{density (in lb/in}^3)$$

Pressure is expressed frequently in grams per square centimeter (g/cm^2), in pounds per square inch (lb/in^2), in atmospheres, and in millimeters of mercury.

Air has weight and therefore exerts a pressure. The atmospheric pressure is due to the weight of the overlying air.

NORMAL ATMOSPHERIC PRESSURE

Normal atmospheric pressure is a defined pressure approximately equal to the average pressure of the atmosphere at sea level. This defined *atmosphere* is equivalent to the pressure exerted by the weight of a column of mercury 760 mm high at 0°C.

$$\text{Pressure} = \text{height of column of mercury} \times \text{density of mercury}$$

$$1 \text{ atmosphere} = 76.0 \text{ cm of Hg} \times 13.595 \text{ g/cm}^3 = 1033 \text{ g/cm}^2$$

$$= 29.9 \text{ in. of Hg} \times 0.491 \text{ lb/in}^3 = 14.7 \text{ lb/in}^2$$

STANDARD CONDITIONS (S.T.P.)

Standard conditions (S.T.P.) denotes a temperature of 0°C (273°K) and normal atmospheric pressure (760 mm of mercury). As both the volume and density of any gas are affected by changes of temperature and pressure, it is customary to reduce all gas volumes to standard conditions for purposes of comparison.

GAS LAWS

At sufficiently low pressures and high temperatures, all gases have been found to obey three simple laws. These laws relate the volume of a gas to the pressure and temperature. A gas which obeys these laws is called an *ideal gas* or *perfect gas*. These laws, which are therefore called the ideal gas laws, are described below. They may be applied only to gases which do not undergo a change in chemical complexity when the temperature or pressure is varied. One exception, for example, is NO_2, which undergoes a dimerization to N_2O_4 at increasing pressures or decreasing temperatures.

BOYLE'S LAW

When the temperature is kept constant, the volume of a given mass of an ideal gas varies inversely with the pressure to which the gas is subjected. In mathematical terms, the product Pressure × Volume of a given mass of gas remains constant. Thus, in comparing the properties of a given mass of an ideal gas under two conditions, which we may call the *initial* and *final* states, we may write the following equations, applicable at constant temperature:

$$(PV)_{initial} = (PV)_{final}$$

or

$$P_1 V_1 = P_2 V_2$$

A given subscript, 1 or 2, refers to a given state of the gas. This law furnishes the most direct test of how well a real gas corresponds to ideal behavior.

CHARLES' LAW

At constant pressure, the volume of a given mass of gas varies directly with the *absolute temperature*. Then, at constant pressure,

$$\left(\frac{V}{T}\right)_{initial} = \left(\frac{V}{T}\right)_{final}$$

or

$$\frac{V_1}{T_1} = \frac{V_2}{T_2}$$

where T_1 and T_2 denote the *absolute* temperatures of the gas at the two states being compared. (Although *absolute temperature* will normally be taken to imply the Kelvin scale in this book, Charles' Law as stated above is equally true for the Kelvin and Rankine scales, so long as the same scale is used for both T_1 and T_2.)

GAY-LUSSAC'S LAW

At constant volume, the pressure of a given mass of gas varies directly with the *absolute temperature*. Then, at constant volume,

$$\frac{P_1}{T_1} = \frac{P_2}{T_2}$$

GENERAL GAS LAW

Any two of the above three gas laws can be employed to derive the general gas law which applies to all possible combinations of changes:

$$\frac{P_1 V_1}{T_1} = \frac{P_2 V_2}{T_2} = \text{a constant}$$

for a given mass of gas used.

Since the majority of gas calculations are concerned with determining a new volume from an old volume, the law is usually given in this form:

$$V_2 = V_1 \times \frac{T_2}{T_1} \times \frac{P_1}{P_2} \quad \text{for a given mass of gas}$$

This expression indicates that the volume of a given mass of gas varies directly with the absolute temperature and inversely with the pressure.

This form conforms closely to logical reasoning. If the old volume (V_1) is being *heated* from T_1 to T_2, the temperature ratio should be set down as a fraction *greater* than 1, as heating causes an increase in volume by expansion. And if the old volume is being *compressed* from P_1 to P_2, the pressure ratio should be a fraction *less* than 1, as compression causes a decrease in volume.

DENSITY OF A GAS

As the volume of a given mass of gas increases, the mass per unit volume (i.e., the density) decreases proportionally. Therefore the density of a gas varies inversely with its volume.

It follows that the density of a gas varies inversely with the absolute temperature and directly with the pressure. This can be written as follows:

$$d_2 \; = \; d_1 \times \frac{T_1}{T_2} \times \frac{P_2}{P_1}$$

DALTON'S LAW OF PARTIAL PRESSURES

The total pressure of a gaseous mixture is equal to the sum of the partial pressures of the components. The partial pressure of a component of a gas mixture is the pressure which that component would exert if it alone occupied the entire volume.

Dalton's law is rigidly accurate only for ideal gases. At pressures of only a few atmospheres or below, gas mixtures may be regarded as ideal gases and this law may be applied in calculations.

COLLECTING GASES OVER WATER

If a gas is collected over a volatile liquid, such as water, a correction is made for the amount of water vapor present with the gas. A gas collected over water is saturated with water vapor, which occupies the total gas volume and exerts a partial pressure. The partial pressure of the water vapor is definite for each temperature and is independent of the nature or pressure of the gas. This definite value of the vapor pressure of water may be found tabulated as a function of temperature in handbooks or in other reference books. The vapor pressure must be subtracted from the total pressure (of the gas + water vapor) in order to obtain the effective partial pressure of the gas being measured. The latter pressure value is used as the actual pressure of the gas collected. Thus

Pressure of gas   =   total pressure − vapor pressure of water

When a gas is collected over mercury, it is not necessary to make a correction for the vapor pressure of mercury, as mercury has a negligible vapor pressure at ordinary temperatures.

DEVIATIONS

The laws discussed above are strictly valid only for ideal gases. Since all gases can be liquefied if they are compressed and cooled sufficiently, all gases become non-ideal at high pressures and low temperatures. The ideal properties are observed at low pressures and high temperatures, conditions far removed from those of the liquid state. For pressures below a few atmospheres practically all gases are sufficiently ideal for the application of the ideal gas laws, with a reliability of a few percent or better.

Solved Problems

AIR PRESSURE DEFINED

7.1. Calculate the difference in pressure between the top and bottom of a vessel **exactly** 76 cm deep when filled at 25°C with (a) water, (b) mercury. Density of mercury at 25° is 13.53 g/cm³, and of water, 0.997 g/cm³.

(a) Pressure = height × density of water = 76 cm × 0.997 g/cm³ = 75.8 g/cm²

(b) Pressure = height × density of mercury = 76 cm × 13.53 g/cm³ = 1028 g/cm²

7.2. How high a column of air would be necessary to cause the barometer to read 76 cm of mercury? Assume that the atmosphere is of uniform density 0.0012 g/cm³. Density of mercury is 13.6 g/cm³.

Air has weight and therefore exerts a pressure. Here the atmospheric pressure, or the pressure exerted by the air, must equal the pressure of the column of mercury.

$$\text{Pressure of mercury} = \text{pressure of air}$$
$$\text{height} \times \text{density of mercury} = \text{height} \times \text{density of air}$$
$$76 \text{ cm} \times 13.6 \text{ g/cm}^3 = \text{height of air} \times 0.0012 \text{ g/cm}^3$$
$$\text{height of air} = \frac{76 \text{ cm} \times 13.6 \text{ g/cm}^3}{0.0012 \text{ g/cm}^3} = 8.6 \times 10^5 \text{ cm} = 8.6 \times 10^3 \text{ m}$$

GAS LAWS

7.3. A mass of oxygen occupies 5.00 liters under a pressure of 740 mm of Hg. Determine the volume of the same mass of gas at standard pressure (760 mm of Hg), the temperature remaining constant.

At constant temperature, the volume of a given mass of gas varies inversely with the pressure to which the gas is subjected.

The *increase* in pressure from 740 mm to 760 mm results in a *decrease* in volume. Hence the original volume (5.00 liters) must be multiplied by a fraction *smaller* than 1, i.e. by $\frac{740 \text{ mm}}{760 \text{ mm}}$ (where the numerator is smaller than the denominator).

$$\text{Volume at 760 mm} = 5.00 \, l \times \frac{740 \text{ mm}}{760 \text{ mm}} = 4.87 \, l$$

7.4. Given the volume of a gas as 200 ml at 800 mm pressure. Calculate the volume of the same mass of gas at 765 mm.

The *decrease* in pressure from 800 mm to 765 mm results in an *increase* in volume. Hence the original volume (200 ml) must be multiplied by a fraction *greater* than 1, i.e. by $\frac{800 \text{ mm}}{765 \text{ mm}}$ (where the numerator is greater than the denominator).

$$\text{Volume at 765 mm} = 200 \text{ ml} \times \frac{800 \text{ mm}}{765 \text{ mm}} = 209 \text{ ml}$$

7.5. A mass of neon occupies 200 ml at 100°C. Find its volume at 0°C, the pressure remaining constant.

At constant pressure, the volume of a given mass of gas varies directly with the absolute temperature. (0°C = 273°K, 100°C = 100 + 273 = 373°K)

The *decrease* in temperature from 373°K to 273°K results in a *decrease* in volume. Hence the original volume (200 ml) must be multiplied by a fraction *smaller* than 1, i.e. by $\frac{273°\text{K}}{373°\text{K}}$.

$$\text{Volume at 0°C} = 200 \text{ ml} \times \frac{273°\text{K}}{373°\text{K}} = 146 \text{ ml}$$

7.6. What volume will 10.0 cubic feet of helium, measured at 12°C, occupy at 36°C? Assume constant pressure. (12°C = 12 + 273 = 285°K, 36°C = 36 + 273 = 309°K)

The *increase* in temperature from 285°K to 309°K results in an *increase* in volume. Hence the original volume must be multiplied by a fraction *greater* than 1, i.e. by $\frac{309°K}{285°K}$.

$$\text{Volume at } 36°C \;=\; 10.0 \text{ ft}^3 \times \frac{309°K}{285°K} \;=\; 10.8 \text{ ft}^3$$

7.7. A steel tank contains carbon dioxide at 27°C and a pressure of 12.0 atmospheres. Determine the internal gas pressure when the tank is heated to 100°C.

At constant volume, the pressure of a given mass of gas varies directly with the absolute temperature. (27°C = 27 + 273 = 300°K, 100°C = 373°K)

The *increase* in temperature from 300°K to 373°K results in an *increase* in pressure. Hence the original pressure must be multiplied by a fraction *greater* than 1, i.e. by $\frac{373°K}{300°K}$.

$$\text{Pressure at } 100°C \;=\; 12.0 \text{ atm} \times \frac{373°K}{300°K} \;=\; 14.9 \text{ atm}$$

7.8. Given 20.0 liters of ammonia at 5°C and 760 mm. Determine its volume at 30°C and 800 mm.

The volume of a given mass of gas varies directly with the absolute temperature and inversely with the pressure. (5°C = 5 + 273 = 278°K, 30°C = 303°K)

The *increase* in temperature from 278°K to 303°K results in an *increase* in volume. The original volume must be multiplied by a fraction *greater* than 1 to correct for the temperature change, i.e. by $\frac{303°K}{278°K}$.

The *increase* in pressure from 760 mm to 800 mm results in a *decrease* in volume. The original volume must be multiplied by a fraction *smaller* than 1 to correct for the pressure change, i.e. by $\frac{760 \text{ mm}}{800 \text{ mm}}$.

$$\text{Final volume} \;=\; 20.0 \, l \times \frac{303°K}{278°K} \times \frac{760 \text{ mm}}{800 \text{ mm}} \;=\; 20.7 \, l$$

Another Method.
Substituting in the General Gas Law equation,

$$V_2 \;=\; V_1 \times \frac{T_2}{T_1} \times \frac{P_1}{P_2} \;=\; 20.0 \, l \times \frac{303°K}{278°K} \times \frac{760 \text{ mm}}{800 \text{ mm}} \;=\; 20.7 \, l$$

Note that all quantities with a given subscript constitute a set of properties of the gas in a particular state. Thus V_1, T_1 and P_1 refer to the gas in its original condition. The final state of the gas is defined by V_2, T_2 and P_2. It should be apparent that the correct answer is independent of the designation of the initial and final states. The initial state could just as well have been characterized by subscripts 2 and the final state by 1. The important point is to use the same subscript for all properties of a particular state of the gas.

7.9. The volume of a quantity of sulfur dioxide at 18°C and 1500 mm is 5.0 cubic feet. Calculate its volume at standard conditions, 0°C and 760 mm. (18°C = 18 + 273 = 291°K, 0°C = 273°K)

The *decrease* in temperature from 291°K to 273°K causes the volume to *decrease*. To correct for the temperature change, the original volume must be multiplied by a fraction *smaller* than 1, i.e. by $\frac{273°K}{291°K}$.

The *decrease* in pressure from 1500 mm to 760 mm causes the volume to *increase*. To correct for the pressure change, the initial volume must be multiplied by a fraction *greater* than 1, i.e. by $\frac{1500 \text{ mm}}{760 \text{ mm}}$.

$$\text{Final volume} = 5.0 \text{ ft}^3 \times \frac{273°\text{K}}{291°\text{K}} \times \frac{1500 \text{ mm}}{760 \text{ mm}} = 9.3 \text{ ft}^3$$

7.10. A mass of hydrogen occupies 874 ft³ at 59°F and 19.8 lb/in². Find its volume at −26°F and 53.6 lb/in².

Rankine temperature (°R) = 460 + Fahrenheit temperature

$$\frac{P_1V_1}{T_1} = \frac{P_2V_2}{T_2}, \quad \frac{19.8 \text{ lb/in}^2 \times 874 \text{ ft}^3}{(59+460)°\text{R}} = \frac{53.6 \text{ lb/in}^2 \times V_2}{(-26+460)°\text{R}}, \quad V_2 = 270 \text{ ft}^3$$

7.11. To how many atmospheres pressure must a liter of gas measured at 1 atmosphere and −20°C be subjected to be compressed to ½ liter when the temperature is 40°C? (−20°C = 253°K, 40°C = 313°K)

The pressure of a given mass of gas varies inversely with the volume and directly with the absolute temperature.

The *decrease* in volume from 1 liter to ½ liter results in an *increase* in pressure. The initial pressure must be multiplied by a fraction *greater* than 1 to correct for the volume change, i.e. by $\frac{1 \text{ liter}}{\frac{1}{2} \text{ liter}}$.

The *increase* in temperature from 253°K to 313°K results in an *increase* in pressure. The original pressure must be multiplied also by another fraction *greater* than 1 to correct for the temperature change, i.e. by $\frac{313°\text{K}}{253°\text{K}}$.

$$\text{Final pressure} = 1 \text{ atm} \times \frac{1 \, l}{\frac{1}{2} \, l} \times \frac{313°\text{K}}{253°\text{K}} = 2.47 \text{ atm}$$

Another Method.

Rearranging the General Gas Law equation, $V_2 = V_1 \times \frac{T_2}{T_1} \times \frac{P_1}{P_2}$, we get

$$P_2 = P_1 \times \frac{V_1}{V_2} \times \frac{T_2}{T_1} = 1 \text{ atm} \times \frac{1 \, l}{\frac{1}{2} \, l} \times \frac{313°\text{K}}{253°\text{K}} = 2.47 \text{ atm}$$

DENSITY OF A GAS

7.12. The density of oxygen is 1.43 grams per liter at S.T.P. (0°C, 760 mm). Determine the density of oxygen at 17°C and 700 mm.

The density of a gas varies inversely with the absolute temperature and directly with the pressure.

The *increase* in temperature from 273°K to 290°K causes a *decrease* in density. To correct for this temperature change, the original density (1.43 g/l) must be multiplied by a fraction *smaller* than 1, i.e. by $\frac{273°\text{K}}{290°\text{K}}$.

The *decrease* in pressure from 760 mm to 700 mm causes a further *decrease* in density. To correct for this pressure change, the initial density must be multiplied also by a fraction *smaller* than 1, i.e. by $\frac{700 \text{ mm}}{760 \text{ mm}}$.

$$\text{Final density} = 1.43 \text{ g/}l \times \frac{273°\text{K}}{290°\text{K}} \times \frac{700 \text{ mm}}{760 \text{ mm}} = 1.24 \text{ g/}l$$

Another Method. $\quad d_2 = d_1 \times \frac{T_1}{T_2} \times \frac{P_2}{P_1} = 1.43 \text{ g/}l \times \frac{273°\text{K}}{290°\text{K}} \times \frac{700 \text{ mm}}{760 \text{ mm}} = 1.24 \text{ g/}l$

7.13. The density of helium is 0.1784 g/l at S.T.P. If a given mass of helium at S.T.P. is allowed to expand to 1.500 times its initial volume by changing the temperature and pressure, compute its resultant density.

The density of a gas varies inversely with the volume. In this case the volume *increases* to 1.500 times its initial value. Hence the density *decreases*, and the initial density must be multiplied by a fraction *smaller* than 1, i.e. by 1/1.500.

$$\text{Resultant density} \;=\; 0.1784 \text{ g/}l \times \frac{1}{1.500} \;=\; 0.1189 \text{ g/}l$$

PARTIAL PRESSURE

7.14. A mixture of gases at 760 mm pressure contains 65.0% nitrogen, 15.0% oxygen, and 20.0% carbon dioxide by volume. What is the partial pressure of each gas in mm?

A fundamental property of gases is that each component of a gas mixture occupies the entire volume of the mixture. The percentage composition by volume refers to the volumes of the separate gases before mixing. Thus 65 volumes of nitrogen, 15 volumes of oxygen, and 20 volumes of carbon dioxide, each at 760 mm pressure, are mixed to give 100 volumes of mixture at 760 mm pressure. The volume of each component is therefore increased by mixing. The pressure must therefore be multiplied by a fraction *smaller* than 1 to correct for the volume change. This fraction is the original volume divided by the final volume, which is the same as the original volume % divided by 100.

$$\text{Partial pressure of N}_2 \;=\; 760 \text{ mm} \times 65/100 \;=\; 494 \text{ mm}$$
$$\text{Partial pressure of O}_2 \;=\; 760 \text{ mm} \times 15/100 \;=\; 114 \text{ mm}$$
$$\text{Partial pressure of CO}_2 \;=\; 760 \text{ mm} \times 20/100 \;=\; 152 \text{ mm}$$

Check: Total pressure = sum of partial pressures = 760 mm

7.15. In a gaseous mixture at 20°C the partial pressures of the components are as follows: hydrogen 200 mm, carbon dioxide 150 mm, methane 320 mm, ethylene 105 mm. What is the total pressure of the mixture and the volume percent of hydrogen?

$$\text{Total pressure of mixture} \;=\; \text{sum of partial pressures}$$
$$=\; (200 + 150 + 320 + 105) \text{ mm} \;=\; 775 \text{ mm}$$

$$\text{Volume fraction of hydrogen} \;=\; \frac{\text{partial pressure of H}_2}{\text{total pressure of mixture}} \;=\; \frac{200 \text{ mm}}{775 \text{ mm}} \;=\; 0.258 \;\doteq\; 25.8\%$$

7.16. A 200 ml flask contained oxygen at 200 mm pressure, and a 300 ml flask contained nitrogen at 100 mm pressure. The two flasks were then connected so that each gas filled their combined volumes. Assuming no change in temperature, what was the partial pressure of each gas in the final mixture and what was the total pressure?

The final total volume was 500 ml.

$$\text{Oxygen:} \qquad P_{\text{final}} \;=\; P_{\text{initial}} \times \frac{V_i}{V_f} \;=\; 200 \text{ mm} \times \frac{200}{500} \;=\; 80 \text{ mm}$$

$$\text{Nitrogen:} \qquad P_f \;=\; P_i \times \frac{V_i}{V_f} \;=\; 100 \times \frac{300}{500} \;=\; 60 \text{ mm}$$

$$\text{Total pressure} \;=\; 80 \text{ mm} + 60 \text{ mm} \;=\; 140 \text{ mm}$$

COLLECTING GASES OVER WATER

7.17. Exactly 100 ml of oxygen is collected over water at 23°C and 800 mm. Compute the standard volume of the dry oxygen. Vapor pressure of water at 23°C is 21.1 mm.

The gas collected is a mixture of oxygen and water vapor. The partial pressure of water vapor in the mixture at 23°C is 21.1 mm. Hence the partial pressure of the oxygen is the total pressure less 21.1 mm.

Pressure of dry oxygen   =   total pressure − vapor pressure of water

=   800 mm − 21.1 mm   =   778.9 mm (or 779 mm)

The *decrease* in temperature from 296°K (23°C) to 273°K (standard temperature) causes the volume to *decrease*. To correct for the temperature change, the initial volume must be multiplied by a fraction *smaller* than 1, i.e. by $\dfrac{273°K}{296°K}$.

The *decrease* in pressure from 779 mm to 760 mm (standard pressure) causes the volume to *increase*. To correct for the pressure change, the initial volume must be multiplied by a fraction *greater* than 1, i.e. by $\dfrac{779\ mm}{760\ mm}$.

$$\text{Final volume}\ =\ 100.0\ ml \times \frac{273°K}{296°K} \times \frac{779\ mm}{760\ mm}\ =\ 94.5\ ml$$

7.18. In a basal metabolism measurement timed at exactly 6 minutes, a patient exhaled 52.5 liters of air, measured over water at 20°C. The vapor pressure of water at 20°C is 17.5 mm. The barometric pressure was 750 mm. The exhaled air analyzed 16.75 volume % oxygen, and the inhaled air 20.32 volume % oxygen, both on a dry basis. Neglecting any solubility of the gases in water and any difference in the total volume of inhaled and exhaled air, calculate the rate of oxygen consumption by the patient in ml (S.T.P.) per minute.

$$\text{Volume of dry air at S.T.P.}\ =\ 52.5\ l \times \frac{273°K}{293°K} \times \frac{(750 - 17.5)\ mm}{760\ mm}\ =\ 47.1\ l$$

$$\text{Rate of oxygen consumption}\ =\ \frac{\text{volume of oxygen (S.T.P.) consumed}}{\text{time in which this volume was consumed}}$$

$$=\ \frac{(0.2032 - 0.1675) \times 47.1\ l}{6\ min}\ =\ 0.280\ l/min\ =\ 280\ ml/min$$

CORRECTION FOR DIFFERENCE IN LEVELS

7.19. A quantity of gas is collected in a graduated tube over mercury. The volume of gas at 20°C is 50.0 ml, and the level of the mercury in the tube is 200 mm above the outside mercury level. The barometer reads 750 mm. Find the volume at S.T.P.

After a gas is collected over a liquid, the receiver is often adjusted so that the level of liquid inside and outside the receiver is the same. When this cannot be done conveniently, it is necessary to correct for the difference in levels.

Since the level of mercury inside the tube is 200 mm higher than outside, the pressure of the gas is 200 mm of Hg less than the atmospheric pressure of 750 mm.

Initial pressure   =   750 mm − 200 mm   =   550 mm;   final pressure   =   760 mm.

Initial temperature   =   20 + 273   =   293°K;   final temperature = 273°K.

$$\text{Final volume}\ =\ 50.0\ ml \times \frac{273°K}{293°K} \times \frac{550\ mm}{760\ mm}\ =\ 33.7\ ml$$

Supplementary Problems

7.20. A mass of oxygen occupies 40.0 cubic feet at 758 mm of mercury. Compute its volume at 635 mm of mercury, temperature remaining constant.   *Ans.* 47.7 ft³

7.21. Ten liters of hydrogen under 1 atmosphere pressure are contained in a cylinder which has a movable piston. The piston is moved in until the same mass of gas occupies 2 liters at the same temperature. Find the pressure in the cylinder.   *Ans.* 5 atm

7.22. A given mass of chlorine occupies 38.0 ml at 20°C. Determine its volume at 45°C, pressure remaining constant.   *Ans.* 41.2 ml

7.23. A quantity of hydrogen is confined in a platinum chamber of constant volume. When the chamber is immersed in a bath of melting ice, the absolute pressure of the gas is 1000 mm of mercury.
(a) What is the centigrade temperature when the pressure manometer reads exactly 100 mm of Hg?
(b) What pressure will be indicated when the chamber is brought to 100°C?
Ans. −246°C, 1366 mm

7.24. Given 1000 cubic feet of helium at 15°C and 763 mm. Calculate the volume at −6°C and 420 mm.
Ans. 1685 ft³

7.25. A mass of gas at 50°C and 785 mm pressure occupies 350 ml. What volume will the gas occupy at S.T.P.?   *Ans.* 306 ml

7.26. If a mass of gas occupies one liter at S.T.P., what volume will it occupy at 300°C and 25 atmospheres pressure?   *Ans.* 0.084 liter

7.27. If a gas occupies 15.7 cubic feet at 60°F and 14.7 lb/in², what volume would it occupy at 100°F and 25 lb/in²?   *Ans.* 9.9 ft³

7.28. Exactly 500 ml of nitrogen is collected over water at 25°C and 755 mm. The gas is saturated with water vapor. Compute the volume of the nitrogen in the dry condition at S.T.P. Vapor pressure of water at 25°C is 23.8 mm.   *Ans.* 441 ml

7.29. A dry gas occupied 127 cc at S.T.P. If this same mass of gas were collected over water at 23°C and a total gas pressure of 745 mm, what volume would it occupy? The vapor pressure of water at 23°C is 21 mm.   *Ans.* 145 cc

7.30. A mass of gas occupies 825 ml at −30°C and 0.556 atmosphere pressure. What is the pressure if the volume becomes 1000 ml and the temperature 20°C?   *Ans.* 0.553 atm

7.31. Calculate the final centigrade temperature required to change 10 liters of helium at 100°K and 0.1 atm to 20 liters at 0.2 atm.   *Ans.* 127°C

7.32. One mole of a gas occupies 22.4 liters at S.T.P. (a) What pressure would be required to compress one mole of oxygen into a five-liter container held at 100°C? (b) What maximum centigrade temperature would be permitted to hold this amount of oxygen in 5 liters if the pressure is not to exceed 3 atmospheres? (c) What capacity would be required to hold this same amount if the conditions of 100°C and 3 atmospheres were fixed?   *Ans.* 6.12 atm, −90°C, 10.2 *l*

7.33. If the density of a certain gas at 30°C and 768 mm is 1.253 g/*l*, find its density at S.T.P.
Ans. 1.376 g/*l*

7.34. A certain container holds 2.55 g of neon at S.T.P. What mass of neon will it hold at 100°C and 10.0 atmospheres pressure?   *Ans.* 18.7 g

7.35. At the top of a mountain the thermometer reads 10°C and the barometer reads 70 cm of mercury. At the bottom of the mountain the temperature is 30°C and the pressure is 76 cm of mercury. Compare the density of the air at the top with that at the bottom.   *Ans.* 0.99 (top) to 1.00 (bottom)

7.36. A volume of 95 ml of nitrous oxide at 27°C is collected in a graduated tube over mercury, the level of mercury inside the tube being 60 mm above the outside mercury level when the barometer reads 750 mm. (*a*) Compute the volume of the same mass of gas at S.T.P. (*b*) What volume would the same mass of gas occupy at 40°C, the barometric pressure being 745 mm and the level of mercury inside the tube 25 mm below that outside?   *Ans.* (*a*) 78 ml, (*b*) 89 ml

7.37. At a certain region of the upper atmosphere, the temperature is estimated to be −100°C and the density just 10^{-9} that of the earth's atmosphere at S.T.P. Assuming the same atmospheric composition, what is the pressure, in mm mercury, of the selected region of the upper atmosphere?
Ans. 4.81×10^{-7} mm

7.38. At 0°C the density of nitrogen at 1 atmosphere is 1.25 g/l. The nitrogen which occupied 1500 cc at S.T.P. was compressed at 0°C to 575 atmospheres and the gas volume was observed to be 3.92 cc, in violation of Boyle's Law. What was the final density of this non-ideal gas?   *Ans.* 478 g/l

7.39. Camphor has been found to undergo a crystalline modification at a temperature of 148°C and a pressure of 3.15×10^4 kg/cm². What is the transition pressure in atmospheres?
Ans. 3.05×10^4 atm

7.40. An abrasive, borazon, is made by heating ordinary boron nitride to 3000°F at one million pounds per square inch. Express the experimental conditions in °C and atmospheres pressure.
Ans. 1649°C, 6.8×10^4 atm

7.41. The respiration of a suspension of yeast cells was measured by observing the decrease of pressure of the gas above the cell suspension. The apparatus was arranged so that the gas was confined to a constant volume, 16.0 cc, and the entire pressure change was caused by uptake of oxygen by the cells. The pressure was measured in a manometer the fluid of which had a density of 1.034 g/cc. The entire apparatus was immersed in a thermostat at 37°C. In a 30-minute observation period the fluid in the open side of the manometer dropped 37 mm. Neglecting the solubility of oxygen in the yeast suspension, compute the rate of oxygen consumption by the cells in cubic millimeters of O_2 (S.T.P.) per hour.   *Ans.* 104 mm³/hr

7.42. A mixture of N_2, NO and NO_2 was analyzed by selective absorption of the oxides of nitrogen. The initial volume of the mixture was 2.74 cc. After treatment with water which absorbed the NO_2 the volume was 2.02 cc. A ferrous sulfate solution was then shaken with the residual gas to absorb the NO, after which the volume was 0.25 cc. All volumes were measured at barometric pressure. Neglecting water vapor, what was the volume percentage of each gas in the original mixture?
Ans. 9.1% N_2, 64.6% NO, 26.3% NO_2

7.43. A 250 ml flask contained krypton at 500 mm. A 450 ml flask contained helium at 950 mm. The contents of the two flasks were mixed by opening a stopcock connecting them. Assuming that all operations were carried out at a uniform constant temperature, calculate the final total pressure and the volume percent of each gas in the resulting mixture. Neglect the volume of the stopcock.
Ans. 789 mm, 22.6% Kr

7.44. The vapor pressure of water at 80°C is 355 mm. A 100 ml vessel contained water-saturated oxygen at 80°C, the total gas pressure being 760 mm. The contents of the vessel were transferred by pumping into a 50 ml vessel at the same temperature. What were the partial pressures of oxygen and of water vapor, and what was the total pressure in the final equilibrated state? Neglect the volume of any water which might condense.
Ans. Water, 355 mm; O_2, 810 mm; total pressure, 1165 mm

Chapter 8

Molecular Weights of Gases

AVOGADRO'S HYPOTHESIS

Avogadro's hypothesis states that: Equal volumes of all gases under the same conditions of temperature and pressure contain the same number of molecules. Thus 1 liter (or ml or cubic foot or other unit of volume) of oxygen contains the same number of molecules as 1 liter (or ml, etc.) of hydrogen or of any other gas.

Avogadro's hypothesis enables us to determine the relative weights of the molecules (molecular weights) of gases. The logic is as follows:

(1) The weight of 1 liter of any gas is the weight of all the molecules in 1 liter of the gas.

(2) One liter of any gas contains the same number of molecules as one liter of any other gas.

(3) Therefore if 1 liter of a gas weighs twice as much as 1 liter of another gas, it is because each molecule of the first gas weighs twice as much as each molecule of the second gas (since the total number of molecules is the same in each case). In this case, the molecular weight of the first gas would be twice that of the second gas.

(4) In general, the *relative* weights of the molecules of all gases can be determined by weighing equivalent volumes of the gases.

For example: At standard conditions, 1 liter of oxygen weighs 1.43 g and 1 liter of carbon monoxide weighs 1.25 g. By Avogadro's hypothesis, 1 liter of carbon monoxide contains the same number of molecules as 1 liter of oxygen. Hence a molecule of carbon monoxide weighs $\frac{1.25}{1.43}$ times as much as a molecule of oxygen. Accordingly, if we take the molecular weight of oxygen as 32, then the molecular weight of carbon monoxide is $\frac{1.25}{1.43} \times 32 = 28$.

The gas density method may be applied so precisely as to be used for atomic weight determinations, particularly for the lighter elements. Even crude gas density experiments may be used, in conjunction with chemical composition data, to establish the molecular weight, and hence molecular formula, of a gaseous compound. For example, a hydride of silicon that has the empirical formula SiH_3 was found to have an approximate gas density at S.T.P. of 2.9 g/l. By comparison with oxygen, whose molecular weight and density are known, the molecular weight of the hydride is then $\frac{2.9}{1.43} \times 32 = 65$. Although the approximate molecular weight might be in error by as much as 10%, it is sufficiently accurate to assign the molecular formula Si_2H_6 (molecular weight = 62.2) and to reject SiH_3 (molecular weight = 31.1), Si_3H_9 (93.3), or higher molecular weight possibilities.

MOLAR VOLUME, OR GRAM-MOLECULAR VOLUME

A molecule of oxygen is known to contain 2 atoms of oxygen. Then the molecular weight of oxygen is twice the atomic weight, 2×16.0, or 32.0.

The density of oxygen is found to be 1.429 g/l at S.T.P. The volume occupied by 1 mole (32.0 g) of oxygen is $\dfrac{32.0 \text{ g}}{1.429 \text{ g}/l}$ = 22.4 liters at S.T.P.

In other words, 22.4 liters of oxygen at S.T.P. contains 1 mole $(6.02 \times 10^{23}$ molecules) of oxygen. By Avogadro's hypothesis, 22.4 liters of any gas at S.T.P. will contain the same number of molecules (6.02×10^{23}) as 22.4 liters of oxygen. Therefore the weight of 22.4 liters of any gas at S.T.P. will be its gram-molecular weight. Thus if 22.4 liters of nitrogen at S.T.P. weighs 28 grams, the molecular weight of nitrogen is 28. The *normal molar volume* of a gas is said to be 22.4 liters, the volume of a mole at S.T.P.

Note. The more exact value for the normal molar volume, 22.414 liters, is the volume of a mole of gas computed to S.T.P. from the observed volume at very low pressures, where all gases have ideal behavior. Because of the deviations from ideal behavior shown by real gases, the actual observed molar volume of a gas at S.T.P. may be slightly different from 22.414 l, either higher or lower. Thus the molar volume of N_2 at S.T.P. is 22.404 l; of H_2, 22.428 l; of CH_4, 22.360 l; of CO_2, 22.262 l; and of NH_3, 22.081 l. In the rest of this chapter, the rounded value 22.4 l will be used for all real gases.

GENERALIZED GAS LAW

A generalization of the gas laws may be regarded as a consequence of Avogadro's hypothesis. If the molar volume is the same for all gases at S.T.P., then the molar volume at any other temperature and pressure is also the same for all ideal gases. This is true because the laws governing volume changes of gases with changes in temperature and pressure are the same for all ideal gases. The mathematical statement of the generalized gas law is

$$PV = nRT$$

where n = number of moles of gas, and R = molar gas constant.

The number of moles, n, can be replaced by g/M, the mass in grams divided by the molecular weight. The density of a gas, d, can then be evaluated directly from the equation $d = \dfrac{g}{V} = \dfrac{PM}{RT}$. Thus at S.T.P. the density of a gas is directly proportional to its molecular weight.

The numerical value of R can be evaluated from the known molar volume of an ideal gas at S.T.P.

$$R = \frac{PV}{nT} = \frac{1 \text{ atm} \times 22.4 \ l}{1 \text{ mole} \times 273^{\circ}\text{K}} = 0.0821 \frac{l \text{ atm}}{^{\circ}\text{K mole}} = 0.0821 \ l \text{ atm} \ ^{\circ}\text{K}^{-1} \text{ mole}^{-1}$$

The student should memorize the generalized gas equation and the numerical value of R. When the value $R = 0.0821 \ l$ atm $^{\circ}\text{K}^{-1}$ mole^{-1} is used, P must be in atmospheres, V in liters, T in degrees Kelvin, and n in moles. (Although R is often expressed in other units, all problems in this book will be solved with the above value. Engineering students may wish to verify the conversions showing that $R = 0.730$ ft^3 atm $^{\circ}R^{-1}$ lb-mole^{-1}, or 1544 ft-lb $^{\circ}R^{-1}$ lb-mole^{-1}.)

DIFFUSION AND EFFUSION OF GASES

If a small amount of a gas, A, is added to one end of a closed tank containing another gas, B, gas A will *diffuse*, or distribute itself uniformly throughout the cylinder in a short time. Different gases have different rates of diffusion. The simplest experiment for characterizing the intrinsic mobility of any one gas is an effusion experiment, in which the gas is allowed to escape through small holes into an evacuated space. In such an experiment the rates of effusion of gases, measured in terms of the number of molecules or moles of gas that escape per unit time, are inversely proportional to the square roots of their densities (Graham's Law).   Then

$$\frac{r_1}{r_2} = \frac{\sqrt{d_2}}{\sqrt{d_1}}$$

where r_1 and d_1 are the rate of effusion and the density of the first gas, and r_2 and d_2 are the rate of effusion and the density of the second gas.

Since the molecular weight of a gas is proportional to its density, the above equation may be written

$$\frac{r_1}{r_2} = \frac{\sqrt{M_2}}{\sqrt{M_1}}$$

where M_1 and M_2 are the molecular weights of the two gases.

GAS VOLUME RELATIONS FROM EQUATIONS

A chemical equation representing the reaction or production of two or more gaseous substances may be used directly to indicate the volumes of the gases participating in the reaction.   The volumes are related to the numbers of molecules indicated in the equation and may be evaluated without reference to the reacting weights of the gases.   For example:

gas	**gas**		**gas**	**gas**
$4\,NH_3$	$+$ $3\,O_2$	$\longrightarrow$	$2\,N_2$	$+$ $6\,H_2O$ (vapor)
4 molecules	3 molecules		2 molecules	6 molecules
4 volumes	3 volumes		2 volumes	6 volumes
4 liters	3 liters		2 liters	6 liters
4 cubic feet	3 cubic feet		2 cubic feet	6 cubic feet

It must be recognized that gas volume calculations require further knowledge: the conditions of temperature and pressure must be stated (or assumed) in order to know which substances exist as gases, and the formulas of these gases must represent the correct number of atoms in the gaseous molecules.   For example, if the reacting and resulting gas volumes in the above equation were measured at atmospheric pressure, but above 100°C, 7 volume units of reactants $(4\,NH_3 + 3\,O_2)$ would produce 8 volume units of products $(2\,N_2 + 6\,H_2O$ vapor); whereas at standard conditions (0°C) the 6 volumes of steam vapor would be condensed to a negligible volume (0.005 in the same units) of liquid water, and the 7 volume units $(4\,NH_3 + 3\,O_2)$ would produce only 2 volume units $(2\,N_2)$ of gaseous products.

Solved Problems

VOLUMES AND MOLECULAR WEIGHTS OF GASES

8.1.   Determine the approximate molecular weight of a gas if 560 ml weighs 1.55 g at S.T.P.

$$M = \frac{gRT}{PV} = \frac{1.55 \text{ g} \times 0.0821 \text{ } l \text{ atm } °K^{-1} \text{ mole}^{-1} \times 273°K}{1 \text{ atm} \times 0.560 \text{ } l} = 62.0 \text{ g/mole}$$

Another Method.

Molecular weight   =   weight of 1 liter at S.T.P. $\times$ number of liters in 1 mole

$$= \frac{1.55 \text{ g}}{0.560 \text{ } l} \times 22.4 \text{ } l/\text{mole} = 62.0 \text{ g/mole}$$

8.2. At 18°C and 765 mm, 1.29 l of a gas weighs 2.71 g. Calculate the approximate molecular weight of the gas.

$$PV \ = \ nRT \quad \text{or} \quad PV \ = \ \frac{g}{M} RT \qquad \text{where} \quad \begin{aligned} g &= \text{number of grams of gas} \\ M &= \text{molecular weight of gas} \end{aligned}$$

$P = \dfrac{765}{760}$ atmospheres, $\quad R = 0.0821 \ l$ atm °K^{-1} mole^{-1}, $\quad T = (18+273)$°K $= 291$°K

$$M \ = \ \frac{gRT}{PV} \ = \ \frac{2.71 \text{ g} \times 0.0821 \ l \text{ atm °K}^{-1} \text{ mole}^{-1} \times 291°\text{K}}{765/760 \text{ atm} \times 1.29 \ l} \ = \ 49.8 \text{ g mole}^{-1}, \quad \text{or} \quad 49.8 \text{ g/mole}$$

Another Method.

Volume of 2.71 g at S.T.P. $\ = \ 1.29 \ l \times \dfrac{765 \text{ mm}}{760 \text{ mm}} \times \dfrac{273°\text{K}}{291°\text{K}} \ = \ 1.22 \ l$

Molecular weight $\ = \ $ weight of 1 l at S.T.P. $\times$ number of liters in 1 mole

$$= \ \frac{2.71 \text{ g}}{1.22 \ l} \times 22.4 \ l/\text{mole} \ = \ 49.8 \text{ g/mole}$$

Note. The student will observe, by comparing this problem with the previous one, that the 22.4 l calculation is faster when data are given at S.T.P., and that the $PV = nRT$ calculation is faster when the data are given for other conditions.

8.3. Determine the volume occupied by 4.0 g of oxygen at S.T.P. Molecular weight of oxygen is 32.

Volume of 4.0 g O_2 at S.T.P. $\ = \ $ number of moles in 4.0 g O_2 $\times$ volume of 1 mole

$$= \ \frac{4.0 \text{ g}}{32 \text{ g/mole}} \times 22.4 \ l/\text{mole} \ = \ 2.8 \ l \ O_2 \text{ at S.T.P.}$$

8.4. What volume would 15.0 g of argon occupy at 90°C and 735 mm?

$$V \ = \ \frac{gRT}{MP} \ = \ \frac{15.0 \text{ g} \times 0.0821 \ l \text{ atm mole}^{-1} \text{ °K}^{-1} \times 363°\text{K}}{39.9 \text{ g mole}^{-1} \times 735/760 \text{ atm}} \ = \ 11.6 \ l$$

Another Method.

At S.T.P. this much argon would occupy the following space:

$$V_{\text{S.T.P.}} \ = \ \frac{15.0}{39.9} \text{ mole} \times 22.4 \ l/\text{mole} \ = \ 8.4 \ l$$

This problem can be completed by the usual procedure for converting from S.T.P. to an arbitrary set of conditions.

$$V_2 \ = \ V_1 \times \frac{T_2}{T_1} \times \frac{P_1}{P_2} \ = \ 8.4 \ l \times \frac{363°\text{K}}{273°\text{K}} \times \frac{760 \text{ mm}}{735 \text{ mm}} \ = \ 11.6 \ l$$

8.5. Compute the approximate density of methane, CH_4, at 20°C and 5.00 atm. The molecular weight of methane is 16.0.

$$\text{Density} \ = \ \frac{g}{V} \ = \ \frac{MP}{RT} \ = \ \frac{16.0 \text{ g mole}^{-1} \times 5.00 \text{ atm}}{0.0821 \ l \text{ atm °K}^{-1} \text{ mole}^{-1} \times 293°\text{K}} \ = \ 3.33 \text{ g/}l$$

Another Method.

$$\text{Density} \ = \ \frac{\text{mass of 1 mole}}{\text{volume of 1 mole}} \ = \ \frac{16.0 \text{ g}}{22.4 \ l \times \dfrac{1.00 \text{ atm}}{5.00 \text{ atm}} \times \dfrac{293°\text{K}}{273°\text{K}}} \ = \ 3.33 \text{ g/}l$$

8.6. If the density of carbon monoxide is 3.17 g/l at -20°C and 2.35 atm, what is its approximate molecular weight?

$$M \ = \ \frac{gRT}{VP} \ = \ \frac{dRT}{P} \ = \ \frac{3.17 \text{ g } l^{-1} \times 0.0821 \ l \text{ atm °K}^{-1} \text{ mole}^{-1} \times 253°\text{K}}{2.35 \text{ atm}} \ = \ 28.0 \text{ g/mole}$$

Another Method.

The density of a gas varies directly with the pressure and inversely with the absolute temperature.

$$\text{Density at S.T.P.} = 3.17 \text{ g}/l \times \frac{1.00 \text{ atm}}{2.35 \text{ atm}} \times \frac{253°\text{K}}{273°\text{K}} = 1.25 \text{ g}/l$$

$$\text{Molecular weight} = \text{weight of 1 } l \text{ at S.T.P.} \times \text{number of liters in 1 mole}$$

$$= 1.25 \text{ g}/l \times 22.4 \text{ }l/\text{mole} = 28.0 \text{ g/mole}$$

8.7. An organic compound had the following analysis: C = 55.8%, H = 7.03%, O = 37.2%. A 1.500 g sample was vaporized and was found to occupy 530 cc at 100°C and 740 mm. What is the molecular formula of the compound?

The approximate molecular weight, calculated from the gas density data, is 89.0. The empirical formula, calculated from the percentage composition data, is C_2H_3O, and the empirical formula weight is 43.0. The exact molecular weight must therefore be $2 \times 43.0 = 86.0$, since this is the only integral multiple of 43.0 which is reasonably close to the approximate molecular weight of 89. The molecule must therefore contain twice the number of atoms in the empirical formula, and the molecular formula must be $C_4H_6O_2$.

Another Method.

Instead of calculating the empirical formula, we can use the composition data to calculate the number of g-at of each element in 89 g of compound.

$$n_C = \frac{.558 \times 89 \text{ g C}}{12.0 \text{ g/g-at}} = 4.1, \quad n_H = \frac{.0703 \times 89 \text{ g H}}{1.01 \text{ g/g-at}} = 6.2, \quad n_O = \frac{.372 \times 89 \text{ g O}}{16.0 \text{ g/g-at}} = 2.1$$

These numbers approximate the numbers of atoms in a molecule, the small deviations from integral values resulting from the approximate nature of the molecular weight. The molecular formula, $C_4H_6O_2$, is obtained without going through the intermediate evaluation of an empirical formula.

8.8. It is found that in 11.2 l at S.T.P. of any gaseous compound of phosphorus there is never less than 15.5 g of phosphorus. Also, this volume of the vapor of phosphorus itself at S.T.P. weighs 62 g. State what conclusions may be drawn from these data with reference to the atomic and molecular weights of phosphorus.

In 22.4 l there is never less than 2×15.5 g $= 31.0$ g of phosphorus. As one atom is the smallest amount of an element that can exist in a molecule, one gram-atom of the element is the smallest amount that can exist in a mole of a compound (22.4 l for gaseous compounds at S.T.P.). Hence the approximate atomic weight of phosphorus is 31.0.

Also, 22.4 l of phosphorus vapor weighs 2×62 g $= 124$ g. Hence the approximate molecular weight of phosphorus is 124, and the molecule contains $124/31 = 4$ atoms (P_4).

8.9. Mercury diffusion pumps are used in the laboratory to produce a high vacuum. Cold traps are generally placed between the pump and the system to be evacuated. These cause the condensation of mercury vapor and prevent mercury from diffusing back into the system. The maximum pressure of mercury that can exist in the system is the vapor pressure of mercury at the temperature of the cold trap. Calculate the number of mercury vapor molecules per cc in a cold trap maintained at −120°C. The vapor pressure of mercury at this temperature is 10^{-16} mm.

$$\text{Moles per cc} = n = \frac{PV}{RT} = \frac{\frac{10^{-16}}{760} \text{ atm} \times 10^{-3} \text{ } l}{0.0821 \text{ } l \text{ atm } °\text{K}^{-1} \text{ mole}^{-1} \times 153°\text{K}} = 1.0 \times 10^{-23} \text{ mole}$$

$$\text{Molecules per cc} = 1.0 \times 10^{-23} \text{ mole} \times 6 \times 10^{23} \text{ molecules/mole} = 6 \text{ molecules}$$

DIFFUSION OF GASES

8.10. Compute the relative rates of effusion of H_2 and CO_2 through a fine pinhole. Molecular weight of $H_2 = 2.0$, of $CO_2 = 44$.

$$\frac{r_{H_2}}{r_{CO_2}} = \sqrt{\frac{M_{CO_2}}{M_{H_2}}} = \sqrt{\frac{44}{2.0}} = \sqrt{22} = 4.7 \quad \text{or} \quad r_{H_2} = 4.7 \times r_{CO_2}$$

COMBINING VOLUME PROBLEMS

8.11. What volume of hydrogen will combine with 12 liters of chlorine to form hydrogen chloride? What volume of hydrogen chloride will be formed? Assume the same temperature and pressure for all gases.

The balanced equation for this reaction is

gas	gas		gas
H_2	$+$ Cl_2	$\longrightarrow$	2 HCl
1 molecule	1 molecule		2 molecules
1 volume	1 volume		2 volumes
1 liter	1 liter		2 liters
12 liters	12 liters		2×12 liters $= 24$ liters

The equation shows that 1 molecule H_2 reacts with 1 molecule Cl_2 to form 2 molecules HCl.

But, by Avogadro's hypothesis, equal numbers of molecules of *gases* under the same conditions of temperature and pressure occupy equal volumes.

Therefore the equation also indicates that 1 volume (liter, cubic foot, or any other unit of volume) of H_2 reacts with 1 volume of Cl_2 to form 2 volumes of HCl (gas).

Thus 12 l of H_2 combines with 12 l of Cl_2 to form $2 \times 12 \, l = 24 \, l$ of HCl.

8.12. What volume of hydrogen will unite with 6 cubic feet of nitrogen to form ammonia? What volume of ammonia will be produced?

gas	gas		gas
N_2	$+$ 3 H_2	$\longrightarrow$	2 NH_3
1 molecule	3 molecules		2 molecules
1 volume	3 volumes		2 volumes
1 ft³	3 ft³		2 ft³
6 ft³	$3 \times 6 = 18$ ft³		$2 \times 6 = 12$ ft³

The equation indicates that

1 molecule N_2 reacts with 3 molecules H_2 to form 2 molecules NH_3

Then   1 volume N_2 reacts with 3 volumes H_2 to form 2 volumes NH_3

and   6 ft³ N_2 reacts with $3 \times 6 = 18$ ft³ H_2 to form $2 \times 6 = 12$ ft³ NH_3

8.13. What volume of O_2 at S.T.P. is required for the complete combustion of one mole of carbon disulfide, CS_2? What volumes of CO_2 and SO_2 at S.T.P. are produced?

liquid	gas		gas		gas
CS_2	$+$ 3 O_2	$\longrightarrow$	CO_2	$+$	2 SO_2
1 mole	3 moles		1 mole		2 moles
	$3 \times 22.4 = 67.2 \, l$		22.4 l		$2 \times 22.4 = 44.8 \, l$

The equation shows that 1 mole CS_2 reacts with 3 moles O_2 to form 1 mole CO_2 and 2 moles SO_2. One mole of a *gas* at S.T.P. occupies 22.4 liters. Therefore:

$$\text{Volume of 3 moles } O_2 \text{ at S.T.P.} = 3 \text{ moles} \times 22.4 \; l/\text{mole} = 67.2 \; l \; O_2$$

$$\text{Volume of 1 mole } CO_2 \text{ at S.T.P.} = 1 \text{ mole} \times 22.4 \; l/\text{mole} = 22.4 \; l \; CO_2$$

$$\text{Volume of 2 moles } SO_2 \text{ at S.T.P.} = 2 \text{ moles} \times 22.4 \; l/\text{mole} = 44.8 \; l \; SO_2$$

Note. CS_2 is a liquid, while O_2, CO_2 and SO_2 are gases. The above *volume* relationship is true only for *gases*. No statement can be made about the volume of 1 mole of CS_2 unless specific information about its density is given.

8.14. How many liters of oxygen, at standard conditions, can be obtained from 100 grams of potassium chlorate?

$$
\begin{array}{cccccc}
\text{solid} & & & & & \text{gas} \\
2 \; KClO_3 & \longrightarrow & 2 \; KCl & + & & 3 \; O_2 \\
\end{array}
$$

2 gfw
= 2 × 122.6 g

3 moles
= 3 × 22.4 l at S.T.P.

Molar Method.

The equation shows that 2 gfw $KClO_3$ gives 3 moles O_2. As in previous chapters, the symbol n will be used for the number of moles, gram-atoms, or gram-formula weights, whichever is appropriate.

$$n_{KClO_3} = \frac{100 \text{ g}}{122.6 \text{ g/gfw}} = 0.816 \text{ gfw } KClO_3$$

$$n_{O_2} = 3/2 \; n_{KClO_3} = 3/2 \times 0.816 = 1.224 \text{ moles } O_2$$

$$\text{Volume of 1.224 moles } O_2 \text{ at S.T.P.} = 1.224 \text{ moles} \times 22.4 \; l/\text{mole} = 27.4 \; l \; O_2$$

Another Method.

The equation shows that 2 gfw $KClO_3$ (2 × 122.6 = 245.2 g) gives 3 molar volumes O_2 (3 × 22.4 l = 67.2 l). Then

$$245.2 \text{ g } KClO_3 \quad \text{gives} \quad 67.2 \; l \; O_2 \quad \text{at S.T.P.}$$

$$1 \text{ g } KClO_3 \quad \text{gives} \quad \frac{67.2}{245.2} \; l \; O_2 \quad \text{at S.T.P.}$$

and

$$100 \text{ g } KClO_3 \quad \text{gives} \quad 100 \times \frac{67.2}{245.2} \; l = 27.4 \; l \; O_2 \quad \text{at S.T.P.}$$

Note. It was not necessary to calculate the *weight* of oxygen formed in either method of solution.

8.15. What volume of oxygen, at 18°C and 750 mm, can be obtained from 100 g of $KClO_3$?

This problem is identical with the previous one, except that the volume of 27.4 liters of O_2 at 0°C and 760 mm must be converted to liters of O_2 at 18°C and 750 mm.

$$\text{Volume at 18°C, 750 mm} = 27.4 \; l \times \frac{(273 + 18)°K}{273°K} \times \frac{760 \text{ mm}}{750 \text{ mm}} = 29.6 \; l$$

Another Method.

$$V = \frac{nRT}{P} = \frac{1.224 \text{ moles} \times 0.0821 \; l \text{ atm mole}^{-1} \; °K^{-1} \times 291°K}{750/760 \text{ atm}} = 29.6 \; l$$

8.16. How many grams of zinc must be dissolved in sulfuric acid in order to obtain 500 ml of hydrogen at 20°C and 770 mm?

$$
\begin{array}{cccccc}
\text{solid} & & & & & \text{gas} \\
Zn & + & H_2SO_4 & \longrightarrow & ZnSO_4 & + & H_2 \\
\end{array}
$$

1 gram-atom
= 65.4 g

1 mole
= 22.4 l at S.T.P.

The number of moles of H_2 may be found by either of the following methods:

$$n_{H_2} = \frac{PV}{RT} = \frac{(770/760) \text{ atm} \times 0.500 \, l}{0.082 \dfrac{l \text{ atm}}{\text{mole deg}} \times 293 \text{ deg}} = 0.0211 \text{ mole } H_2$$

or
$$V_{\text{S.T.P.}} = 500 \text{ ml} \times \frac{273°\text{K}}{(273+20)°\text{K}} \times \frac{770 \text{ mm}}{760 \text{ mm}} = 472 \text{ ml} = 0.472 \, l$$

$$n_{H_2} = \frac{0.472 \, l}{22.4 \, l/\text{mole}} = 0.0211 \text{ mole } H_2$$

The equation shows that 1 mole H_2 requires 1 gram-atom Zn. Then 0.0211 mole H_2 requires 0.0211 gram-atom Zn. (Atomic weight of Zn is 65.4).

Grams of Zn in 0.0211 g-atom $= 0.0211$ g-atom $\times 65.4$ g/g-atom $= 1.38$ g Zn

8.17. A natural gas sample contains 84% (by volume) CH_4, 10% C_2H_6, 3% C_3H_8, and 3% N_2. If a series of catalytic reactions could be used for converting all the carbon atoms of the gas into butadiene, C_4H_6, with 100% efficiency, how much butadiene could be prepared from 100 g of the natural gas?

Molecular weights: $CH_4 = 16$, $C_2H_6 = 30$, $C_3H_8 = 44$, $N_2 = 28$, $C_4H_6 = 54$. The volume percent of a gas mixture is the same as the mole percent.

$$\begin{aligned} 100 \text{ moles mixture} &= 84 \text{ moles } CH_4 + 10 \text{ moles } C_2H_6 + 3 \text{ moles } C_3H_8 + 3 \text{ moles } N_2 \\ &= 84 \times 16 \text{ g } CH_4 + 10 \times 30 \text{ g } C_2H_6 + 3 \times 44 \text{ g } C_3H_8 + 3 \times 28 \text{ g } N_2 \\ &= 1860 \text{ g natural gas} \end{aligned}$$

Gram-atoms C in 100 moles mixture $= 84 \times 1 + 10 \times 2 + 3 \times 3 = 113$ gram-atoms C.

Since   4 g-atoms C give 1 mole (54 g) C_4H_6,

113 g-atoms C give 113/4 moles $C_4H_6 = 113/4$ moles $\times 54$ g/mole $= 1530$ g C_4H_6.

Then   1860 g natural gas yields 1530 g C_4H_6

and   100 g natural gas yields $\dfrac{1530}{1860} \times 100$ g $= 82$ g C_4H_6

Supplementary Problems

8.18. If 200 ml of a gas weighs 0.268 g at S.T.P., what is its molecular weight?   *Ans.* 30.0

8.19. Compute the volume of 11 g of nitrous oxide, N_2O, at S.T.P.   *Ans.* 5.6 l

8.20. What volume will 1.216 g of SO_2 gas occupy at 18°C and 755 mm?   *Ans.* 456 ml

8.21. A 1.225 g mass of a volatile liquid is vaporized, giving 400 ml of vapor when measured over water at 30°C and 770 mm. The vapor pressure of water at 30°C is 32 mm. What is the molecular weight of the substance?   *Ans.* 78.4

8.22. Compute the weight of one liter of ammonia gas, NH_3, at S.T.P.   *Ans.* 0.76 g/l

8.23. Determine the density of H_2S gas at 27°C and 2.00 atmospheres.   *Ans.* 2.77 g/l

8.24. Find the molecular weight of a gas whose density at 40°C and 785 mm is 1.286 g/l.   *Ans.* 32.0

8.25. What weight of hydrogen at S.T.P. could be contained in a vessel that holds 4.0 g of oxygen at S.T.P.?   *Ans.* 0.25 g

8.26. One of the methods for estimating the temperature of the center of the sun is based on the ideal gas law. If the center is assumed to consist of gases whose average molecular weight is 2.0, and if the density and pressure are 1.4 g/cm^3 and 1.3×10^9 atmospheres respectively, calculate the temperature.   *Ans.* 2.3×10^7 °K

8.27. An electronic vacuum tube was sealed off during manufacture at a pressure of 1.2×10^{-5} mm at 27°C. Its volume is 100 ml. Compute the number of gas molecules remaining in the tube.
Ans. 4×10^{13} molecules

8.28. An empty steel gas tank with valve weighed 125 lb. Its capacity is 1.5 ft^3. When the tank was filled with oxygen to 2000 lb/in^2 at 25°C, what percent of the total weight of the full tank was O_2? Assume the validity of the ideal gas laws.   *Ans.* 12%

8.29. Pure oxygen gas is not necessarily the most compact source of oxygen for confined fuel systems because of the weight of the cylinder necessary to confine the gas. Other compact sources are hydrogen peroxide and lithium peroxide. The oxygen-yielding reactions are:

$$2\,H_2O_2 \longrightarrow 2\,H_2O + O_2 \qquad \text{and} \qquad 2\,Li_2O_2 \longrightarrow 2\,Li_2O + O_2$$

Rate (*a*) 65% (by weight) H_2O_2 and (*b*) pure Li_2O_2 in terms of % of total weight which is "available" oxygen. Neglect the weights of the containers. Compare with the previous problem.
Ans. (*a*) 31%, (*b*) 35%

8.30. An iron meteorite was analyzed for its isotopic argon content. The amount of Ar^{36} was 2.00×10^{-4} cc (S.T.P.) per kg of meteorite. If each Ar^{36} atom had been formed by a single cosmic event, how many such events must there have been per gram of meteorite?   *Ans.* 5.4×10^{12}

8.31. Three volatile compounds of a certain element have gaseous densities calculated back to S.T.P. as follows: 6.75, 9.56 and 10.08 g/l. The three compounds contain 96.0%, 33.9% and 96.4% of the element in question, respectively. What is the most probable atomic weight of the element?
Ans. 72.6, although the data do not exclude 72.6/*n*, where *n* is a positive integer.

8.32. Chemical absorbers can be used to remove exhaled CO_2 of space travelers in short space flights. Li_2O is one of the most efficient in terms of absorbing capacity per unit weight. If the reaction is

$$Li_2O + CO_2 \longrightarrow Li_2CO_3$$

what is the absorption efficiency of pure Li_2O in ft^3 CO_2 (S.T.P.) capacity per pound?
Ans. 12.0 ft^3/lb

8.33. Argon gas liberated from crushed meteorites does not have the same isotopic composition as atmospheric argon. The gas density of a particular sample of meteoritic Ar was found to be 1.481 g/l at 27°C and 740 mm. What is the average atomic weight of this sample of argon?   *Ans.* 37.5

8.34. Exactly 500 ml of a gas at S.T.P. weighs 0.581 g. The composition of the gas is as follows: C = 92.24%, H = 7.76%. Derive its molecular formula.   *Ans.* C_2H_2

8.35. A hydrocarbon has the following composition: C = 82.66%, H = 17.34%. The density of the vapor is 0.2308 g/l at 30°C and 75 mm. Determine its molecular weight and its molecular formula.
Ans. 58.1, C_4H_{10}

8.36. How many grams of oxygen are contained in 10.5 liters of oxygen measured over water at 25°C and 740 mm? Vapor pressure of water at 25°C is 24 mm.   *Ans.* 12.9 g

8.37. The density of NO was very carefully determined to be 0.2579 g/l at a temperature and pressure at which oxygen's measured density was 0.2749 g/l. On the basis of this information and the known atomic weight of oxygen, calculate the atomic weight of nitrogen.   *Ans.* 14.02

8.38. (*a*) Assuming that air consists of 79 mole % N_2, 20 % O_2 and 1 % Ar, compute the average molecular weight of air. (*b*) What is the density of air at 25°C and 1 atm?   *Ans.* 28.9, 1.18 g/l

8.39. An empty flask open to the air weighed 24.173 g. The flask was filled with the vapor of an organic liquid and was sealed off at the barometric pressure at 100°C. At room temperature the flask then weighed 25.002 g. The flask was then opened, filled with water at room temperature, and weighed 176 g. The barometric reading was 725 mm. All weighings were done at the temperature of the room, 25°C. What is the molecular weight of the organic vapor? Allow for the weight of air in the flask during the initial weighing, using the result from Problem 8.38.   *Ans.* 211

8.40. A 50 ml sample of a hydrogen-oxygen mixture was placed in a gas buret at 18°C and confined at barometric pressure. A spark was passed through the sample so that the formation of water could go to completion. The resulting pure gas had a volume of 10 ml at barometric pressure. What was the initial mole percentage of hydrogen in the mixture (a) if the residual gas after sparking was hydrogen, (b) if the residual gas was oxygen?   *Ans.* (a) 73%, (b) 53%

8.41. How much water vapor is contained in a cubic room 10 ft along an edge if the relative humidity is 50% and the temperature is 80°F? The vapor pressure of water at 80°F is 26.2 mm. The relative humidity expresses the partial pressure of water as percentage of the water vapor pressure.
Ans. 0.79 lb

8.42. Uranium isotopes have been separated by taking advantage of the different rates of diffusion of the two isotopic forms of UF_6. One form contains uranium of atomic weight 238, and the other of atomic weight 235. What are the relative rates of diffusion of these two molecules under ideal conditions?
Ans. UF_6 with U^{235} is faster by a factor of 1.004

8.43. The pressure in a vessel that contained pure oxygen dropped from 2000 mm to 1500 mm in 47 minutes when the oxygen leaked through a small hole. When the same vessel was filled with another gas, the pressure dropped from 2000 mm to 1500 mm in 74 minutes. What is the molecular weight of the second gas?   *Ans.* 79

8.44. A large cylinder of helium filled at 2000 lb/in² had a small thin orifice through which helium escaped at the rate of 3.4 millimoles per hour. How long would it take for 10 millimoles of CO to leak through a similar orifice if the CO were confined at the same pressure?   *Ans.* 7.8 hr

8.45. Ethane gas, C_2H_6, burns in air as indicated by the equation: $2\,C_2H_6 + 7\,O_2 \longrightarrow 4\,CO_2 + 6\,H_2O$. Determine the number of
(a) moles of CO_2 and of H_2O formed when 1 mole of C_2H_6 is burned.
(b) liters of O_2 required to burn 1 liter of C_2H_6.
(c) liters of CO_2 formed when 25 liters of C_2H_6 is burned.
(d) liters (S.T.P.) of CO_2 formed when 1 mole of C_2H_6 is burned.
(e) moles of CO_2 formed when 25 liters (S.T.P.) of C_2H_6 is burned.
(f) grams of CO_2 formed when 25 liters (S.T.P.) of C_2H_6 is burned.
Ans. (a) 2 moles CO_2, 3 moles H_2O;   (b) 3.5 *l*;   (c) 50 *l*;   (d) 44.8 *l*;   (e) 2.23 moles;   (f) 98.2 g

8.46. In order to economize on the oxygen supply in space ships, it has been suggested that the oxygen in exhaled CO_2 be converted to water by a reduction with hydrogen. The CO_2 output per astronaut has been estimated as 2.25 pounds per 24 hour day. An experimental catalytic converter reduces CO_2 at a rate of 500 cc (S.T.P.) per minute. What fraction of the time would such a converter have to operate in order to keep up with the CO_2 output of one astronaut?   *Ans.* 72%

8.47. How many grams of $KClO_3$ are needed to prepare 18 liters of oxygen which is collected over water at 22°C and 760 mm? Vapor pressure of water at 22°C is 19.8 mm.   *Ans.* 59.2 g

8.48. What volume of hydrochloric acid solution, of specific gravity 1.18 and containing 35.0% by weight of HCl, must be allowed to react with zinc in order to liberate 4.68 g of hydrogen?   *Ans.* 410 ml

8.49. Fifty grams of aluminum is to be treated with a 10% excess of H_2SO_4. The chemical equation for the reaction is: $2\,Al + 3\,H_2SO_4 \longrightarrow Al_2(SO_4)_3 + 3\,H_2$.
(a) What volume of concentrated sulfuric acid, of specific gravity 1.80 and containing 96.5% H_2SO_4 by weight, must be taken? (b) What volume of hydrogen would be collected over water at 20°C and 785 mm? Vapor pressure of water at 20°C is 17.5 mm.   *Ans.* 173 ml acid, 66.2 *l* H_2

8.50. A commercial grade of calcium carbide, CaC_2, on reacting with water liberated 195 ml of acetylene, C_2H_2, from a 0.712 g sample. The gas was measured over water at 15°C and 748 mm. Vapor pressure of water at 15°C is 13 mm. $CaC_2 + 2\,H_2O \longrightarrow Ca(OH)_2 + C_2H_2$. Determine:

(a) the volume of the acetylene at standard conditions,

(b) the weight of pure CaC_2 needed to give this volume,

(c) the percent by weight of CaC_2 in the commercial sample.

Ans. 179 ml, 0.512 g, 71.9%

8.51. What volume of hydrogen sulfide, at standard conditions, is required to precipitate completely PbS from 500 ml of a solution containing 1.20 g $Pb(NO_3)_2$ in 100 ml of solution?   *Ans.* 406 ml

8.52. In an Oklahoma oil field, 41,000 gallons of hydrochloric acid, containing 1.34 lb of pure HCl per gallon, were pumped into a well to remove binding $CaCO_3$ deposits.

(a) How many liters of CO_2 at 20°C and 1 atmosphere were thereby formed?

(b) How many kilograms of dry ice (CO_2) might have been made from this gas if 60% of it were recovered?

Ans. $8.2 \times 10^6\ l$, 9.0×10^3 kg

8.53. A catalytic process is known for converting *n*-heptane, C_7H_{16}, into toluene, C_7H_8. Hydrogen gas is a by-product. What volume at S.T.P. of hydrogen must be removed when 75 g of toluene are formed? *Ans.* 73 *l*

8.54. A 100 g sample of zinc which is 95% pure is treated with hydrochloric acid. What volume of hydrogen is produced (a) at S.T.P., (b) at 30°C and 768 mm?   *Ans.* 32.5 *l*, 35.7 *l*

8.55. The strength of hydrogen peroxide solutions is often expressed as "volume concentration", which is defined as the number of cc of oxygen (S.T.P.) formed by the decomposition of 1 cc of the peroxide solution according to the equation $2\,H_2O_2 \longrightarrow 2\,H_2O + O_2$. What is the "volume concentration" of an 85% by weight peroxide solution whose density is 1.401 g/cc?   *Ans.* 392

8.56. Calculate (a) the weight of MnO_2 and (b) the volume of hydrochloric acid, of specific gravity 1.12 and containing 40.0% HCl by weight, needed to produce 2.00 *l* of chlorine gas at standard conditions, according to the reaction: $MnO_2 + 4\,HCl \longrightarrow MnCl_2 + 2\,H_2O + Cl_2$.

Ans. 7.76 g MnO_2, 29.1 ml solution

Chapter 9

Structure of Matter

VALENCE

The formulas of the chemical compounds are no accident. There is an NaCl, but no $NaCl_2$; there is a CaF_2, but no CaF. On the other hand, certain pairs of elements form two, or even more, different compounds, e.g. Cu_2O, CuO, N_2O, NO, NO_2. A few sets of simple rules given here help to systematize the formulas of many simple compounds. A more extensive study of chemistry is needed to classify the more complex substances. The power of an atom to combine with others is called *valence*, and we may distinguish between the two main types, *ionic valence* and *covalence*.

IONIC VALENCE

Ionic valence describes the combination between oppositely charged particles or *ions*. The principal forces are the classical electrical forces operating between any two charged particles. Some of the more common ions are listed in Table 9-1 below. The charges of the elementary ions can be understood in terms of the electronic structure of atoms. A textbook should be consulted for a full discussion. If we accept the ionic charges as chemical facts, we can easily write the empirical formulas for ionic compounds to conform to the requirement that the compound as a whole is neutral. The charge of an ion is often called the *ionic valence* since it determines the number of opposite charges the ion can combine with to form a neutral compound.

The names of the elementary *cations* (positive ions) listed in the left-hand column of Table 9-1 below are simply the names of the elements, e.g. calcium ion. The names of the *anions* (negative ions) listed in the left-hand column are derived from the names of the elements by replacing the last (or last two) syllable(s) of the element names by the suffix *-ide*, e.g. chloride, oxide. The significance of the ionic radius for the elementary ions will be discussed later in this chapter. When two cations exist for the same element, the newer internationally accepted Stock convention is to write the charge per atom in parentheses after the name of the metal. There is still some remnant of the older custom of differentiating the two states by use of the *-ous* suffix for the lower of the two charge states and *-ic* for the higher. Distinctive suffixes are used for the complex oxy-anions in a partly systematic way, *-ate* for the most common or most stable and *-ite* for the ion containing less oxygen. The prefix *per-* is added to an *-ate* name to indicate even more oxygen, and the *hypo-* prefix is added to an *-ite* name to indicate a lower oxygen content than in the *-ite* ion.

Table 9-1

SOME COMMON IONS			
Ion	Ionic Radius	Ion	Name
H^+		Cu^+	copper (I), or cuprous
Li^+	0.60 Å	Cu^{2+}	copper (II), or cupric
Na^+	0.95 Å	Fe^{2+}	iron (II), or ferrous
K^+	1.33 Å	Fe^{3+}	iron (III), or ferric
Cs^+	1.69 Å	Cr^{2+}	chromium (II), or chromous
Ag^+	1.26 Å	Cr^{3+}	chromium (III), or chromic
Mg^{2+}	0.65 Å	Hg_2^{2+}	mercury (I), or mercurous
Ca^{2+}	0.99 Å	Hg^{2+}	mercury (II), or mercuric
Sr^{2+}	1.13 Å	NH_4^+	ammonium
Ba^{2+}	1.35 Å	OH^-	hydroxide
Zn^{2+}	0.74 Å	HCO_3^-	bicarbonate, or hydrogen carbonate
Cd^{2+}	0.97 Å	CO_3^{2-}	carbonate
Ni^{2+}	0.69 Å	NO_3^-	nitrate
Al^{3+}	0.50 Å	NO_2^-	nitrite
H^-		PO_4^{3-}	(ortho)phosphate
F^-	1.36 Å	SO_4^{2-}	sulfate
Cl^-	1.81 Å	SO_3^{2-}	sulfite
Br^-	1.95 Å	ClO_4^-	perchlorate
I^-	2.16 Å	ClO_3^-	chlorate
O^{2-}	1.40 Å	ClO_2^-	chlorite
S^{2-}	1.84 Å	ClO^-	hypochlorite
		$Cr_2O_7^{2-}$	dichromate
		CrO_4^{2-}	chromate
		MnO_4^-	permanganate

COVALENCE

The covalent force between atoms sharing two or more "bonding" electrons is related to the *delocalization*, or smearing out, of the electrons over a wider region of space in the bond than they occupy in the separated atoms. A stylized scheme has been developed for representing the electron distribution in covalent molecules and ions by structural formulas such as the following.

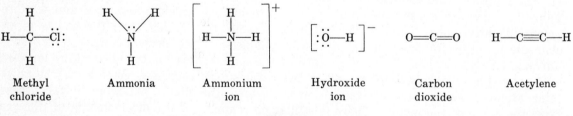

| Methyl chloride | Ammonia | Ammonium ion | Hydroxide ion | Carbon dioxide | Acetylene |

Fig. 9-1

In these formulas, a *line* between two atoms represents a *pair* of shared electrons and a *dot* represents an unshared electron. *Two lines* constitute a *double bond* of four shared electrons, and *three lines* constitute a *triple bond* of six shared electrons. The total number of electrons shown in such a molecular structure is equal to the sum of the number of outer shell, or *valence*, electrons in the free atoms: 1 for H, 4 for C, 5 for N, 6 for O, and

7 for Cl. For an ionic structure, one additional electron must be added to this sum for each unit of negative charge in the whole ion, as in OH^-, and one electron must be subtracted from the sum for each unit of positive charge on the ion, as in NH_4^+. The number of pairs of electrons shared by an atom is called its *covalence*.

The structures above conform to the *octet* rule, according to which a total of eight electrons, either shared or unshared, should be in the region of each atom beyond the first period of the periodic table. For hydrogen this number is two. The covalence of hydrogen is always one. The covalence of oxygen is practically always one or two. The covalence of carbon is four in almost all its stable compounds. Thus each carbon is expected to form either four single bonds, two double bonds, or a single and a triple bond. It is expected that all shared electrons will be counted twice in applying the octet rule, once for each atom participating in the bond. Although the octet rule is not a rigid rule of chemical bonding, it is obeyed for C, N, O, and F in almost all of their compounds. The rule breaks down commonly for elements in the third and higher periods of the periodic table.

It often happens that a single structural formula involving only pairs of shared electrons is inadequate to represent a substance correctly. In such cases, several such diagrams must be written and the true structure is said to be a *resonance* hybrid of the several diagrams. Consider the following two examples:

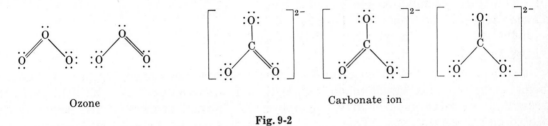

Ozone   Carbonate ion

Fig. 9-2

Experiment has shown that the two terminal oxygens in ozone are equivalent, that is, they are equidistant from the central oxygen. If only one of the resonance diagrams were written, it would appear that one of the terminal oxygens is bonded more strongly to the central oxygen (by a double bond) than is the other (by a single bond) and that the more strongly bonded atom should be closer to the central atom. The hybrid of the two ozone structures gives equal weight to the extra bonding of the two terminal oxygen atoms. Similarly, the three resonance structures of carbonate are needed to account for the experimental fact that all three oxygens are equidistant from the central carbon.

The total bond strength of a substance for which resonance structures are written is greater than would be expected if there were only one formal structure. This *additional* stabilization is called *resonance energy*. This fact is an extension of the principal source of covalent bond energy, the delocalization of electrons about the two atoms forming the bond. As a result of resonance in ozone, for example, the electrons constituting the second pair of the double bond are delocalized around the 3 oxygen atoms. The writing of two or more resonance structures is a way of overcoming some of the inadequacies of a single valence bond structure. We shall continue to use the valence bond formalism in this book, although an alternative procedure of *molecular orbitals* is probably more suited to the quantitative treatment of chemical binding. The molecular orbital approach, requiring advanced mathematical procedures, treats the valence electrons as being spread out over the whole molecule instead of being localized in particular bonds.

FORMAL CHARGE

Although a molecule as a whole is electrically neutral, it is of some interest to know whether there are local charges which can be identified with particular parts of a molecule, the algebraic sum of which would equal zero. In an ion, the algebraic sum would equal the charge of the ion as a whole. The determination of such detailed charge distributions is a fascinating current problem for research chemists. One approximate, but very useful, method of apportioning charges within a molecule or ion is the scheme of *formal charges*. In this scheme the shared electrons in a covalent bond are divided equally between the two atoms forming the bond. Unshared valence electrons on an atom are assigned exclusively to that atom. Each atom is then assigned a formal charge depending on the difference between the number of valence electrons assigned to it in the structure and the number of valence electrons possessed by that same atom in the free state. These charges may be written near the atoms on the structural diagrams, as shown for single resonance structures of ozone and carbonate. The central oxygen in ozone is assigned just 5 electrons (2 in the unshared pair plus half of the three pairs in the bonds); this atom, being one

electron short of the complement of six in a free oxygen atom, is thus assigned a formal charge of +1. The terminal oxygen connected by a single bond is assigned 7 (6 in the unshared pairs plus half of one pair in the bond); having one electron more than a neutral oxygen atom, this atom is assigned a formal charge of −1. The other terminal oxygen has no formal charge because 6 electrons are assigned to it (4 in the unshared pairs plus half of the two shared pairs).

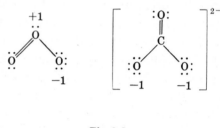

Fig. 9-3

A rough rule useful in choosing one electronic structure over another is that large formal charges, of magnitude much greater than 1, are avoided if possible. In particular, structures are to be avoided in which appreciable formal charges of the same sign are located on adjacent atoms. (This rule can be made even more useful by allowing for the unequal sharing of electrons in a covalent bond, as is done in more advanced treatments of molecular structure.)

BOND ANGLES

The structural formulas as usually written are not intended to reflect the actual distances or angles within the molecule. Such information, however, is often available. The relative angles between covalent bonds in a molecule are determined by the orbitals occupied by the bonding electrons. A few of the simpler rules can be stated here.

1. The three p orbitals in the valence shell make 90° angles with respect to each other. In SbH$_3$, for example, the three Sb—H bonds are practically perpendicular to each other and are presumed to occupy the three valence p orbitals of Sb.

2. Angles larger than 90° are made by bonds which occupy mixtures, or *hybrids*, of s and p orbitals. In H$_2$O vapor, the angle between the two O—H bonds is about 105°, showing the mixing of s and p orbitals.

3. Four equivalent bonds pointing from a central atom to the four corners of a regular tetrahedron can be made by mixing one s orbital with three p orbitals. This is the situation in CH$_4$, SiCl$_4$ and in many other compounds of the Group IV elements. A regular tetrahedron is shown in Fig. 9-4. In CH$_4$, as an example of this type of structure, the C is located at the center (not shown in the diagram) and an H at each corner. Any two CH bonds (not shown in the diagram) make an angle of 109°28′.

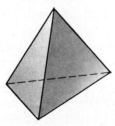

Fig. 9-4

4. Bond angles of 120° can be achieved by hybridization of one *s* orbital with two *p* orbitals. This is the case in BCl_3 and in C_2H_4. This angle insures that all six atoms in C_2H_4 lie in the same plane, as shown in Fig. 9-5.

Fig. 9-5   Fig. 9-6

5. A bond angle of 180° can result from the hybridization of one *s* orbital with one *p* orbital, as in C_2H_2 (see Fig. 9-6) or in the high temperature form of $BeCl_2$.

6. Bond angles much below 90° usually introduce considerable strain into a molecule.

COVALENT RADII

Considerable systematization of experimental bond lengths can be achieved by postulating that each single covalent bond distance, measured between centers of the bonding atoms, is the sum of the *covalent radii* of the two atoms. A table of such radii of the various atoms then becomes a useful compact source of quick and fairly reliable estimates of bond lengths. Such a compilation is shown in Table 9-2. (This procedure is not perfectly accurate, because atoms are not really hard spheres of unchangeable radius.)

Table 9-2

SINGLE BOND COVALENT RADII			
C	0.77 Å	O	0.66 Å
Si	1.17 Å	S	1.04 Å
N	0.70 Å	F	0.64 Å
P	1.10 Å	Cl	0.99 Å
Sb	1.41 Å	I	1.33 Å

Double bonds are shorter. From precisely measured carbon—carbon bond lengths in H_3C—CH_3, H_2C=CH_2 and HC≡CH, of 1.54 Å, 1.33 Å and 1.20 Å, a rule of thumb has been developed to allow a 0.21 Å shortening for any double bond and 0.34 Å shortening for any triple bond. In the case of resonance, a bond length is intermediate between the values it would have in the separate resonance structures. Some applications of Table 9-2 are shown below.

Substance	Bond	Predicted Length	Observed Length
CH_3Cl	C—Cl	$r_C + r_{Cl} =$ $0.77 + 0.99 = 1.76$ Å	1.77 Å
$(CH_3)_2O$	C—O	$r_C + r_O =$ $0.77 + 0.66 = 1.43$ Å	1.43 Å
H_2CO	C=O	$r_C + r_O - 0.21 =$ $1.43 - 0.21 = 1.22$ Å	1.22 Å
ICN	C≡N	$r_C + r_N - 0.34 =$ $0.77 + 0.70 - 0.34 = 1.13$ Å	1.16 Å

ISOMERS

There are many examples of two or more substances which have the same numbers of atoms of each element per molecule but differ in the spatial arrangement of the atoms. Such substances, known as *isomers,* are different substances and they may differ in their physical properties (like melting point, boiling point, and density) and in their chemical properties (reactivities toward other substances). The rules of valence can be used in many cases to predict the occurrence of isomerism. For example, two different structures can be drawn for C_4H_{10}, both conforming to the octet rule for carbon. In *n*-butane,

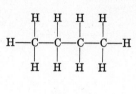

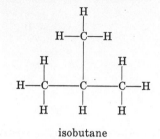

n-butane

isobutane

Fig. 9-7

two of the carbons are bonded to one carbon each and two are bonded to two each. In isobutane, three of the carbons are bonded to one carbon each and the fourth to three carbons. This difference in the description of the binding partners is called *structural isomerism*. *n*-butane melts at $-135°$ and boils at $0°$, while isobutane melts at $-145°$ and boils at $-10°$. Within any one molecule there is a continuous, practically unrestricted, rotation of the atoms about any C—C single bond, so that all structures which on paper differ only by the angular position of an atom or group of atoms are really the same, as shown in Fig. 9-8.

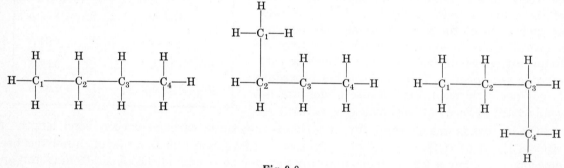

Fig. 9-8

These diagrams are all representations of the same substance, *n*-butane. In each, we can number the carbon atoms 1 to 4 as we move from one end of the molecule to the other through successive C—C bonds. We see that the kinds of neighbors of a given numbered carbon atom are the same in all three diagrams.

A different situation arises with compounds containing a carbon—carbon double bond. The extra strength of the double bond restricts the motion of groups with respect to this bond, so that diagrams showing different angular positions represent different substances. Consider the following structures:

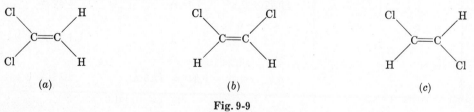

(a)   (b)   (c)

Fig. 9-9

Structures (a) and (b) are obviously structural isomers of each other, as are (a) and (c), since in (a) both chlorines are attached to the same carbon and in (b) or (c) one chlorine is attached to each carbon. Structures (b) and (c) are called *geometrical isomers* because they differ in the spatial arrangement of their atoms but not in the listing of the mere numbers of atoms of each kind to which each atom is bonded. All three structures are therefore different substances.   An analogous situation occurs in compounds with square symmetry.   The following two compounds are geometrical isomers.

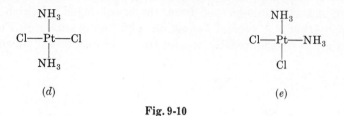

Fig. 9-10

Because of the rigidity of the square planar arrangement of the Pt, N and Cl atoms, the two forms are distinctive.   For example, the distance between the two chlorine atoms in (d) is greater than in (e).   If this compound had been tetrahedral with Pt in the center, isomerism would not have occurred because any corner of a regular tetrahedron is equidistant to the other three corners.

CRYSTALS

A crystalline substance is an ordered array of its component atoms, ions, or molecules. The arrangement of the simpler particles in the array is called a *lattice*.   Every lattice is a three-dimensional stacking of identical building blocks called *unit cells*.   The properties of the entire crystal, including its overall symmetry, can be understood in terms of the unit cell.

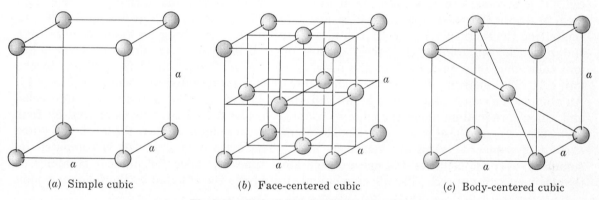

(a) Simple cubic   (b) Face-centered cubic   (c) Body-centered cubic

Fig. 9-11.   Unit Cells of Cubic Symmetry

The most symmetrical crystals have cubic lattices.   Within the cubic system, three kinds of elementary lattices are possible; the *simple cubic*, the *face-centered cubic*, and the *body-centered cubic*.   The corresponding unit cells are represented in Fig. 9-11.   The length of the cube edge is designated by a.   A substance made up of identical atoms crystallizing in the simple cubic lattice would have an atom located at each of the corners of the cubic unit cell, as indicated in Fig. 9-11(a).   (Actually no such crystalline substance exists.)   In a face-centered lattice an atom would be located at each corner of the unit cell and at the center of each of the six faces, as shown in Fig. 9-11(b).   In the body-centered lattice an atom would be located at each corner and one at the center of the unit cell, as shown in Fig. 9-11(c).   In all cases, the crystal could be thought of as a three-dimensional stack of unit cell cubes, packed face to face so as to fill all of the space occupied by the crystal.   Crystal classes less symmetrical than the cubic have unit cells which may be thought of as more or less distorted cubes, the opposite faces of which are parallel to each other.   The name of such a general geometric shape is a *parallelepiped*.   The cube is a very special parallelepiped in which each face is a square, and each face is perpendicular to two other faces.   A brick or cardboard box is a more general parallelepiped in which each face is a rectangle, usually with unequal sides.   In the most general parallelepiped the faces of the unit cell need not even be rectangles but are simply parallelograms.   Crystals with hexagonal symmetry, like snow (ice), have unit cells that are prisms with a vertical axis perpendicular to a rhombus-shaped base, the equal edges of which are at 60° and 120° with respect to each other.   A typical hexagonal unit cell is shown in Fig. 9-12.   The length of the rhombus edge is designated by a, and the height of the unit cell by c.

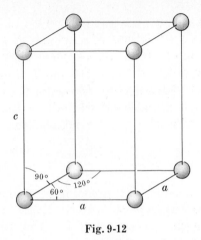

Fig. 9-12

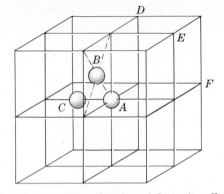

Stacking of cubes showing eight unit cells. Only a few representative atoms are shown.

Fig. 9-13

The bulk density of a perfect crystal can be computed from the properties of the unit cell.   It is necessary to apportion the mass of the crystal among the various unit cells, and then to divide the mass apportioned to one unit cell by the volume of the unit cell.   In computing the mass of a unit cell, it is necessary to avoid assigning an atom to more than one unit cell.   An atom at the corner of one unit cell is also at the corner of seven other unit cells; thus only one-eighth of the mass of any corner atom should be assigned to one unit cell.   This sharing of a corner by eight unit cells is represented by atom A in Fig. 9-13. An atom at the center of a face of a unit cell, like B in Fig. 9-13, is shared by two unit cells, and a non-corner atom along the edge of a unit cell, like C in Fig. 9-13, is shared by four unit cells.   In all these cases, only inside atoms are considered since those at the outer surface of a macro-crystal are relatively rare.   The mass of a unit cell is computed by summing, over all elements, the mass of one atom times the allocated number of atoms of that element per unit cell.   The allocated number of atoms is computed by the following sum:

$\frac{1}{8}$(number of atoms on corners of the unit cell)

$+$ $\frac{1}{4}$(number of non-corner atoms on unit cell edges)

$+$ $\frac{1}{2}$(number of non-edge atoms on the faces of the unit cell)

$+$ number of interior atoms within the unit cell

CRYSTAL FORCES

The strength of the forces holding crystalline substances together varies considerably. In *molecular crystals*, like CO_2 and benzene, each molecule is almost independent of all the others and retains practically the same internal geometry (bond lengths and angles) that it has in the gaseous or liquid state.   The crystal is held together by the relatively weak van der Waals forces, and the melting point is never very high.   For substances that can undergo intermolecular hydrogen binding, the intermolecular forces in the crystal can be great enough to impose a noticeable change in the molecular geometry.   Water is an example; the angle between the two O—H bonds in the vapor is about 105° but becomes 109°28′ in the crystal to conform to the crystalline, rather than molecular, spatial requirements.   In *metals*, a special type of crystalline force comes into play, characterized by a non-directional nature.   Fixed bond angles do not play a major role in metals, and the stablest crystal structures for the elementary metals are those with the densest packing. In the *covalent crystals*, like diamond or silicon carbide, the crystal is held together by a three-dimensional network of covalent bonds, the mutual angles of which are determined largely by the valence requirements of the individual atoms.

The attractive forces operating in *ionic crystals* are mostly electrostatic in nature, the classical attraction between oppositely charged particles. To avoid the repulsion between similarly charged particles, ionic substances crystallize in structures in which a positive and a negative ion can come within touching distance while ions of like charge are kept away from each other. In fact, the dimensions of most simple purely ionic crystals can be understood by assuming a fixed *ionic radius* for each ion, valid for all compounds of that ion, and the nearest cation-anion distance equal to the sum of the ionic radii of the cation and anion. Radii for some elementary ions are listed in Table 9-1.

Solved Problems

FORMULAS

9.1. Write formulas for the following ionic compounds:

(*a*) barium oxide, (*b*) aluminum chloride, (*c*) magnesium orthophosphate.

(*a*) The formula is BaO since the $+2$ charge on one barium ion just balances the -2 charge on the oxide ion.

(*b*) Three chloride ions, -1 charge on each, are needed to balance the $+3$ charge of one aluminum ion. The formula is $AlCl_3$.

(*c*) Since neither 2 (the positive charge on magnesium ion) nor 3 (the negative charge on orthophosphate) is an integral multiple of the other, the smallest number must be found which is a multiple of both in a manner analogous to finding the least common denominator. This number is 6. The formula, $Mg_3(PO_4)_2$, shows 6 units of positive charge (3×2) and 6 units of negative charge (2×3) per empirical formula unit.

9.2. Name the following compounds: (*a*) Mg_3P_2, (*b*) $Hg_2(NO_3)_2$, (*c*) NH_4TcO_4.

(*a*) Although the ion of phosphorus is not included in Table 9-1, it must be an anion, since magnesium forms only a cation. Application of the usual formula for naming binary compounds gives the name, magnesium phosphide.

(*b*) Since the charge in $2\,NO_3^-$ (nitrate ions) is -2, the total cationic charge (Hg_2) must be $+2$. Since the average charge per Hg is $+1$, the name is mercury (I) nitrate (or mercurous nitrate).

(*c*) The charge on the anion must be -1 in order to balance the $+1$ charge on the ammonium ion. Since Tc belongs to the same group of the periodic table as Mn, TcO_4^- is analogous to MnO_4^-, permanganate. The name is thus ammonium pertechnetate.

9.3. Determine the charges of the complex ions in parentheses [or brackets] in the following formulas:

(*a*) $Na_2(MnO_4)$ (*c*) $K_2H_3(IO_6)$ (*e*) $Ca_3(CoF_6)_2$ (*g*) $(UO_2)Cl_2$

(*b*) $H_4[Fe(CN_6)]$ (*d*) $Na_2(B_4O_7)$ (*f*) $Mg_3(BO_3)_2$ (*h*) $(SbO)_2SO_4$

(*a*) The charge of (MnO_4) must balance that of $2\,Na^+$; i.e., $(MnO_4)^{2-}$. (This ion is called *manganate* and is different from permanganate.)

(*b*) The ion in brackets must balance the charge of $4\,H^+$; i.e., $[Fe(CN)_6]^{4-}$.

(c) The charge of (IO_6) must balance the charge of $2\,K^+$ and $3\,H^+$; i.e., $(IO_6)^{5-}$.

(d) The charge of (B_4O_7) must balance the charge of $2\,Na^+$; i.e., $(B_4O_7)^{2-}$.

(e) The charge of $2\,(CoF_6)$ ions must balance the charge of $3\,Ca^{2+}$; i.e., $(CoF_6)^{3-}$.

(f) The charge of $2\,(BO_3)$ ions must balance the charge of $3\,Mg^{2+}$; i.e., $(BO_3)^{3-}$.

(g) The charge of the (UO_2) ion must balance the charge of $2\,Cl^-$; i.e., $(UO_2)^{2+}$.

(h) The charge of $2\,(SbO)$ ions must balance the charge of SO_4^{2-}; i.e., $(SbO)^+$.

9.4. The formula of calcium pyrophosphate is $Ca_2P_2O_7$. Determine the formulas of sodium pyrophosphate and ferric pyrophosphate.

The charge of the pyrophosphate ion must be -4 to balance the charge of $2\,Ca^{2+}$. We can then write $Na_4P_2O_7$ and $Fe_4(P_2O_7)_3$.

9.5. Write all possible octet structural formulas for   (a) CH_4O, (b) C_2H_3F, (c) N_3^-.

(a) Since the valence level of hydrogen is saturated with two electrons, each hydrogen can form only one covalent bond. Thus hydrogen cannot serve as a bridge between C and O. The only possibility of providing 4 bonds to the C is to have 3 H's and the O bonded directly to it. Only one structure is possible, as shown in Fig. 9-14. Note that the total number (14) of valence electrons in the structure is the sum of the numbers of valence electrons in the free component atoms: 4 (in C) + 6 (in O) + 4 (in 4 H).

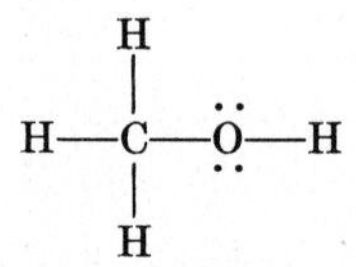

Fig. 9-14

(b) The only way of providing 4 bonds for each C within the limitation of 18 valence electrons is to have a carbon—carbon double bond, as in Fig. 9-15. The reader should convince himself by trial and error that no other structure is possible.

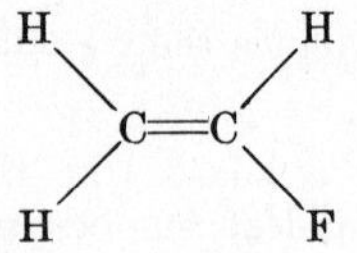

Fig. 9-15

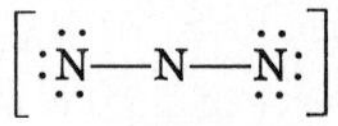

:N—N—N:

Incorrect Structure

Fig. 9-16

(c) The total number of available valence electrons is 16 (5 in each of the 3 free N atoms plus 1 arising from the net ionic charge). From this it can easily be seen that a linear structure without multiple bonds would not satisfy the octet rule. For example, in Fig. 9-16, the available electrons provide octets for the terminal nitrogen atoms but 4 less than an octet for the central one. The only way to have the same number of electrons provide octets for more atoms is to reduce the number of unshared electrons by 4 and increase the number of shared electrons accordingly, i.e. by forming either two double bonds or one triple bond. The possible linear structures are shown in Fig. 9-17.

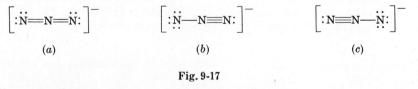

(a)   (b)   (c)

Fig. 9-17

Two resonance structures are written with the triple bond because there is no reason why one of the terminal nitrogens in an isolated azide ion should be different from the other. A different type of structure, involving a ring, also satisfies the octet rule. This structure (Fig. 9-18) is ruled out because the bond angles of 60° or less demanded by the structure require considerable strain from the normal angles of bonding.

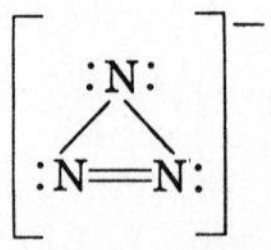

Fig. 9-18

9.6. Experimentally the azide ion, N_3^-, is found to be linear, with each adjacent nitrogen—nitrogen distance equal to 1.16 Å. (*a*) Evaluate the formal charge at each nitrogen in each of the 3 linear octet structures depicted in Fig. 9-17. (*b*) Predict the relative importance of these 3 resonance structures in N_3^-.

(*a*) In structure (*a*) of Fig. 9-17, the central N is assigned $\frac{1}{2}$ of the 4 shared pairs, or 4 electrons. This number is one less than in a free N atom, and thus this atom has a formal charge of +1. Each terminal N is assigned 4 unshared electrons plus $\frac{1}{2}$ of the 2 shared pairs, or 6 altogether. The formal charge is thus −1. The net charge of the ion, −1, is the sum of $2(-1) + 1$.

In structures (*b*) and (*c*) of Fig. 9-17, the central N again has 4 assigned electrons with a resulting formal charge of +1. The terminal triply bonded N has 2 plus $\frac{1}{2}$ of 3 pairs, or a total of 5, with a formal charge of 0. The singly bonded terminal N has 6 plus $\frac{1}{2}$ of 1 pair, or a total of 7, with a formal charge of −2. The net charge of the ion, −1, is the sum of +1 and −2.

(*b*) For structure (*a*) the nitrogen—nitrogen bond distance is predicted to be $(0.70 + 0.70)$ minus the double bond shortening of 0.21 Å, or 1.19 Å. The observed bond length, 1.16 Å, is just slightly shorter, possibly because of a small contribution from structures (*b*) and (*c*), which would give additional shortening because of the triple bond.

9.7. The sulfate ion is tetrahedral with four equal S—O distances of 1.49 Å. Draw a reasonable structural formula consistent with these facts.

It is possible to place the 32 valence electrons (6 for each of the 5 Group VI atoms plus 2 for the net negative charge) in an octet structure having only single bonds. There are two objections to this formula (Fig. 9-19). (1) The predicted bond distance, $r_S + r_O = 1.04 + 0.66 = 1.70$, is much too high. (2) The calculated formal charge on the sulfur, +2, is rather high. An alternative is to

Fig. 9-19   Fig. 9-20

write resonance structures containing double bonds. A structure like that in Fig. 9-20 places zero formal charge on the sulfur and −1 on each of the singly bonded oxygens. The shrinkage in bond length due to double bond formation helps to account for the low observed bond distance. Other resonance structures with alternate locations of the double bonds are, of course, implied. Structures like this with an expanded valence level beyond the octet are generally considered to involve *d* orbitals of the central atom. This is the reason why second period elements (C, N, O, F) do not form compounds requiring more than 8 valence electrons per atom; there is no 2*d* sub-shell.

9.8. Draw all the octet resonance structures of benzene, C_6H_6, and of naphthalene, $C_{10}H_8$. Benzene is known to have hexagonal symmetry and the carbon framework of naphthalene consists of two coplanar fused hexagons. Invoke bonding only between adjacent atoms.

Benzene.

Fig. 9-21

The hydrogens are distributed uniformly, one on each carbon, to conform with the hexagonal symmetry. This leaves three carbon—carbon bonds to be formed at each carbon to bring its total covalence to four. Alternating single and double bonds form the only scheme for writing formulas within all of the above restrictions. Hydrocarbons containing planar rings in which every ring carbon forms one double bond and two single bonds are called *aromatic* hydrocarbons. A short-hand notation has been developed for writing aromatic structures by the use of polygons. A carbon atom is assumed to lie at each corner of the polygons. Carbon—carbon bonds are written in the polygons but carbon—hydrogen bonds are not explicitly written. The same two structures shown above can be written as follows.

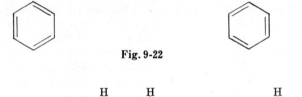

Fig. 9-22

Naphthalene.

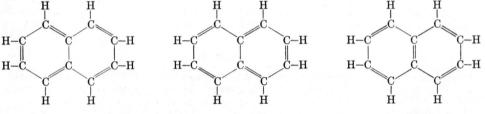

Fig. 9-23

The coplanarity of naphthalene indicates its aromatic character. Note that the two carbons at the fusion of the two rings reach their covalence of four without bonding to hydrogen. The short-hand notation of these structures is as follows:

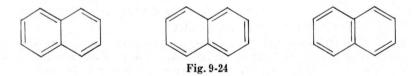

Fig. 9-24

A carbon—hydrogen bond is assumed at those carbons possessing only three bonds within the carbon skeleton.

9.9. Write all the structural isomeric formulas for C_4H_9Cl.

The molecular composition resembles butane, C_4H_{10}, with the exception that one H is substituted by Cl. The formulas can be found by picking all the differently located hydrogens in the two isomeric formulas of C_4H_{10} (Fig. 9-7).

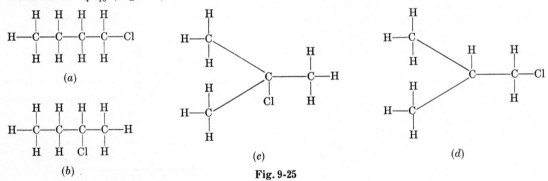

Fig. 9-25

Note that the two end carbons of *n*-butane are identical and the two interior carbons are identical. Thus a chlorine substituted on the left-hand carbon would have given the same compound as (a), only viewed from a different end. Similarly, substitution on the carbon next to the left would have given a compound identical with (b). In *iso*-butane, the three terminal carbons are identical. Because of the free rotation around the C—C single bond axis, it is immaterial which of the several hydrogens attached to a given C is substituted; all positions become averaged in time anyway because of the free rotation.

9.10. Write formulas for all the structural and geometric isomers of C_4H_8.

If the four carbons are in a row, there must be one double bond in order to satisfy the tetra-covalence of carbon. The double bond occurs either in the center of the molecule or toward an end. In the former case, two geometrical isomers occur with different positions of the terminal carbons with respect to the double bond; in the latter case, two structural isomers occur differing in the extent of branching within the skeleton. Additional possibilities are ring structures without double bonds.

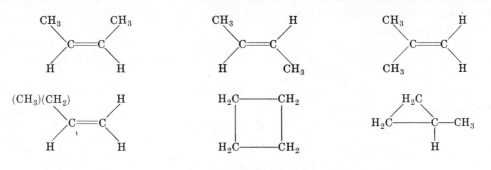

Fig. 9-26

Note the short-hand notation of grouping 2 or 3 hydrogens with the carbon to which they are bonded. It is understood, of course, that each hydrogen is bonded to the carbon of its group and that the carbon of the group is bonded to the next carbon or to the carbon of the adjoining group or groups.

MOLECULAR GEOMETRY

9.11. The C—C single bond distance is 1.54 Å. What is the distance between the terminal carbons in propane, C_3H_8? Assume that the 4 bonds of any carbon atom are pointed toward the corners of a regular tetrahedron.

The two terminal carbons may be thought of as lying on two corners of a regular tetrahedron at the center of which lies the middle carbon atom. The problem is one of finding the edge of a regular tetrahedron if the distance from the center to a corner is known.

Geometrical Method.

A simple way of drawing a regular tetrahedron is to select alternating corners of a cube and to connect each of the selected corners with each of the other three, as in Figure 9-27.

To prove that the resulting four-sided figure is indeed a regular tetrahedron, we note that each edge is the diagonal of one of the faces of the cube. Since all faces of a cube are equal, all face diagonals must be equal. Therefore the four sides of the inscribed polygon are all equal equilateral triangles and the figure must be regular.

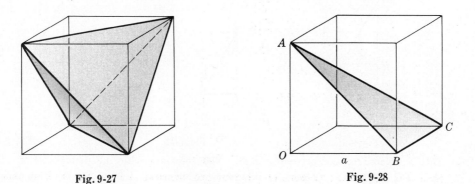

Fig. 9-27   Fig. 9-28

In Fig. 9-28, the face diagonal AB is the hypotenuse of right triangle OAB. Let $OA = OB = a$, the side of the cube. Then $(AB)^2 = (OA)^2 + (OB)^2 = 2a^2$ and $AB = a\sqrt{2}$.

We will also be interested in evaluating the length of a cube diagonal, AC. The shaded triangle ABC is a right triangle with right angle ABC. In the shaded triangle, $(AC)^2 = (BC)^2 + (AB)^2 = a^2 + 2a^2 = 3a^2$ and $AC = a\sqrt{3}$.

In Fig. 9-27, it is obvious that the center of the tetrahedron is the center of the cube, since the center of a cube is equidistant from all its corners. The center of a cube lies at the center of any diagonal running through the cube, like AC in Fig. 9-28. In our propane problem, the distance between adjacent carbon atoms, 1.54 Å, is one-half of the cube diagonal, $\frac{1}{2}a\sqrt{3}$. The distance between the terminal carbons is the face diagonal, $a\sqrt{2}$.

Then $\frac{1}{2}a\sqrt{3} = 1.54$ Å or $a = (3.08/\sqrt{3})$ Å, and the distance between the terminal carbons in propane is $a\sqrt{2} = 3.08\sqrt{2/3} = 2.51$ Å.

Trigonometric Method.

Construct isosceles triangle ABC in Fig. 9-29 with $\angle ACB = 109°28'$, the central tetrahedral angle. Draw a line from C perpendicular to AB; this line bisects $\angle ACB$ and bisects AB at D. Thus $\angle ACD = \frac{1}{2}(109°28') = 54°44'$.

Point C corresponds to the position of the central carbon in propane; AC and BC correspond to the two C—C single bonds, 1.54 Å each, making an angle of 109°28'. Then

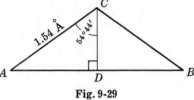

$$AB = 2\,AD = 2(AC \sin 54°44')$$
$$= 2(1.54 \text{ Å})(.816) = 2.51 \text{ Å}$$

Fig. 9-29

9.12. Compute the chlorine to chlorine distances in each of the isomeric $C_2H_2Cl_2$ compounds shown in Fig. 9-9, Page 74. Use the table of covalent radii and the average double bond shortening for estimating bond distances.

The starting point of the solution is the set of diagrams below, based on the 120° angle in double bond compounds.

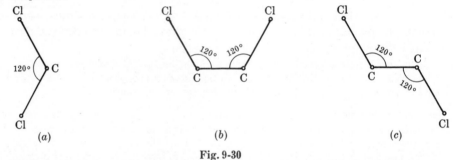

Fig. 9-30

(a) From Table 9-2, Page 73, we estimate the C—Cl bond length to be 1.76 Å. This is the hypotenuse AB of a 60° right triangle, ABD in Fig. 9-31, whose side AD is half the non-bonded chlorine—chlorine distance AC. Then $AC = 2\,AD = 2(1.76 \sin 60°) = 3.04$ Å.

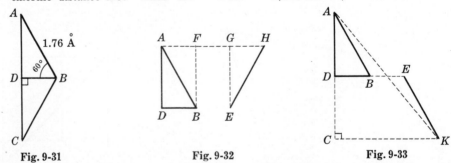

Fig. 9-31   Fig. 9-32   Fig. 9-33

(b) This is related to the preceding case. The chlorine—chlorine distance (Fig. 9-32) is $AH = AF + FG + GH$. $AF = GH = BD = 1.76 \cos 60° = 0.88$ Å. FG is the C—C double bond distance, 1.33 Å. Thus $AH = 0.88 + 1.33 + 0.88 = 3.09$ Å.

(c) This can be solved by reference to parts (a) and (b). The chlorine—chlorine distance is AK, the hypotenuse of right triangle ACK. The legs of this triangle ACK (Fig. 9-33) are AC and CK. AC was evaluated in (a) to be 3.04 Å. CK equals AH, evaluated in (b) to be 3.09 Å. Then

$$AK = \sqrt{(3.04)^2 + (3.09)^2} = 4.33 \text{ Å}.$$

CRYSTALS

9.13. Metallic gold crystallizes in the face-centered cubic lattice. The length of the cubic unit cell [Fig. 9-11(b)], a, is 4.070 Å.

(a) What is the closest distance between gold atoms?

(b) How many "nearest neighbors" does each gold atom have at the distance calculated in (a)?

(c) What is the density of gold?

(a) Consider a corner gold atom in Fig. 9-11(b), Page 75. The closest distance to another corner atom is a, 4.070 Å. The distance to an atom at the center of a face is one-half the diagonal of that face. In Problem 9.11 we saw that the diagonal of a face is $a\sqrt{2}$. Half the diagonal is thus $\frac{1}{2}a\sqrt{2} = \frac{1}{2}(4.070)\sqrt{2} = 2.878$ Å. This distance is shorter than the distance between adjacent corners and is therefore the closest distance between atoms.

(b) The problem is to find how many face-centers are equidistant from a corner atom. Point A in Fig. 9-13, Page 76, may be taken as the reference corner atom. On that same figure, B is one of the face-center points at the nearest distance to A. In plane ABD in the figure there are three other points equally close to A: the centers of the squares in the upper right, lower left, and lower right quadrants of the plane, measured around A. Plane ACE, parallel to the plane of the paper, also has points in the centers of each of the squares in the four quadrants around A. Also, plane ACF, perpendicular to the plane of the paper, has points in the centers of each of the squares in the four quadrants around A. Thus there are 12 nearest neighbors in all.

   The same result would have been obtained by counting the nearest neighbors around B, a face-centered point.

(c) To compute the density, we must count the number of gold atoms that can be uniquely assigned to one unit cell. This number is $\frac{1}{8}$ times the number of occupied corners in the unit cell plus $\frac{1}{2}$ times the number of occupied face-centers. Since a cube has 8 corners and 6 faces, the number is $8/8 + 6/2 = 4$. The volume of the unit cell is a^3. The density is then the mass of 4 gold atoms divided by a^3.

$$\text{Density} = \frac{4 \text{ at} \times 197.0 \frac{u}{\text{at}} \times \frac{1}{6.023 \times 10^{23}} \frac{g}{u}}{(4.070 \times 10^{-8} \text{ cm})^3} = 19.4 \text{ g/cc}$$

   The reverse of this type of calculation can be used for a precise determination of Avogadro's number. In such a case, both the lattice dimension and the density must be known precisely.

9.14. CsCl crystallizes in a cubic structure that has a Cs^+ at each corner and a Cl^- at the center of the unit cell. Use the ionic radii listed in Table 9-1, Page 70, to predict the lattice constant a and compare with the value of a calculated from the observed density of CsCl, 3.97 g/cc.

   Let us assume that the closest Cs^+ and Cl^- distance is the sum of the ionic radii of Cs^+ and Cl^-, or $(1.69 + 1.81)$ Å $= 3.50$ Å. This is the distance from the center of the unit cell to a corner, or one-half the cube diagonal. The cube diagonal was found to be $a\sqrt{3}$ in Problem 9.11. Then

$$\tfrac{1}{2}a\sqrt{3} = 3.50 \text{ Å} \qquad \text{and} \qquad a = 2(3.50)/\sqrt{3} \text{ Å} = 4.04 \text{ Å}$$

   The density can be used if we count the number of ions of each type per unit cell. The number of assigned Cs^+ ions per unit cell is one-eighth of the number of corner Cs^+ ions, or $\frac{1}{8}(8) = 1$. The only Cl^- in the unit cell is the center Cl^-, so that the assigned number of chloride ions is also 1. (This type of assignment of ions or atoms in a compound must always agree with the empirical formula of the compound, as the 1 : 1 ratio in this problem does.) The assigned mass per unit cell is thus that of 1 formula unit of CsCl, $\frac{132.9 + 35.5}{6.02 \times 10^{23}}$ g $= 2.797 \times 10^{-22}$ g.

$$\text{Volume of unit cell} = a^3 = \frac{\text{mass}}{\text{density}} = \frac{2.797 \times 10^{-22} \text{ g}}{3.97 \text{ g/cc}} = 70.4 \times 10^{-24} \text{ cc}, \qquad \text{from which}$$

$$a = \sqrt[3]{70.4 \times 10^{-24} \text{ cm}^3} = 4.13 \times 10^{-8} \text{ cm} = 4.13 \text{ Å}$$

   The value based on the experimental density is to be considered the more reliable since it is based on a measured property of CsCl, while the ionic radii are based on averages over many different compounds.

9.15. Ice crystallizes in a hexagonal lattice. At the low temperature at which the structure was determined the lattice constants were $a = 4.53$ Å and $c = 7.41$ Å (Fig. 9-12, Page 76). How many H_2O molecules are contained in a unit cell?

The volume V of the unit cell in Fig. 9-12 is

$$V = \text{area of rhombus base} \times \text{height } c$$

$$= (a^2 \sin 60°)c = (4.53 \text{ Å})^2(.866)(7.41 \text{ Å}) = 132 \text{ Å}^3 = 132 \times 10^{-24} \text{ cm}^3$$

Although the density of ice at the experimental temperature is not stated, it could not be very different from the value at $0°C$, 0.92 g/cc.

$$\text{Mass of unit cell} = V \times \text{density} = (132 \times 10^{-24} \text{ cm}^3)(0.92 \text{ g/cm}^3)(0.602 \times 10^{24} \, u/g)$$

$$= 73 \, u, \text{ i.e. 73 in atomic weight units.}$$

This is close to 4 times the molecular weight of H_2O; we conclude that there are 4 molecules of H_2O per unit cell. The discrepancy between $73\,u$ and the actual mass of 4 molecules, $72\,u$, is undoubtedly due to the uncertainty in the density at the experimental temperature.

Supplementary Problems

FORMULAS

9.16. Determine the ionic valences of the groups in parentheses in the following formulas: (a) $Ca(C_2O_4)$, (b) $Ca(C_2H_3O_2)_2$, (c) $Mg_3(AsO_3)_2$, (d) $(MoO)Cl_3$, (e) $(CrO_2)F_2$, (f) $(PuO_2)Br$, (g) $(PaO)_2S_3$.
   Ans. (a) -2, (b) -1, (c) -3, (d) $+3$, (e) $+2$, (f) $+1$, (g) $+3$

9.17. Write formulas for the following ionic compounds: (a) lithium hydride, (b) calcium bromate, (c) chromium (II) oxide, (d) thorium (IV) perchlorate, (e) nickel orthophosphate, (f) zinc sulfate.
   Ans. (a) LiH, (b) $Ca(BrO_3)_2$, (c) CrO, (d) $Th(ClO_4)_4$, (e) $Ni_3(PO_4)_2$, (f) $ZnSO_4$.

9.18. Name the following compounds: (a) $Mg(IO)_2$, (b) $Fe_2(SO_4)_3$, (c) $CaMnO_4$, (d) $KReO_4$, (e) $CaWO_4$, (f) $CoCO_3$.
   Ans. (a) magnesium hypoiodite, (b) iron (III) sulfate or ferric sulfate, (c) calcium manganate, (d) potassium perrhenate, (e) calcium tungstate, (f) cobalt (II) carbonate.

9.19. The formula of potassium arsenate is K_3AsO_4. The formula of potassium ferrocyanide is $K_4Fe(CN)_6$. Write the formulas of (a) calcium arsenate, (b) ferric arsenate, (c) barium ferrocyanide, (d) aluminum ferrocyanide.   Ans. (a) $Ca_3(AsO_4)_2$, (b) $FeAsO_4$, (c) $Ba_2Fe(CN)_6$, (d) $Al_4[Fe(CN)_6]_3$

9.20. Draw octet structural formulas for each of the following: (a) C_2HCl, (b) C_2H_6O, (c) C_2H_4O, (d) NH_3O, (e) NO_2^- (both oxygens terminal), (f) N_2O_4 (all oxygens terminal), (g) OF_2.

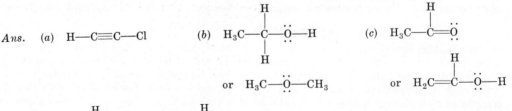

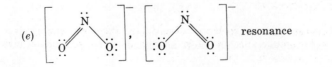

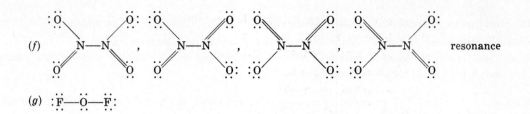

(f)   ⋯   ,   ⋯   ,   ⋯   ,   ⋯   resonance

(g) :F̈—Ö—F̈:

9.21. Complete the following structures by adding unshared electron pairs when necessary. Evaluate the formal charges where they differ from zero.

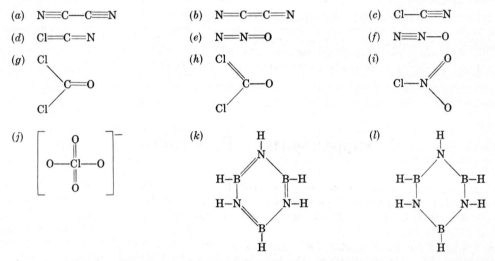

(a) N≡C—C≡N

(b) N=C=C=N

(c) Cl—C≡N

(d) Cl=C=N

(e) N=N=O

(f) N≡N—O

(g) Cl, C=O, Cl

(h) Cl, C—O, Cl

(i) Cl—N with O's

(j) [O—Cl—O with O's]⁻

(k) borazine ring

(l) borazine ring

Ans. (a) All zero; (b) +1 on one N (which does not have an octet), −1 on the other; (c) all zero; (d) +1 on Cl, −1 on N; (e) −1 on terminal N, +1 on central N; (f) +1 on central N, −1 on O; (g) all zero; (h) +1 on doubly bonded Cl, −1 on O; (i) +1 on N, −1 on singly bonded O; (j) +1 on Cl, −1 on each singly bonded O; (k) +1 on each N, −1 on each B; (l) all zero.

9.22. The chlorine to oxygen bond distance in ClO_4^- is 1.44 Å. What do you conclude about the valence bond structures for this ion?
Ans. There must be considerable double bond character in the bonds.

9.23. How many structural isomers can be drawn for each of the following: (a) C_5H_{12}, (b) C_3H_7Cl, (c) $C_3H_6Cl_2$, (d) $C_4H_8Cl_2$, (e) $C_5H_{11}Cl$, (f) C_6H_{14}, (g) C_7H_{16}?
Ans. (a) 3, (b) 2, (c) 4, (d) 9, (e) 8, (f) 5, (g) 9

9.24. How many resonance structures can be written for each of the following aromatic hydrocarbons:

(a) anthracene, 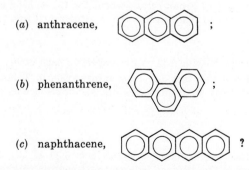 ;

(b) phenanthrene, ;

(c) naphthacene, ?

Consider double bonds only between adjacent carbons. (The circle inside the hexagon is a short-hand notation used to designate an aromatic ring without having to write out all the valence bond structures.)   *Ans.* (a) 4, (b) 5, (c) 5

9.25. The structure of 1,3-butadiene is often written as $H_2C\!=\!CH\!-\!CH\!=\!CH_2$. The distance between the central carbon atoms is 1.46 Å. Comment on the adequacy of the assigned structure.

Ans. There must be non-octet resonance structures involving double bonding between the central carbons, such as $\overset{+}{C}H_2\!-\!CH\!=\!CH\!-\!\overset{..}{C}H_2{}^-$.

9.26. How many structural and geometrical isomers can be written for the following without counting ring compounds: (a) C_3H_5Cl, (b) $C_3H_4Cl_2$, (c) C_4H_7Cl, (d) C_5H_{10}?

Ans. (a) 4, (b) 7, (c) 11, (d) 6

MOLECULAR GEOMETRY

9.27. The platinum—chlorine distance has been found to be 2.32 Å in several crystalline compounds. If this value applies to both of the compounds shown in Fig. 9-10, Page 75, what is the chlorine—chlorine distance in (a) structure (d), (b) structure (e)?   Ans. (a) 4.64 Å, (b) 3.28 Å

9.28. What is the length of a polymer molecule containing 1001 carbon atoms singly bonded in a line if the molecule could be stretched to its maximum length consistent with maintenance of the normal tetrahedral angle within any C—C—C group?   Ans. 1.26×10^3 Å

9.29. A plant virus was examined by the electron microscope and was found to consist of uniform cylindrical particles 150 Å in diameter and 3000 Å long. The virus has a specific volume of 0.73 cc/g. If the virus particle is considered to be one molecule, what is its molecular weight?

Ans. 4.4×10^7

9.30. Assuming the additivity of covalent radii in the C—Cl bond, what would be the chlorine—chlorine distances in each of the three dichlorobenzenes? Assume that the ring is a regular hexagon and that each C—Cl bond lies on a line through the center of the hexagon. The distance between adjacent carbons is 1.40 Å.

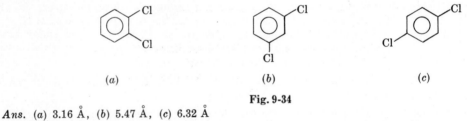

(a)   (b)   (c)

Fig. 9-34

Ans. (a) 3.16 Å, (b) 5.47 Å, (c) 6.32 Å

9.31. Estimate the length and width of the carbon skeleton in anthracene (see Problem 9-24(a) for a diagram). Assume hexagonal rings with equal carbon—carbon distances of 1.40 Å.

Ans. 7.3 Å long, 2.8 Å wide

9.32. There are two structural isomers of $C_2H_4Cl_2$. (a) In the isomer in which both chlorines are attached to the same carbon, find the chlorine—chlorine distance. (b) In the other isomer find the minimum and maximum distances between the two chlorines as one CH_2Cl group rotates about the C—C bond as an axis. Assume tetrahedral angles and additivity of covalent bond radii. (Hint. Refer to Prob. 9-11.)

Ans. (a) 2.87 Å; (b) minimum 3.57 Å, maximum 4.58 Å

9.33. BBr_3 is a symmetrical planar molecule, all B—Br bonds lying 120° with respect to each other. The distance between Br atoms is found to be 3.24 Å. From this fact and from information in Table 9-2, Page 73, estimate the covalent radius of boron, assuming that the bonds are all single bonds.

Ans. 0.73 Å

CRYSTALS

9.34. Potassium crystallizes in a body-centered cubic lattice, with a unit cell length $a = 5.20$ Å. (a) What is the distance between nearest neighbors? (b) What is the distance between next-nearest neighbors? (c) How many nearest neighbors does each K atom have? (d) How many next-nearest neighbors does each K have? (e) What is the calculated density of crystalline K?

Ans. (a) 4.50 Å, (b) 5.20 Å, (c) 8, (d) 6, (e) 0.924 g/cc

9.35. A common crystal form for many metals is the hexagonal close-packed lattice. In terms of Fig. 9-12, Page 76, $c = 1.633a$. There is an atom at each corner of the unit cell and another atom which can be located by moving one-third the distance along the diagonal of the rhombus base, starting at the lower left-hand corner, and moving perpendicularly by $\frac{1}{2}c$. Mg crystallizes in this lattice and has a density of 1.74 g/cc. (a) What is the volume of the unit cell? (b) What is a? (c) What is the distance between nearest neighbors? (d) How many nearest neighbors does each atom have?
Ans. (a) 46.4 Å^3, (b) 3.20 Å, (c) 3.20 Å, (d) 12

9.36. The NaCl lattice has a cubic unit cell shown in Fig. 9-35. KBr crystallizes in this lattice. (a) How many K$^+$ ions and how many Br$^-$ ions are in each unit cell? (b) Assuming the additivity of ionic radii, what is a? (c) Calculate the density of a perfect KBr crystal.
Ans. (a) 4 each, (b) 6.56 Å, (c) 2.80 g/cc

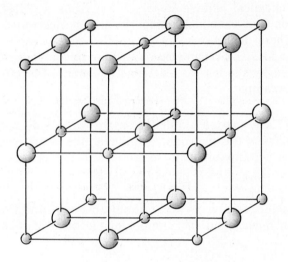

Unit Cell of NaCl-type Lattice.
Small circles represent cation positions.
Large circles represent anion positions.

Fig. 9-35

9.37. MgS and CaS both crystallize in the NaCl-type lattice (Fig. 9-35). From ionic radii listed in Table 9-1, Page 70, what conclusion can you draw about anion-cation contact in these crystals?
Ans. Ca^{2+} and S^{2-} can be in contact, but Mg^{2+} and S^{2-} cannot. In MgS, if Mg^{2+} and S^{2-} were in contact there would not be enough room for the sulfide ions along the diagonal of a square constituting one quarter of a unit cell cube face.

9.38. Each rubidium halide crystallizing in the NaCl-type lattice has a unit cell length a 0.30 Å greater than the corresponding potassium salt of the same halogen. What is the ionic radius of Rb$^+$ computed from these data?   *Ans.* 1.48 Å

9.39. Iron crystallizes in several modifications. At about 1184°, the body-centered cubic α form undergoes a transition to the face-centered cubic γ form. Assuming that the distance between nearest neighbors is the same in the two forms at the transition temperature, calculate the ratio of the density of γ-iron to that of α-iron at the transition temperature.   *Ans.* 1.09

9.40. The ZnS zinc blende structure is cubic. The unit cell may be described as a face-centered zinc ion sub-lattice with sulfide ions in the centers of alternating sub-cubes made by partitioning the main cube into 8 equal parts. (a) How many nearest neighbors does each Zn^{2+} have? (b) How many nearest neighbors does each S^{2-} have? (c) What angle is made by the lines connecting any S^{2-} to any two of its nearest neighbors?   *Ans.* (a) 4, (b) 4, (c) 109°28′

Chapter 10

Oxidation-Reduction

OXIDATION-REDUCTION REACTIONS

Oxidation is a chemical change in which electrons are lost by an atom or group of atoms, and reduction is a chemical change in which electrons are gained by an atom or group of atoms. These definitions can be applied most simply in the case of elementary substances and their ions. A transformation that converts a neutral atom to a positive ion must be accompanied by the loss of electrons and must, therefore, be an oxidation. Consider the following example:

$$Fe \longrightarrow Fe^{2+} + 2e$$

Electrons (symbol e) are written explicitly on the right side and provide equality of total charge on the two sides of the equation. Similarly, the transformation of a neutral element to an anion must be accompanied by electron gain and is classified as a reduction, as in the following case:

$$Cl_2 + 2e \longrightarrow 2\,Cl^-$$

Oxidation and reduction always occur simultaneously, and the total number of electrons lost in the oxidation must equal the number of electrons gained in the reduction.

OXIDATION STATE

It is not immediately apparent from the ionic charge alone whether a compound substance is undergoing oxidation or reduction. For example, MnO_2 reacts with hydrochloric acid to produce, among other things, chlorine gas and the Mn^{2+} ion. It is obvious that neutral chlorine is produced by *oxidation* of Cl^-. We infer, therefore, that MnO_2 is undergoing *reduction* in spite of the fact that a neutral substance, MnO_2, is converted partly to a cationic substance, Mn^{2+}. In another example, arsenious acid, H_3AsO_3, reacts with I_2 to form, among other things, the arsenate ion $HAsO_4^{2-}$ and the iodide ion, I^-. Since the iodine is *reduced* (neutral halogen to anion), the arsenious acid must be *oxidized*, in spite of the conversion of neutral H_3AsO_3 to anionic $HAsO_4^{2-}$.

Oxidation state is a concept that is helpful in diagnosing quickly the state of oxidation or reduction of particular atoms in such compound species as MnO_2, H_3AsO_3, and $HAsO_4^{2-}$. Oxidation state of an atom in some chemical combination is the arbitrary electrical charge assigned to that atom according to a prescribed set of rules. Other expressions often used to refer to oxidation state are *oxidation number* or *valence state*. In what follows, Roman numerals will be used to designate oxidation state.

It should be made clear that oxidation state is not the same as *formal charge*. Formal charge is based on an attempted mapping out of the real charge distribution of a molecule or ion among the constituent atoms in accordance with detailed knowledge of structure and electronic binding. Oxidation state, however, is a simpler assignment that does not require information about such electronic variables as single or multiple bonding and octet *versus* non-octet structures. Oxidation state is computed directly from the compositional formula itself, such as MnO_2 or H_3AsO_3. The rules for oxidation state assignment follow.

(1)   In ionic binary compounds the oxidation state is equated to the charge per atom. $CdCl_2$ is an ionic compound and may be designated $Cd^{++}(Cl^-)_2$ to show its ionic character. The cadmium possesses a true charge of $+2$ and an oxidation state of $+II$. Each chloride ion possesses a true charge of -1 and an oxidation state of $-I$. In Hg_2Cl_2, each mercury in Hg_2^{2+} has an average charge of $+1$ and an oxidation state of $+I$. The chlorine in Cl^- is again $-I$.

(2)   In covalent, or non-ionic compounds, the electrons involved in bond formation are not completely transferred from one element to the other, but are shared more or less equally by the bonding atoms. For purposes of oxidation state computation, however, it is conventional to assign each bonding electron to some particular atom. If the atoms are of the same kind, half of the bonding electrons are assigned to each of the two atoms. If the atoms are different, all electrons in the bond are arbitrarily assigned to that atom which has the greater electronegativity, or attractiveness for electrons. The most electronegative elements in order of decreasing electronegativity are F, O, N, Cl. C is more electronegative than H. Metals are less electronegative than nonmetals. Modern textbooks should be consulted for the complete order of electronegativities of the elements. The above definitions lead to the following rules as corollaries.

(a)   The oxidation state of a free and uncombined element is zero.

   Examples.
   Hg in Hg, H in H_2, O in O_2, S in S_8, etc.

   In H_2, of the two electrons in the molecule, one is assigned to each hydrogen atom. A hydrogen with one electron is the same as a neutral free hydrogen atom. Thus the oxidation state is zero.

(b)   The oxidation state of hydrogen in compounds is usually $+I$, except in the case of the metallic hydrides, where it is $-I$.

   Examples.
   In NH_3 the nitrogen atom is bonded directly to each of the three hydrogen atoms. Since nitrogen is more electronegative than hydrogen, all the bonding electrons are assigned to nitrogen. Each hydrogen is thus left with zero assigned electrons, one less than a free hydrogen atom. Thus the hydrogen has an apparent charge, or an oxidation state, of $+I$. The arbitrary nature of this charge designation is apparent in NH_3, in which the true charge separation between the nitrogen and the hydrogens is very slight. NH_3 never ionizes in water, for example, to produce hydrogen ions.

   In CaH_2, on the contrary, each hydrogen, being more electronegative than calcium, is assigned two electrons, one more than a free hydrogen atom. The oxidation state of hydrogen is therefore $-I$.

(c)   The oxidation state of oxygen in compounds is usually $-II$, except in peroxides where it is $-I$ or in fluorine compounds where it may be positive.

(d)   The algebraic sum of the positive and negative oxidation states of all atoms in a compound is equal to zero.

(e)   The algebraic sum of the positive and negative oxidation states of all atoms in an ion is equal to the charge of the ion.

   Examples.
   In H_2SiO_3 the oxidation state of Si is $+IV$ in order that Rule 2d may be observed.

   In PO_4^{-3}, the oxidation state of P is $+V$ in order that Rule 2e may be observed: $+V + 4(-II) = -3$. Only the phosphate ion as a whole has a true charge, or ionic valence. Each atom within this complex ion has only an arbitrarily assigned charge or oxidation state.

(3)   Many elements have only one oxidation state (in addition to zero for the uncombined element). Others have several oxidation states. During a course in college chemistry the student should learn and remember the more common oxidation states of the elements.

(4)   The definitions of oxidation and reduction may be generalized as follows: *The increase of oxidation state is oxidation and the decrease of oxidation state is reduction.* In the reduction of MnO_2 to Mn^{2+}, for example, the oxidation state of manganese is changed from $+IV$ to $+II$; in the oxidation of H_3AsO_3 to $HAsO_4^{2-}$, the oxidation state of arsenic is changed from $+III$ to $+V$.

IONIC NOTATION FOR EQUATIONS

In writing oxidation-reduction equations, care should be taken to write formulas only for compounds or ions that have true chemical existence, like MnO_2, H_3AsO_3, $HAsO_4^{2-}$. Mn(IV), As(III), and As(V) are not true species and should not be written into equations.

When the ionic convention is to be used for writing formulas, the following rules are customarily observed.

(1)   Ionic substances are written in the ionic form only if the ions are separated from each other in the reaction medium. Thus, NaCl would be the conventional notation for reactions involving solid salt because the ions in the solid are not separated from the crystal. Salt reactions in solution, however, would be indicated by Na^+ and Cl^-, or by either of these ions alone if only the sodium *or* the chlorine undergoes a change in oxidation state. Fairly insoluble salts, like CuS, are always written in the neutral form.

(2)   Partially ionized substances are written in the ionic form only if the extent of ionization is appreciable (about 20% or more). Thus water, which is ionized to the extent of less than one part in a hundred million, is written as H_2O. Strong acids, like hydrochloric acid and nitric acid, may be written in the ionized form; but weak acids, like nitrous, acetic, and sulfurous acids, are always written in the molecular form. Ammonia, a weak base, is written NH_3. Strong bases, like sodium hydroxide, are written in the ionized form Na^+ and OH^-, for water solutions.

(3)   Some complex ions are so stable that one or more of the groups from which they are formed do not exist in appreciable amounts outside the complex. In such a case the formula for the entire complex is written. Thus, the ferricyanide ion is always written as $Fe(CN)_6^{3-}$, never as separate ferric and cyanide ions. $Cu(NH_3)_4^{++}$ is the notation for the common blue complex ion formed by cupric salts in ammonia solutions.

(4)   A mixed convention will be employed in this chapter as an aid to indicate whether a given compound may be written in the ionized form. Thus, $Ba^{++}(NO_3^-)_2$ should indicate to the student that barium nitrate has the overall composition of $Ba(NO_3)_2$, that it is soluble and ionized, and that either the barium ion, Ba^{++}, or the nitrate ion, NO_3^-, may be written separately if desired. The absence of an ionic notation, as in As_2S_5, means that only the neutral formula of the whole compound should be used. This mixed convention will not be uniformly used in subsequent chapters. Experienced chemists may write $Ba(NO_3)_2$ for a reaction in aqueous solution, recognizing the ionic character of this salt even without an explicit ionic notation.

BALANCING OXIDATION-REDUCTION EQUATIONS

The principles of oxidation-reduction are the basis of two simple systematic methods for balancing these equations. If all the products of reaction are known, the balancing may be done directly either by the *ion-electron partial method* or by the *oxidation-state method*. After the student has acquired more experience he will be able to predict some or all of the principal products on the basis of such rules as the following.

(1)   If a free halogen is reduced, the product of reduction must be the halogenide ion (charge $= -1$).

(2) If a metal that has only one positive valence is oxidized, the oxidation state of the product is obvious.

(3) Reductions of concentrated nitric acid lead to NO_2, whereas the reduction of dilute nitric acid may lead to NO, N_2, NH_4^+, or other products, depending on the nature of the reducing agent and the extent of the dilution.

(4) Permanganate ion, MnO_4^-, is reduced to Mn^{++} in distinctly acid solution, as is MnO_2. The reduction product of permanganate in neutral or alkaline solution may be $MnO(OH)$, MnO_2, or MnO_4^{--}.

(5) If a peroxide is reduced, the product of reduction must contain oxygen in the $-II$ oxidation state, as in H_2O or OH^-. If a peroxide is oxidized, molecular oxygen is formed.

(6) Dichromate, $Cr_2O_7^{--}$, is reduced in acid solution to Cr^{+++}.

Ion-electron Partial Method of Balancing Oxidation-reduction Equations.

(1) Write a *skeleton equation* that includes those reactants and products that contain the elements undergoing a change of oxidation state.

(2) Write a partial skeleton equation for the oxidizing agent, with the element undergoing a reduction in oxidation state on each side of the equation. The element should not be written as a free atom or ion unless it really exists as such. It should be written as part of a real molecular or ionic species.

(3) Write another partial skeleton equation for the reducing agent, with the element undergoing an increase in oxidation state on each side.

(4) Balance each partial equation as to number of atoms of each element. In neutral or acid solution, H_2O and H^+ may be added for balancing oxygen and hydrogen atoms. The oxygen atoms are balanced first. For each excess oxygen atom on one side of an equation balance is secured by adding one H_2O to the *other* side. Then H^+ is used to balance hydrogens. Note that O_2 and H_2 are not used to balance the oxygen and hydrogen atoms unless they are known to be principal participants in the reaction.

If the solution is alkaline, OH^- may be used. For each excess oxygen on one side of an equation balance is secured by adding one H_2O to the *same* side and $2\,OH^-$ to the *other* side. If hydrogen is still unbalanced after this is done, balance is secured by adding one OH^- for each excess hydrogen on the *same* side as the excess and one H_2O on the *other* side. If both oxygen and hydrogen are in excess on the same side of the skeleton equation, an OH^- can be written on the other side for each paired excess of H and O.

(5) If an element undergoing a change of oxidation state is complexed in one of its states with some other element, balance the complexing groups with a species of that element in the same oxidation state occurring in the complex.

(6) Balance each partial equation as to the number of charges by adding electrons to either the left or the right side of the equation. If the rules have been followed carefully it will be found that electrons must be added to the left in the partial equation for the oxidizing agent and to the right in the partial for the reducing agent.

(7) Multiply each partial equation by a number chosen so that the total number of electrons lost by the reducing agent equals the number of electrons gained by the oxidizing agent.

(8) Add the two partial equations resulting from the multiplications. In the sum equation, cancel any terms common to both sides. All electrons should cancel.

(9) For an understanding of the nature of the reaction, step (8) can be considered the last step. For calculations involving the masses of reactants or products, transform the ionic equation of step (8) into a molecular equation. This is done by adding to each side of the equation equal numbers of the ions which do not undergo electron transfer but which are present together with the reactive components in neutral chemical substances. Proper pairs of ions may be combined to give a molecular formula.

(10) Check the final equation by counting the number of atoms of each element on each side of the equation and by computing the net charge on each side.

Oxidation-state Method of Balancing Oxidation-reduction Equations.

(1) Write a skeleton equation that includes those reactants and products that contain the elements undergoing a change of oxidation state.

(2) Determine the *change* in oxidation state which some element in the oxidizing agent undergoes. The number of electrons gained is equal to this change times the number of atoms undergoing the change.

(3) Determine the same for some element in the reducing agent.

(4) Multiply each principal formula by such numbers as to make the total number of electrons lost by the reducing agent equal to the number of electrons gained by the oxidizing agent.

(5) By inspection supply the proper coefficients for the rest of the equation.

(6) Check the final equation by counting the number of atoms of each element on both sides of the equation.

Solved Problems

FORMULAS AND OXIDATION STATE

10.1. Given that the oxidation state of hydrogen is +I, of oxygen −II, and of fluorine −I. Determine the oxidation state of the other elements in the following compounds: (a) PH_3, (b) H_2S, (c) CrF_3, (d) H_2SO_4, (e) H_2SO_3, (f) Al_2O_3.

(a) H_3 represents an oxidation state sum of +3 (+1 for each of the three hydrogens). Then the oxidation state of P must be −III, since the algebraic sum of the oxidation states of all atoms in a compound must equal zero.

(b) Oxidation state sum of H_2 is +2; then oxidation state of S = −II.

(c) Oxidation state sum of F_3 is −3; then oxidation state of Cr = +III.

(d) Oxidation state sum of H_2 is +2 and of O_4 is −8, or a total of +2 − 8 = −6; then oxidation state of S = +VI. The sulfate radical, SO_4, has an ionic valence of −2, since the positive ionic valence of H_2 (i.e., 2 H^+) is +2.

(e) Oxidation state sum of H_2 is +2 and of O_3 is −6, or a total of +2 − 6 = −4; then oxidation state of S = +IV. The sulfite radical, SO_3, has an ionic valence of −2, since the ionic charge of H_2 is +2.

(f) Oxidation state sum of O_3 is −6; then oxidation state sum of Al_2 is +6, and oxidation state of Al = $\frac{1}{2}(+6)$ = +III.

BALANCING OXIDATION-REDUCTION EQUATIONS

10.2. Balance the following oxidation-reduction equation.

$$H^+NO_3^- + H_2S \longrightarrow NO + S + H_2O$$

Ion-electron Method.

(1) The skeleton equation is given above.

(2) The oxidizing agent is nitrate ion, NO_3^-, since it contains the element, N, which undergoes a decrease in oxidation state. The reducing agent is H_2S, since it contains the element, S, which undergoes an increase in oxidation state. (S^{--} might have been selected as the reducing agent, but H_2S is preferable because of the very slight degree of ionization of the acid in nitric acid solution. Only a very small fraction exists as S^{--}.)

(3) The partial skeleton equation for the oxidizing agent is

$$NO_3^- \longrightarrow NO$$

(4) The partial skeleton equation for the reducing agent is

$$H_2S \longrightarrow S$$

(5) *a.* In the first partial equation, $2\,H_2O$ must be added to the right side in order to balance the oxygen atoms. Then $4\,H^+$ must be added to the left side to balance the H atoms.

$$4\,H^+ + NO_3^- \longrightarrow NO + 2\,H_2O$$

b. The second partial equation can be balanced by adding $2\,H^+$ to the right side.

$$H_2S \longrightarrow S + 2\,H^+$$

(6) *a.* In equation 5a above, the net charge on the left is $+4 - 1 = +3$, and on the right it is 0. Therefore 3 electrons must be added to the left side.

$$4\,H^+ + NO_3^- + 3e \longrightarrow NO + 2\,H_2O$$

b. In equation 5b, the net charge on the left is 0, and on the right it is $+2$. Hence 2 electrons must be added to the right side.

$$H_2S \longrightarrow S + 2\,H^+ + 2e$$

(7) Equation 6a must be multiplied by 2, and equation 6b by 3.

a. $$8\,H^+ + 2\,NO_3^- + 6e \longrightarrow 2\,NO + 4\,H_2O$$

b. $$3\,H_2S \longrightarrow 3\,S + 6\,H^+ + 6e$$

(8) Addition of equations 7a and 7b results in

$$8\,H^+ + 2\,NO_3^- + 3\,H_2S + 6e \longrightarrow 2\,NO + 4\,H_2O + 3\,S + 6\,H^+ + 6e$$

Since $6\,H^+$ and $6e$ are common to both sides, they may be canceled.

$$2\,H^+ + 2\,NO_3^- + 3\,H_2S \longrightarrow 2\,NO + 4\,H_2O + 3\,S$$

This form of the equation shows all the reactant ions and compounds in the proper form.

(9) If we want to know how much HNO_3 is required, we merely combine H^+ with NO_3^-.

$$2\,H^+NO_3^- + 3\,H_2S \longrightarrow 2\,NO + 4\,H_2O + 3\,S$$

Oxidation-state Method.

(1) Note that the oxidation state of N changes from $+V$ in NO_3^- to $+II$ in NO.

(2) The oxidation state of S changes from $-II$ in H_2S to 0 in S.

(3) Electron balance *diagrams* can be written as follows. (These are not *equations*.)

(*1*) $$N(+V) + 3e \longrightarrow N(+II)$$

(*2*) $$S(-II) \longrightarrow S(0) + 2e$$

(4) In order that the number of electrons lost shall equal the number gained, we must multiply diagram (*1*) by 2, and (*2*) by 3.

$$2\,N(+V) + 6e \longrightarrow 2\,N(+II)$$

$$3\,S(-II) \longrightarrow 3\,S(0) + 6e$$

Hence the coefficient of $H^+NO_3^-$ and of NO is 2, and of H_2S and of S is 3. Part of the skeleton equation can now be filled in.

$$2\,H^+NO_3^- + 3\,H_2S \longrightarrow 2\,NO + 3\,S$$

(5) The 8 atoms of H on the left (2 from $H^+NO_3^-$ plus 6 from H_2S) must form $4\,H_2O$ on the right. The final and complete equation is

$$2\,H^+NO_3^- + 3\,H_2S \longrightarrow 2\,NO + 3\,S + 4\,H_2O$$

Note that the oxygen atoms were balanced automatically without special attention.

The first partly filled skeleton equation could have been written in terms of NO_3^- rather than $H^+NO_3^-$. In this as well as the subsequent solved problems, the neutral compound notation will be used in the oxidation state method for variety, and the ionic notation in the ion-electron method.

10.3. Balance the following oxidation-reduction equation.

$$K^+MnO_4^- + K^+Cl^- + (H^+)_2SO_4^{--} \longrightarrow Mn^{++}SO_4^{--} + (K^+)_2SO_4^{--} + H_2O + Cl_2$$

Ion-electron Method.

The skeleton partial equations may be written

(a) $\qquad\qquad\qquad MnO_4^- \longrightarrow Mn^{++}$

(b) $\qquad\qquad\qquad Cl^- \longrightarrow Cl_2$

Partial (a) requires $4\,H_2O$ on the right to balance the oxygen atoms; then $8\,H^+$ on the left to balance the hydrogen. Partial (b) balances in routine fashion.

(a) $\qquad\qquad 8\,H^+ + MnO_4^- \longrightarrow Mn^{++} + 4\,H_2O$

(b) $\qquad\qquad 2\,Cl^- \longrightarrow Cl_2$

The net charge on the left of Partial (a) is $+8 - 1 = +7$, and on the right it is $+2$; therefore 5 electrons must be added to the left. In Partial (b) the net charge on the left is -2, and on the right it is 0; therefore 2 electrons must be added to the right.

(a) $\qquad\qquad 8\,H^+ + MnO_4^- + 5e \longrightarrow Mn^{++} + 4\,H_2O$

(b) $\qquad\qquad 2\,Cl^- \longrightarrow Cl_2 + 2e$

The multiplying factors are seen to be 2 and 5 respectively.

$$16\,H^+ + 2\,MnO_4^- + 10e \longrightarrow 2\,Mn^{++} + 8\,H_2O$$
$$\underline{\qquad\qquad 10\,Cl^- \longrightarrow 5\,Cl_2 + 10e \qquad\qquad}$$
$$16\,H^+ + 2\,MnO_4^- + 10\,Cl^- \longrightarrow 2\,Mn^{++} + 8\,H_2O + 5\,Cl_2$$

Since MnO_4^- was added as $KMnO_4$, $2\,MnO_4^-$ introduce $2\,K^+$ to the left side of the equation; and since K^+ does not react, the same number will appear on the right side. Since Cl^- was added as KCl, the $10\,Cl^-$ introduce $10\,K^+$ to each side of the equation. Since H^+ was added as H_2SO_4, $16\,H^+$ introduce $8\,SO_4^{--}$ to each side of the equation. Then

$$16\,H^+ + 8\,SO_4^{--} + 2\,K^+ + 2\,MnO_4^- + 10\,K^+ + 10\,Cl^- \longrightarrow 2\,Mn^{++} + 8\,H_2O + 5\,Cl_2$$
$$+ 12\,K^+ + 8\,SO_4^{--}$$

Pairs of ions may be grouped on the left side to show that the chemical reagents were H_2SO_4, $KMnO_4$, and KCl. Pairs of ions may be grouped on the right to show that evaporation of the solution, after reaction has taken place, would cause the crystallization of $MnSO_4$ and K_2SO_4.

$$8\,(H^+)_2SO_4^{--} + 2\,K^+MnO_4^- + 10\,K^+Cl^- \longrightarrow 2\,Mn^{++}SO_4^{--} + 6\,(K^+)_2SO_4^{--}$$
$$+ 5\,Cl_2 + 8\,H_2O$$

Oxidation-state Method.

Mn undergoes a change in oxidation state from $+VII$ in MnO_4^- to $+II$ in Mn^{++}. Cl undergoes a change in oxidation state from $-I$ in Cl^- to 0 in Cl_2. The electron balance diagrams are

(1) $\qquad\qquad Mn(+VII) + 5e \longrightarrow Mn(+II)$

(2) $\qquad\qquad 2\,Cl(-I) \longrightarrow 2\,Cl(0) + 2e$

Diagram (*2*) was written in terms of 2 Cl atoms because these atoms occur in pairs in the product Cl_2. The multiplying factors are 2 and 5, just as in the previous method.

$$2 \text{ Mn}(+VII) + 10e \longrightarrow 2 \text{ Mn}(+II)$$

$$10 \text{ Cl}(-I) \longrightarrow 10 \text{ Cl}(0) + 10e$$

Hence the coefficient of $K^+MnO_4^-$ and of $Mn^{++}SO_4^{--}$ is 2, of K^+Cl^- is 10, of Cl_2 is 5 ($\frac{1}{2} \times 10$).

$$2 \text{ K}^+MnO_4^- + 10 \text{ K}^+Cl^- \longrightarrow 2 \text{ Mn}^{++}SO_4^{--} + 5 \text{ Cl}_2 \qquad \text{(Incomplete)}$$

So far no provision has been made for the H_2O, $(H^+)_2SO_4^{--}$, and $(K^+)_2SO_4^{--}$. The 8 atoms of oxygen from $2 \text{ K}^+MnO_4^-$ form 8 H_2O. For 8 H_2O we need 16 atoms of hydrogen which can be furnished by $8 \text{ (H}^+)_2SO_4^{--}$. The 12 atoms of K ($10 \text{ K}^+Cl^- + 2 \text{ K}^+MnO_4^-$) yield $6 \text{ (K}^+)_2SO_4^{--}$. Note that all the oxygen in the oxidizing agent is converted to water. The sulfate radical retains its identity throughout the reaction.

$$2 \text{ K}^+MnO_4^- + 10 \text{ K}^+Cl^- + 8 \text{ (H}^+)_2SO_4^{--} \longrightarrow 2 \text{ Mn}^{++}SO_4^{--} + 5 \text{ Cl}_2$$
$$+ 6 \text{ (K}^+)_2SO_4^{--} + 8 \text{ H}_2O$$

10.4. Balance the following oxidation-reduction equation.

$$(K^+)_2Cr_2O_7^{--} + H^+Cl^- \longrightarrow K^+Cl^- + Cr^{+++}(Cl^-)_3 + H_2O + Cl_2$$

Ion-electron Method.

The balancing of the partial equation for the oxidizing agent proceeds as follows.

$$Cr_2O_7^{--} \longrightarrow Cr^{+++}$$
$$Cr_2O_7^{--} \longrightarrow 2 \text{ Cr}^{+++}$$
$$Cr_2O_7^{--} \longrightarrow 2 \text{ Cr}^{+++} + 7 \text{ H}_2O$$
$$14 \text{ H}^+ + Cr_2O_7^{--} \longrightarrow 2 \text{ Cr}^{+++} + 7 \text{ H}_2O$$
$$14 \text{ H}^+ + Cr_2O_7^{--} + 6e \longrightarrow 2 \text{ Cr}^{+++} + 7 \text{ H}_2O \qquad \text{(Balanced)}$$

The balancing of the partial equation for the reducing agent is as follows.

$$Cl^- \longrightarrow Cl_2$$
$$2 \text{ Cl}^- \longrightarrow Cl_2$$
$$2 \text{ Cl}^- \longrightarrow Cl_2 + 2e \qquad \text{(Balanced)}$$

The overall equation is

$$1 \times [14 \text{ H}^+ + Cr_2O_7^{--} + 6e \longrightarrow 2 \text{ Cr}^{+++} + 7 \text{ H}_2O]$$
$$3 \times [\qquad\qquad 2 \text{ Cl}^- \longrightarrow Cl_2 + 2e \qquad\qquad]$$
$$\overline{14 \text{ H}^+ + Cr_2O_7^{--} + 6 \text{ Cl}^- \longrightarrow 2 \text{ Cr}^{+++} + 7 \text{ H}_2O + 3 \text{ Cl}_2}$$

The 14 H^+ were added as 14 H^+Cl^-, and 6 of the 14 chloride ions were oxidized. To each side of the equation 8 more Cl^- can be added to represent those Cl^- which were not oxidized. Similarly, 2 K^+ may be added to each side to show that $Cr_2O_7^{--}$ came from $(K^+)_2Cr_2O_7^{--}$.

$$14 \text{ H}^+ + 6 \text{ Cl}^- + 8 \text{ Cl}^- + 2 \text{ K}^+ + Cr_2O_7^{--} \longrightarrow 2 \text{ Cr}^{+++} + 2 \text{ K}^+ + 8 \text{ Cl}^- + 3 \text{ Cl}_2 + 7 \text{ H}_2O$$
$$14 \text{ H}^+Cl^- + (K^+)_2Cr_2O_7^{--} \longrightarrow 2 \text{ Cr}^{+++}(Cl^-)_3 + 2 \text{ K}^+Cl^- + 3 \text{ Cl}_2 + 7 \text{ H}_2O$$

Oxidation-state Method.

The electron balance diagrams are written in terms of 2 atoms of Cr and 2 atoms of Cl because of the appearance of pairs of atoms of these kinds in $(K^+)_2Cr_2O_7^{--}$ and Cl_2.

For the oxidizing agent:
$$2 \text{ Cr}(+VI) \longrightarrow 2 \text{ Cr}(+III)$$
$$2 \text{ Cr}(+VI) + 6e \longrightarrow 2 \text{ Cr}(+III)$$

For the reducing agent:
$$2 \text{ Cl}(-I) \longrightarrow 2 \text{ Cl}(0)$$
$$2 \text{ Cl}(-I) \longrightarrow 2 \text{ Cl}(0) + 2e$$

Multiplying: $1 \times [2\ Cr(+VI)\ +\ 6e\ \longrightarrow\ 2\ Cr(+III)\ ]$

$3 \times [\qquad\qquad 2\ Cl(-I)\ \longrightarrow\ 2\ Cl(0)\ +\ 2e]$

Summing: $2\ Cr(+VI)\ +\ 6\ Cl(-I)\ \longrightarrow\ 2\ Cr(+III)\ +\ 6\ Cl(0)$

Hence, $(K^+)_2 Cr_2 O_7{}^{--}\ +\ 6\ H^+Cl^-\ \longrightarrow\ 2\ Cr^{+++}(Cl^-)_3\ +\ 3\ Cl_2$ (Incomplete)

The equation is still unbalanced, as no provision has been made for the K^+Cl^-, H_2O, or the H^+Cl^- which acts as an acid (as opposed to the H^+Cl^- which acts as reducing agent).

By inspection we can see that the 7 atoms of oxygen in $(K^+)_2 Cr_2 O_7{}^{--}$ form $7\ H_2O$. For $7\ H_2O$ we need 14 atoms of H which can be furnished by $14\ H^+Cl^-$. Since 6 of the chloride ions were oxidized to Cl_2, the remaining 8 $(14-6)$ should appear on the right as K^+Cl^- or $Cr^{+++}(Cl^-)_3$. Moreover, $1\ (K^+)_2 Cr_2 O_7{}^{--}$ yields $2\ K^+Cl^-$. Thus the coefficient of H^+Cl^- is 14, of H_2O is 7, and of K^+Cl^- is 2.

$(K^+)_2 Cr_2 O_7{}^{--}\ +\ 14\ H^+Cl^-\ \longrightarrow\ 2\ Cr^{+++}(Cl^-)_3\ +\ 3\ Cl_2\ +\ 7\ H_2O\ +\ 2\ K^+Cl^-$ (Balanced)

Note that here again all the oxygen in the oxidizing agent is converted to water.

10.5. Balance the following oxidation-reduction equation by the oxidation-state method.

$$FeS_2\ +\ O_2\ \longrightarrow\ Fe_2O_3\ +\ SO_2$$

The two special features of this problem are that both the iron and sulfur in FeS_2 undergo a change of oxidation state and that the reduction product of oxygen gas occurs in combination with both iron and sulfur. The electron balance diagram for oxidation must contain Fe and S atoms in the ratio of 1 to 2, since this is the ratio in which they are oxidized. The two diagrams, with their multiplying factors, are

$4 \times [Fe(+II)\ +\ 2\ S(-I)\ \longrightarrow\ Fe(+III)\ +\ 2\ S(+IV)\ +\ 11e]$

$11 \times [\qquad 2\ O(zero)\ +\ 4e\ \longrightarrow\ 2\ O(-II)\qquad\qquad\qquad]$

$4\ Fe(+II)\ +\ 8\ S(-I)\ +\ 22\ O(zero)\ \longrightarrow\ 4\ Fe(+III)\ +\ 8\ S(+IV)\ +\ 22\ O(-II)$

Hence, $4\ FeS_2\ +\ 11\ O_2\ \longrightarrow\ 2\ Fe_2O_3\ +\ 8\ SO_2$ (Balanced)

10.6. Balance the following oxidation-reduction equation by the ion-electron method.

$$Zn\ +\ Na^+NO_3{}^-\ +\ Na^+OH^-\ \longrightarrow\ (Na^+)_2 ZnO_2{}^{--}\ +\ NH_3\ +\ H_2O$$

The skeleton partial for the oxidizing agent is

$$NO_3{}^-\ \longrightarrow\ NH_3$$

In alkaline solutions, each excess oxygen atom is balanced by adding one H_2O to the same side of the equation and $2\ OH^-$ to the opposite side. Each excess hydrogen atom is balanced by adding $1\ OH^-$ to the same side and $1\ H_2O$ to the opposite side. In this example we must add $3\ H_2O$ to the left and $6\ OH^-$ to the right to balance the excess oxygen of the $NO_3{}^-$. Also, we must add $3\ OH^-$ to the right and $3\ H_2O$ to the left to balance the excess hydrogen of the NH_3. In all, $9\ OH^-$ must be added to the right and $6\ H_2O$ to the left.

$NO_3{}^-\ +\ 6\ H_2O\ \longrightarrow\ NH_3\ +\ 9\ OH^-$

$NO_3{}^-\ +\ 6\ H_2O\ +\ 8e\ \longrightarrow\ NH_3\ +\ 9\ OH^-$ (Balanced)

The balancing of the partial equation for the reducing agent is as follows.

$Zn\ \longrightarrow\ ZnO_2{}^{--}$

$4\ OH^-\ +\ Zn\ \longrightarrow\ ZnO_2{}^{--}\ +\ 2\ H_2O$

$4\ OH^-\ +\ Zn\ \longrightarrow\ ZnO_2{}^{--}\ +\ 2\ H_2O\ +\ 2e$ (Balanced)

The overall equation is

$1 \times [NO_3{}^-\ +\ 6\ H_2O\ +\ 8e\ \longrightarrow\ NH_3\ +\ 9\ OH^-\qquad\qquad]$

$4 \times [\qquad 4\ OH^-\ +\ Zn\ \longrightarrow\ ZnO_2{}^{--}\ +\ 2\ H_2O\ +\ 2e]$

$NO_3{}^-\ +\ 6\ H_2O\ +\ 4\ Zn\ +\ 16\ OH^-\ \longrightarrow\ NH_3\ +\ 9\ OH^-\ +\ 4\ ZnO_2{}^{--}\ +\ 8\ H_2O$

Canceling, $NO_3^- + 4 Zn + 7 OH^- \longrightarrow NH_3 + 4 ZnO_2^{--} + 2 H_2O$

Adding ions and combining,

$$Na^+NO_3^- + 4 Zn + 7 Na^+OH^- \longrightarrow NH_3 + 4 (Na^+)_2ZnO_2^{--} + 2 H_2O$$

10.7. Balance the following equation by the ion-electron method.

$$HgS + H^+Cl^- + H^+NO_3^- \longrightarrow (H^+)_2HgCl_4^{--} + NO + S + H_2O$$

The partial equation for the oxidizing agent is balanced by the straightforward procedure.

$$4 H^+ + NO_3^- + 3e \longrightarrow NO + 2 H_2O$$

The skeleton for the reducing agent is

$$HgS \longrightarrow S$$

The above skeleton contains both reduced and oxidized forms of sulfur, the only element undergoing a change in oxidation state. The unbalance is not one of hydrogen or oxygen atoms but of mercury. According to the overall equation, the form in which mercury exists among the products is the ion $HgCl_4^{--}$. If such an ion is added to the right to balance the mercury, then chloride ions must be added to the left to balance the chlorine. In general, it is always allowable to add ions necessary for complex formation when such an addition does not require the introduction of new oxidation states [Rule (5)].

$$HgS + 4 Cl^- \longrightarrow S + HgCl_4^{--} + 2e \qquad \text{(Balanced)}$$

The overall equation is

$$2 \times [4 H^+ + NO_3^- + 3e \longrightarrow NO + 2 H_2O \qquad]$$
$$3 \times [\qquad HgS + 4 Cl^- \longrightarrow S + HgCl_4^{--} + 2e]$$
$$\overline{8 H^+ + 2 NO_3^- + 3 HgS + 12 Cl^- \longrightarrow 2 NO + 4 H_2O + 3 S + 3 HgCl_4^{--}}$$

Adding ions and combining,

$$3 HgS + 2 H^+NO_3^- + 12 H^+Cl^- \longrightarrow 3 S + 3 (H^+)_2HgCl_4^{--} + 2 NO + 4 H_2O$$

10.8. Complete and balance the following skeleton equation for a reaction in acid solution.

$$H_2O_2 + MnO_4^- + \qquad \longrightarrow$$

The products must be deduced from chemical experience. MnO_4^- contains manganese in its highest oxidation state. Hence, if reaction occurs at all, MnO_4^- is reduced. Since the solution is acid the reduction product will be Mn^{++}. The H_2O_2 must therefore act as a reducing agent in this reaction, and its only possible oxidation product is O_2. The usual procedure may be followed from this point, leading to the following solution.

$$2 \times [MnO_4^- + 8 H^+ + 5e \longrightarrow Mn^{++} + 4 H_2O \quad]$$
$$5 \times [\qquad H_2O_2 \longrightarrow O_2 + 2 H^+ + 2e]$$
$$\overline{2 MnO_4^- + 16 H^+ + 5 H_2O_2 \longrightarrow 2 Mn^{++} + 8 H_2O + 5 O_2 + 10 H^+}$$

After canceling,
$$2 MnO_4^- + 6 H^+ + 5 H_2O_2 \longrightarrow 2 Mn^{++} + 8 H_2O + 5 O_2$$

The above equation is as definite a statement as can be made within the given overall skeleton. If we are to write an equation in terms of neutral substances, we are free to decide which salt of MnO_4^- and which acid to use. If we use $K^+MnO_4^-$ and $(H^+)_2SO_4^{--}$, we obtain

$$2 K^+MnO_4^- + 3 (H^+)_2SO_4^{--} + 5 H_2O_2 \longrightarrow 2 Mn^{++}SO_4^{--} + 5 O_2 + (K^+)_2SO_4^{--} + 8 H_2O$$

COMPARISON OF THE TWO METHODS OF BALANCING

Both methods lead to the correct form of the balanced equation. The ion-electron method has two advantages:

(1) It differentiates those components of a system which react from those which do not. In the previous problem any permanganate and any strong acid could have been used, not necessarily $K^+MnO_4^-$ and $(H^+)_2SO_4^{--}$. The use of complete formulas for neutral substances is useful only for the calculation of weight relations. Since permanganate ion cannot be weighed on a balance, it is necessary to choose one particular permanganate, like $K^+MnO_4^-$, and weigh an amount of $K^+MnO_4^-$ that will give the correct weight of MnO_4^-.

(2) The second advantage of the ion-electron method is that the half-reactions of the partial equations can actually be made to occur independently. Most oxidation-reduction reactions can be carried out as galvanic cell processes for producing an electrical potential. This can be done by placing the reducing agent and oxidizing agent in separate vessels and making electrical connections between the two. It has been found that the reaction taking place in each beaker corresponds exactly to a partial equation written according to the rules for the ion-electron method.

Some chemists prefer to use the ion-electron method for oxidation-reduction reactions carried out in dilute aqueous solutions, where free ions have more or less independent existence, and to use the oxidation-state method for oxidation-reduction reactions between solid chemicals or for reactions in concentrated acid media.

STOICHIOMETRY IN OXIDATION-REDUCTION

10.9. In Problem 10-3, how much Cl_2 could be reduced by the reaction of 100 g $KMnO_4$?

This weight-weight problem is solved like all other weight-weight problems if the balanced molecular equation is used, with reference to the fact that 2 gfw $KMnO_4$ can produce 5 moles Cl_2.

The problem can also be solved directly from the balanced ionic equation showing that 2 MnO_4^- ions yield 5 molecules of Cl_2. We need add only the chemical equivalence of 2 MnO_4^- with 2 $KMnO_4$.

In either case, the solution is

$$x \text{ g } Cl_2 = 100 \text{ g } KMnO_4 \times \frac{1 \text{ gfw } KMnO_4}{158 \text{ g } KMnO_4} \times \frac{5 \text{ moles } Cl_2}{2 \text{ gfw } KMnO_4} \times \frac{70.9 \text{ g } Cl_2}{1 \text{ mole } Cl_2} = 112 \text{ g } Cl_2$$

Supplementary Problems

Write balanced ionic and molecular equations for the following.

10.10. $CuS + H^+NO_3^-$ (dilute) $\longrightarrow$ $Cu^{++}(NO_3^-)_2 + S + H_2O + NO$

10.11. $K^+MnO_4^- + H^+Cl^- \longrightarrow K^+Cl^- + Mn^{++}(Cl^-)_2 + H_2O + Cl_2$

10.12. $Fe^{++}(Cl^-)_2 + H_2O_2 + H^+Cl^- \longrightarrow Fe^{+++}(Cl^-)_3 + H_2O$

10.13. $As_2S_5 + H^+NO_3^-$ (conc.) $\longrightarrow$ $H_3AsO_4 + H^+HSO_4^- + H_2O + NO_2$

10.14. $Cu + H^+NO_3^-$ (conc.) $\longrightarrow$ $Cu^{++}(NO_3^-)_2 + H_2O + NO_2$

10.15. $Cu + H^+NO_3^-$ (dilute) $\longrightarrow$ $Cu^{++}(NO_3^-)_2 + H_2O + NO$

10.16. $Zn + H^+NO_3^-$ (dilute) $\longrightarrow$ $Zn^{++}(NO_3^-)_2 + H_2O + NH_4^+NO_3^-$

10.17. $(Na^+)_2C_2O_4^{--} + K^+MnO_4^- + (H^+)_2SO_4^{--} \longrightarrow (K^+)_2SO_4^{--} + (Na^+)_2SO_4^{--} + H_2O$
$$+ Mn^{++}SO_4^{--} + CO_2$$

10.18. $CdS + I_2 + H^+Cl^- \longrightarrow Cd^{++}(Cl^-)_2 + H^+I^- + S$

10.19. $MnO + PbO_2 + H^+NO_3^- \longrightarrow H^+MnO_4^- + Pb^{++}(NO_3^-)_2 + H_2O$

10.20. $Cr^{+++}(I^-)_3 + K^+OH^- + Cl_2 \longrightarrow (K^+)_2CrO_4^{--} + K^+IO_4^- + K^+Cl^- + H_2O$

Note that both the chromium and the iodide are oxidized in this reaction.

10.21. $(Na^+)_2HAsO_3^{--} + K^+BrO_3^- + H^+Cl^- \longrightarrow Na^+Cl^- + K^+Br^- + H_3AsO_4$

10.22. $(Na^+)_2TeO_3^{--} + Na^+I^- + H^+Cl^- \longrightarrow Na^+Cl^- + Te + H_2O + I_2$

10.23. $U^{++++}(SO_4^{--})_2 + K^+MnO_4^- + H_2O \longrightarrow (H^+)_2SO_4^{--} + (K^+)_2SO_4^{--}$
$+ Mn^{++}SO_4^{--} + UO_2^{++}SO_4^{--}$

10.24. $I_2 + (Na^+)_2S_2O_3^{--} \longrightarrow (Na^+)_2S_4O_6^{--} + Na^+I^-$

10.25. $Ca^{++}(OCl^-)_2 + K^+I^- + H^+Cl^- \longrightarrow I_2 + Ca^{++}(Cl^-)_2 + H_2O + K^+Cl^-$

10.26. $Bi_2O_3 + Na^+OH^- + Na^+OCl^- \longrightarrow Na^+BiO_3^- + Na^+Cl^- + H_2O$

10.27. $(K^+)_3Fe(CN)_6^{---} + Cr_2O_3 + K^+OH^- \longrightarrow (K^+)_4Fe(CN)_6^{----} + (K^+)_2CrO_4^{--} + H_2O$

10.28. $H^+NO_3^- + H^+I^- \longrightarrow NO + I_2 + H_2O$

10.29. $Mn^{++}SO_4^{--} + (NH_4^+)_2S_2O_8^{--} + H_2O \longrightarrow MnO_2 + (H^+)_2SO_4^{--} + (NH_4^+)_2SO_4^{--}$

10.30. $(K^+)_2Cr_2O_7^{--} + Sn^{++}(Cl^-)_2 + H^+Cl^- \longrightarrow Cr^{+++}(Cl^-)_3 + SnCl_4 + K^+Cl^- + H_2O$

10.31. $Co^{++}(Cl^-)_2 + (Na^+)_2O_2^{--} + Na^+OH^- + H_2O \longrightarrow Co(OH)_3 + Na^+Cl^-$

10.32. $Cu(NH_3)_4^{++}(Cl^-)_2 + K^+CN^- + H_2O \longrightarrow NH_3 + NH_4^+Cl^- + (K^+)_2Cu(CN)_3^{--}$
$+ K^+CNO^- + K^+Cl^-$

10.33. $Sb_2O_3 + K^+IO_3^- + H^+Cl^- + H_2O \longrightarrow HSb(OH)_6 + K^+Cl^- + ICl$

10.34. $Ag + K^+CN^- + O_2 + H_2O \longrightarrow K^+Ag(CN)_2^- + K^+OH^-$

10.35. $WO_3 + Sn^{++}(Cl^-)_2 + H^+Cl^- \longrightarrow W_3O_8 + (H^+)_2SnCl_6^{--} + H_2O$

10.36. $Co^{++}(Cl^-)_2 + K^+NO_2^- + H^+C_2H_3O_2^- \longrightarrow (K^+)_3Co(NO_2)_6^{---} + NO + K^+C_2H_3O_2^-$
$+ K^+Cl^- + H_2O$

10.37. $V(OH)_4^+Cl^- + Fe^{++}(Cl^-)_2 + H^+Cl^- \longrightarrow VO^{++}(Cl^-)_2 + Fe^{+++}(Cl^-)_3 + H_2O$

Balance the following equations by the oxidation-state method.

10.38. $NH_3 + O_2 \longrightarrow NO + H_2O$

10.39. $CuO + NH_3 \longrightarrow N_2 + H_2O + Cu$

10.40. $KClO_3 + H_2SO_4 \longrightarrow KHSO_4 + O_2 + ClO_2 + H_2O$

10.41. $Sn + HNO_3 \longrightarrow SnO_2 + NO_2 + H_2O$

10.42. $I_2 + HNO_3 \longrightarrow HIO_3 + NO_2 + H_2O$

10.43. $KI + H_2SO_4 \longrightarrow K_2SO_4 + I_2 + H_2S + H_2O$

10.44. $KBr + H_2SO_4 \longrightarrow K_2SO_4 + Br_2 + SO_2 + H_2O$

10.45. $Cr_2O_3 + Na_2CO_3 + KNO_3 \longrightarrow Na_2CrO_4 + CO_2 + KNO_2$

10.46. $P_2H_4 \longrightarrow PH_3 + P_4H_2$

10.47. $Ca_3(PO_4)_2 + SiO_2 + C \longrightarrow CaSiO_3 + P_4 + CO$

Complete and balance the following skeleton equations by the ion-electron partial method.

10.48. $I^- + NO_2^- \longrightarrow I_2 + NO$ (acid solution)

10.49. $Au + CN^- + O_2 \longrightarrow Au(CN)_4^-$ (aqueous solution)

10.50. $MnO_4^- \longrightarrow MnO_4^{--} + O_2$ (alkaline solution)

10.51. $P \longrightarrow PH_3 + H_2PO_2^-$ (alkaline solution)

10.52. $Zn + As_2O_3 \longrightarrow AsH_3$ (acid solution)

10.53. $Zn + ReO_4^- \longrightarrow Re^-$ (acid solution)

10.54. $ClO_2 + O_2^{--} \longrightarrow ClO_2^-$ (alkaline solution)

10.55. $Cl_2 + IO_3^- \longrightarrow IO_4^-$ (alkaline solution)

10.56. $V \longrightarrow HV_6O_{17}^{---} + H_2$ (alkaline solution)

Chapter 11

Equivalent Weight

INTRODUCTION

Some stoichiometric problems can be simplified by using the equivalent weight instead of the atomic weight, formula weight, or molecular weight. The equivalent weight is that fraction of the formula (or atomic or molecular) weight which corresponds to one defined unit of chemical reaction. This principle is useful in two types of reaction, acid-base and oxidation-reduction. The definitions of the reaction units in these and related cases are given in the following sections.

ACIDS AND BASES

The defined unit of reaction is the following neutralization reaction.

$$H^+ + OH^- \longrightarrow H_2O$$

(1) The equivalent weight of an acid is that fraction of the formula weight which contains or can supply for reaction one acid H^+. A *gram-equivalent* (*g-eq*) is that weight which contains or can supply for reaction one mole of H^+.

Examples.

The equivalent weights of HCl and $HC_2H_3O_2$ are the same as their molecular weights, since each contains one acidic hydrogen per molecule. A g-eq of each of these molecules is the same as a mole. The equivalent weight of H_2SO_4 is usually half the molecular weight and a g-eq is half a mole, since both hydrogens are replaceable in most reactions of dilute sulfuric acid. A g-eq of H_3PO_4 may be 1 mole, 1/2 mole, or 1/3 mole, depending on whether 1, 2, or 3 hydrogen atoms per molecule are replaced in a particular reaction. A g-eq of H_3BO_3 is always one mole, since only one hydrogen is replaceable in neutralization reactions. The equivalent weight of SO_3 is one-half the molecular weight, since SO_3 can react with water to give $2\,H^+$.

$$SO_3 + H_2O \longrightarrow 2\,H^+ + SO_4^{2-}$$

There are no simple rules for predicting how many hydrogens of an acid will be replaced in a given neutralization. That is why the chemical properties of each substance must be studied.

(2) The equivalent weight of a base is that fraction of the formula weight which contains or can supply one OH^-, or can react with one H^+.

Examples.

The equivalent weights of NaOH, NH_3 (which can react with water to give $NH_4^+ + OH^-$), $Mg(OH)_2$, and $Al(OH)_3$, are equal to 1/1, 1/1, 1/2, and 1/3 of their formula weights, respectively.

(3) The equivalent weight of a salt can be defined in terms of the use of the salt as an acid or base. The equivalent weight may be variable depending on whether the formula unit is being used for 1, 2, or 3 units of the reference neutralization reaction.

Examples.

Salt	Typical Reaction	Equivalent Weight
NH_4Cl	$NH_4^+ + OH^- \longrightarrow NH_3 + H_2O$	FW
$NaHCO_3$	$HCO_3^- + OH^- \longrightarrow CO_3^{2-} + H_2O$	FW
$NaHCO_3$	$HCO_3^- + H^+ \longrightarrow CO_2 + H_2O$	FW
Na_2CO_3	$CO_3^{2-} + 2\,H^+ \longrightarrow CO_2 + H_2O$	$\frac{1}{2} \times$ FW
Na_2CO_3	$CO_3^{2-} + H^+ \longrightarrow HCO_3^-$	FW
NaH_2PO_4	$H_2PO_4^- + OH^- \longrightarrow HPO_4^{2-} + H_2O$	FW
NaH_2PO_4	$H_2PO_4^- + 2\,OH^- \longrightarrow PO_4^{3-} + 2\,H_2O$	$\frac{1}{2} \times$ FW

OXIDIZING AND REDUCING AGENTS

The equivalent weight of an oxidizing or reducing agent for a particular reaction is equal to its formula weight divided by the total number of electrons gained or lost when the reaction of that formula unit occurs. Thus, the equivalent weight of an oxidizing or reducing agent

$$= \frac{\text{formula weight of oxidizing or reducing substance}}{\text{number of electrons gained or lost}}$$

A given oxidizing or reducing agent may have more than one equivalent weight, depending on the reaction for which it is used.

A special case is that of two elements reacting to form a binary compound. For each element the number of electrons transferred per atom is the oxidation state in the compound. The equivalent weight then becomes the atomic weight divided by the oxidation state.

CHEMICAL REACTION CALCULATIONS BY METHOD OF EQUIVALENT WEIGHTS

Equivalent weights are so defined that equal numbers of gram-equivalents of two substances react exactly with each other. This is true for neutralization because one H^+ neutralizes one OH^-, and for oxidation-reduction because the number of electrons lost by the reducing agent equals the number of electrons gained by the oxidizing agent. This principle makes it possible to solve problems involving combining weights without writing out a balanced equation.

Solved Problems

ACIDS AND BASES

11.1. Find the equivalent weight of $HClO_4 \cdot H_2O$ and of H_2SO_4, assuming complete neutralization.

The equivalent weight of an acid is that fraction of the formula weight containing one replaceable hydrogen.

One formula unit of $HClO_4 \cdot H_2O$ contains 1 replaceable hydrogen, and one molecule of H_2SO_4 contains 2 replaceable hydrogens. Hence,

$$\text{EW of } HClO_4 \cdot H_2O = \text{FW} = 118.5$$
$$\text{EW of } H_2SO_4 = \tfrac{1}{2}\text{MW} = \tfrac{1}{2} \times 98.08 = 49.04$$

11.2. Calculate the equivalent weight of KOH, $Ca(OH)_2$, $Al(OH)_3$, Fe_2O_3.

The equivalent weight of a base is that fraction of the formula weight supplying one replaceable OH^-.

One formula unit of KOH contains 1 OH^-; 1 formula unit of $Ca(OH)_2$ contains 2 OH^- ions; 1 formula unit of $Al(OH)_3$ contains 3 OH^- ions. Hence,

$$EW \text{ of KOH} \quad = \quad FW \quad = \quad 56.11$$
$$EW \text{ of } Ca(OH)_2 \quad = \quad \tfrac{1}{2} FW \quad = \quad \tfrac{1}{2} \times 74.10 \quad = \quad 37.05$$
$$EW \text{ of } Al(OH)_3 \quad = \quad \tfrac{1}{3} FW \quad = \quad \tfrac{1}{3} \times 78.00 \quad = \quad 26.00$$

The EW of an oxide is that fraction of the FW containing one-half an oxide ion. The reason is: (a) soluble oxides form hydroxides in water by the reaction

$$O^{2-} + H_2O \longrightarrow 2\,OH^-$$

and (b) insoluble oxides can be neutralized by acids by the following stoichiometry,

$$O^{2-} + 2\,H^+ \longrightarrow H_2O$$

Hence, $EW \text{ of } Fe_2O_3 = \tfrac{1}{6} FW = \tfrac{1}{6} \times 159.70 = 26.62$

11.3. What is the equivalent weight of K_3PO_4 for reactions with HCl if the end product of the reaction is: (a) HPO_4^{2-}, (b) $H_2PO_4^-$, (c) H_3PO_4?

Formula weight of $K_3PO_4 = 212.3$.

(a) The reaction is a one-proton reaction; thus $EW = FW = 212.3$.

$$PO_4^{3-} + H^+ \longrightarrow HPO_4^{2-}$$

(b) The reaction is a two-proton reaction; thus $EW = \tfrac{1}{2} FW = 106.2$.

$$PO_4^{3-} + 2\,H^+ \longrightarrow H_2PO_4^-$$

(c) The reaction is a three-proton reaction; thus $EW = \tfrac{1}{3} FW = 70.8$.

$$PO_4^{3-} + 3\,H^+ \longrightarrow H_3PO_4$$

11.4. Determine the equivalent weight of H_2SO_4 in each of the following reactions:

(a) $NaOH + H_2SO_4 \longrightarrow NaHSO_4 + H_2O$, (b) $2\,NaOH + H_2SO_4 \longrightarrow Na_2SO_4 + 2\,H_2O$.

(a) Here only one hydrogen of H_2SO_4 is replaced to form the acid salt $NaHSO_4$. Hence in this reaction the equivalent weight of H_2SO_4 = molecular weight = 98.08.

(b) Here both hydrogens are replaced, forming the normal salt Na_2SO_4. Hence in this reaction the equivalent weight of H_2SO_4 = 1/2 × molecular weight = 49.04.

11.5. What weight of Na_2CO_3 is required to neutralize 4.89 g of HCl? The reaction is

$$2\,HCl + Na_2CO_3 \longrightarrow 2\,NaCl + H_2O + CO_2$$

1 g-eq HCl = 1 mole = 36.46 g. 1 g-eq Na_2CO_3 = $\tfrac{1}{2}$ gfw = $\tfrac{1}{2}(105.99\text{ g})$ = 53.00 g.

Equivalent weight may be considered to be a property expressed in units of grams per g-eq. The number of g-eq of Na_2CO_3 must equal the number of g-eq of HCl.

$$\text{Number of g-eq in 4.89 g HCl} = \frac{4.89\text{ g}}{36.46\text{ g/g-eq}} = 0.1341\text{ g-eq HCl}$$

Therefore 0.1341 g-eq of Na_2CO_3 is needed.

Weight of Na_2CO_3 needed = 0.1341 g-eq × 53.00 g/g-eq = 7.11 g Na_2CO_3

Another Method.

Since 1 g-eq HCl (36.46 g) requires 1 g-eq Na_2CO_3 (53.00 g), then

$$1\text{ g HCl requires } \frac{53.00}{36.46}\text{ g } Na_2CO_3$$

and $4.89\text{ g HCl requires } 4.89 \times \dfrac{53.00}{36.46}\text{ g } = 7.11\text{ g } Na_2CO_3$

OXIDIZING AND REDUCING AGENTS

11.6. (a) In a given reaction MnO_2, in HCl solution, is reduced to $MnCl_2$. What is the equivalent weight of MnO_2 in this reaction? (b) In a given reaction HNO_3 is reduced to NO_2. What is the equivalent weight of HNO_3 in this reaction?

(a) In MnO_2 the oxidation state of Mn is +IV; in $MnCl_2$ it is +II. The number of electrons gained per manganese is 2, the change in oxidation state. This number can also be found by writing the ion-electron partial equation

$$MnO_2 + 4\,H^+ + 2e \longrightarrow Mn^{2+} + 2\,H_2O$$

$$\text{Equivalent weight of } MnO_2 = \frac{\text{formula weight}}{\text{change in oxidation state}} = \frac{86.94}{2} = 43.47$$

Note that Mn is the only element in MnO_2 whose oxidation state changes during the reaction.

(b) In HNO_3 the oxidation state of N is +V; in NO_2 it is +IV. The change in oxidation state is 1.

$$\text{Equivalent weight of } HNO_3 = \frac{\text{formula weight}}{\text{change in oxidation state}} = \frac{63.02}{1} = 63.02$$

11.7. The following is an unbalanced equation for a reaction in which $KMnO_4$ acts as an oxidizing agent: $KMnO_4 + HCl \longrightarrow MnCl_2 + KCl + Cl_2 + H_2O$. Determine the equivalent weight of $KMnO_4$ as an oxidizing agent in acid solution.

The oxidation state of Mn in $KMnO_4$ is +VII; in $MnCl_2$ it is +II. The change in oxidation state is 5.

$$\text{Equivalent weight of } KMnO_4 = \frac{\text{formula weight}}{\text{change in oxidation state}} = \frac{158.04}{5} = 31.61$$

11.8. The following is an unbalanced equation for an oxidation-reduction reaction:

$$FeSO_4 + HNO_3 + H_2SO_4 \longrightarrow Fe_2(SO_4)_3 + NO + H_2O$$

Determine the equivalent weight of (a) HNO_3 (which acts here as an oxidizing agent) and (b) $FeSO_4$ (which acts here as a reducing agent).

(a) The oxidation state of N in HNO_3 is +V; in NO it is +II. The change in oxidation state is 3.

$$\text{Equivalent weight of } HNO_3 = \frac{\text{formula weight}}{\text{change in oxidation state}} = \frac{63.02}{3} = 21.01$$

(b) The oxidation state of Fe in $FeSO_4$ is +II; in $Fe_2(SO_4)_3$ it is +III. The change in oxidation state is 1. Hence, equivalent weight of $FeSO_4$ = formula weight = 151.91.

11.9. Determine the equivalent weight of (a) $K_2Cr_2O_7$ (oxidizing agent) and (b) H_2S (reducing agent) in the reaction represented by the following unbalanced equation:

$$K_2Cr_2O_7 + H_2S + H_2SO_4 \longrightarrow Cr_2(SO_4)_3 + K_2SO_4 + S + H_2O$$

(c) What weight $K_2Cr_2O_7$ is needed to complete the oxidation of 100 g H_2S in this reaction?

(a) In $K_2Cr_2O_7$ the oxidation state of Cr is +VI, and in $Cr_2(SO_4)_3$ it is +III. Each Cr atom thus undergoes a decrease of 3 units of oxidation state. The number of electrons gained by the two Cr atoms per formula unit of $K_2Cr_2O_7$ is thus $2 \times (6-3) = 6$. This result can be obtained also from the ion-electron partial equation:

$$Cr_2O_7{}^{2-} + 14\,H^+ + 6e \longrightarrow 2\,Cr^{3+} + 7\,H_2O$$

$$\text{Equivalent weight of } K_2Cr_2O_7 = \frac{\text{formula weight}}{\text{number of electrons gained}} = \frac{294.19}{6} = 49.03$$

(b) The oxidation state of S in H_2S is −II; in S it is 0. The change in oxidation state is 2.

$$\text{Equivalent weight of } H_2S = 34.08/2 = 17.04$$

(c)   Number of g-eq of H_2S = $\dfrac{100 \text{ g } H_2S}{17.04 \text{ g/g-eq}}$ = 5.87 g-eq H_2S

Number of g-eq of $K_2Cr_2O_7$ needed = number of g-eq of H_2S = 5.87 g-eq

Weight of $K_2Cr_2O_7$ needed = 5.87 g-eq $\times$ 49.03 g $K_2Cr_2O_7$/g-eq = 287.8 g $K_2Cr_2O_7$

11.10. A 5.00 gram sample of an acid was dissolved in water, and an excess of zinc was added. The weight of the evolved hydrogen was 0.0672 g. (*a*) Calculate the equivalent weight of the acid. (*b*) What weight of NaOH would 9.00 g of this acid neutralize?

Since one H^+ is a standard unit of reaction for both acid-base and oxidation-reduction reactions, the number of g-eq of H_2 liberated must equal the number of g-eq of the acid.

(*a*)   Number of gram-equivalents of H_2 = $\dfrac{0.0672 \text{ g } H_2}{1.008 \text{ g } H_2/\text{g-eq}}$ = 0.0667 g-eq

Equivalent weight of acid = $\dfrac{5.00 \text{ g}}{0.0667 \text{ g-eq}}$ = 75.0 g/g-eq

(*b*)   Number of gram-equivalents of acid = $\dfrac{9.00 \text{ g}}{75.0 \text{ g/g-eq}}$ = 0.120 g-eq

Therefore 0.120 g-eq of NaOH is required.

Weight of NaOH required = 0.120 g-eq $\times$ 40.0 g NaOH/g-eq = 4.80 g NaOH

EQUIVALENT WEIGHT OF AN ELEMENT

11.11. A 3.245 gram mass of cadmium was dissolved in dilute acid, and 0.0582 gram of hydrogen was evolved. Calculate the equivalent weight and resulting oxidation state of cadmium.

The equivalent weight of hydrogen in all such reactions is the same as the atomic weight (1.008), because two electrons are absorbed per mole of hydrogen.

$$2 \, H^+ + 2e \longrightarrow H_2$$

First Method.

Since 0.0582 g of H is displaced by 3.245 g of Cd, then

$$1 \text{ g of H is displaced by } \dfrac{3.245}{0.0582} \text{ g of Cd}$$

and 1.008 g of H is displaced by $1.008 \times \dfrac{3.245}{0.0582} \text{ g}$ = 56.2 g Cd

Hence the equivalent weight of Cd is 56.2.

Ionic valence = oxidation state = $\dfrac{\text{atomic weight}}{\text{equivalent weight}}$ = $\dfrac{112.4}{56.2}$ = 2

Another Method.

Number of g-eq H_2 = $\dfrac{0.0582 \text{ g } H_2}{1.008 \text{ g } H_2/\text{g-eq}}$ = 0.0578 g-eq

Equivalent weight of Cd = $\dfrac{3.245 \text{ g Cd}}{0.0578 \text{ g-eq}}$ = 56.2 g Cd/g-eq

11.12. An oxide of molybdenum contains 70.58% by weight molybdenum. Determine the equivalent weight and oxidation state of molybdenum in this compound.

The equivalent weight of oxygen in all oxides is one-half its atomic weight, since the oxidation state of oxygen in oxides (but not in peroxides) is $-$II.

Percent of oxygen in molybdenum oxide = 100.00% $-$ 70.58% = 29.42% by weight.

Since 29.42 parts of O unites with 70.58 parts of Mo, then

$$1 \text{ part of O unites with } \frac{70.58}{29.42} \text{ parts of Mo,}$$

and   8.000 parts of O unites with $8.000 \times \frac{70.58}{29.42} = 19.19$ parts of Mo

Hence the equivalent weight of Mo is 19.19. Oxidation state $= \frac{\text{atomic weight}}{\text{equivalent weight}} = \frac{95.94}{19.19} = 5.$

Another Method.

$$\text{Equivalent weight of Mo } = \frac{70.58 \text{ g Mo}}{\dfrac{29.42 \text{ g O}_2}{8.000 \text{ g O}_2/\text{g-eq}}} = 19.19 \text{ g Mo/g-eq}$$

11.13. Given that the equivalent weight of silver is 107.87. Calculate the equivalent weight of magnesium if 0.3636 g of it displaces 3.225 g of silver from its salts.

1 g-eq of an element will react with or displace 1 g-eq of another element. To find the equivalent weight of magnesium we must find how many grams of Mg displace a g-eq of Ag (107.87 g).

Since   3.225 g of Ag is equivalent to 0.3636 g of Mg, then

$$1 \text{ g of Ag is equivalent to } \frac{0.3636}{3.225} \text{ g of Mg,}$$

and   107.87 g of Ag is equivalent to $107.87 \times \frac{0.3636}{3.225}$ g Mg $= 12.16$ g Mg

Hence the equivalent weight of Mg is 12.16.

Another Method.

$$\text{Equivalent weight of Mg } = \frac{0.3636 \text{ g Mg}}{\dfrac{3.225 \text{ g Ag}}{107.87 \text{ g Ag/g-eq}}} = 12.16 \text{ g Mg/g-eq}$$

Supplementary Problems

ACIDS AND BASES

11.14. Calculate the equivalent weight of each of the following acids and bases, assuming complete neutralization: HBr, H_2SO_3, H_3PO_4, LiOH, $Zn(OH)_2$.
Ans. 80.92, 41.04, 32.67, 23.95, 49.69

11.15. What is the equivalent weight of H_2S in each of the following reactions?
(a) $NaOH + H_2S \longrightarrow NaHS + H_2O$ *Ans.* 34.08
(b) $2 NaOH + H_2S \longrightarrow Na_2S + 2 H_2O$ *Ans.* 17.04

11.16. Determine the equivalent weight of $H_4P_2O_7$ when it is used to make
(a) $Na_2H_2P_2O_7$, (b) $Na_3HP_2O_7$, (c) $Na_4P_2O_7$.   *Ans.* (a) 88.99, (b) 59.33, (c) 44.49

11.17. What weight of LiOH is necessary to neutralize 75.0 grams of H_3PO_4 if one of the three hydrogens remains unneutralized? Use the method of equivalent weights.   *Ans.* 36.7 g

11.18. What is the equivalent weight of an acid, 1.248 g of which requires exactly 0.2475 g of KOH for neutralization.   *Ans.* 283 g/g-eq

11.19. Crystalline potassium hydrogen phthalate, $KHC_8H_4O_4$, is often used as a standard acid because it can be purified easily and weighed. There is one ionizable H^+ per formula unit. How many milliequivalents are contained in 0.7325 g of this salt? 1 milliequivalent (meq) = .001 g-eq.
Ans. 3.587

OXIDIZING AND REDUCING AGENTS

11.20. Determine the equivalent weight of each of the underlined oxidizers or reducers in each of the following unbalanced oxidation-reduction reactions.

   (a) $\underline{MnO_2} + \underline{FeSO_4} + H_2SO_4 \longrightarrow MnSO_4 + Fe_2(SO_4)_3 + H_2O$ *Ans.* 43.47, 151.91

   (b) $\underline{KMnO_4} + \underline{H_2C_2O_4} \cdot 2\,H_2O + H_2SO_4 \longrightarrow MnSO_4 + K_2SO_4 + CO_2 + H_2O$ *Ans.* 31.61, 63.03

   (c) $\underline{K_2Cr_2O_7} + \underline{SnCl_2} + HCl \longrightarrow CrCl_3 + SnCl_4 + KCl + H_2O$ *Ans.* 49.03, 94.80

   (d) $\underline{Ce(SO_4)_2} + \underline{Hg_2SO_4} \longrightarrow Ce_2(SO_4)_3 + HgSO_4$ *Ans.* 332.25, 248.62

11.21. In reaction (b) of the above problem, compute the weight of $KMnO_4$ necessary to oxidize 0.3572 grams $H_2C_2O_4 \cdot 2\,H_2O$. In reaction (c) of the same problem, compute the weight of $K_2Cr_2O_7$ necessary to oxidize 0.2000 grams $SnCl_2$.   *Ans.* 0.1791 g $KMnO_4$, 0.1034 g $K_2Cr_2O_7$

11.22. When Fe^{3+} oxidizes I^- to I_2, the product of reduction is Fe^{2+}. How many grams of iodine can be formed by the action of 13.20 grams of $FeCl_3$ on an excess of KI?   *Ans.* 10.33 g

EQUIVALENT WEIGHT OF AN ELEMENT

11.23. Analysis shows that 16.20 g of an oxide of iron contains 11.33 g of iron. What are the equivalent weight and oxidation state of iron in this oxide?   *Ans.* 18.61, 3

11.24. When 1.532 g of vanadium was dissolved in dilute acid, 0.0909 g of hydrogen was evolved. Find the equivalent weight and resulting oxidation state of vanadium.   *Ans.* 16.99, 3

11.25. Determine the equivalent weight and oxidation state of tin in (a) the oxide which contains 88.12% tin, (b) the oxide which contains 78.77% tin.   *Ans.* 59.35, 2; (b) 29.68, 4

11.26. A determination of the equivalent weight of sulfur was carried out by synthesizing silver sulfide. 20.9810 g of silver was heated in an atmosphere of sulfur vapor to form 24.0975 g of silver sulfide. If the equivalent weight of silver is known to be 107.870, what is the experimental value of the equivalent weight of sulfur?   *Ans.* 16.023

11.27. When a 0.482 gram sample of a metal was added to an excess of dilute sulfuric acid, 120 cc of dry hydrogen gas (S.T.P.) was evolved. Compute the equivalent weight of the metal.   *Ans.* 45.0

11.28. An isotope of curium was made by a nuclear reaction. The product was roasted in air and the resulting curium oxide was found to contain 11.59% oxygen. What is the equivalent weight of curium in this compound? What is the mass number of this curium isotope?   *Ans.* 61.0, 244

Chapter 12

Expressing Concentrations of Solutions

SOLUTE AND SOLVENT

In a solution of one substance in another substance, the dissolved substance is called the *solute*. The substance in which the solute is dissolved is called the *solvent*. When the relative amount of one substance in a solution is much greater than that of the other, the substance present in greater amount is generally regarded as the solvent. When the relative amounts of the two substances are of the same order of magnitude, it becomes difficult, in fact arbitrary, to specify which substance is the solvent.

EXPRESSING CONCENTRATIONS IN PHYSICAL UNITS

When physical units are employed, the concentrations of solutions are generally expressed in the following ways.

(1) By the weight of solute per unit volume of solution.

> **Example.**
> 20 grams of KCl per liter of solution.

(2) By the percentage composition, or the number of grams of solute per 100 grams of solution.

> **Example.**
> A 10% aqueous NaCl solution contains 10 grams of NaCl per 100 grams of solution. Ten grams of NaCl is mixed with 90 grams of water to form 100 grams of solution.

(3) By the weight of solute per weight of solvent.

> **Example.**
> 5.2 grams of NaCl in 100 grams of water.

EXPRESSING CONCENTRATIONS IN CHEMICAL UNITS

(1) The *molarity* of a solution is the number of moles of the solute contained in one liter of solution.

> **Example.**
> A 0.5 molar (0.5 M) solution of H_2SO_4 contains 49.04 grams of H_2SO_4 per liter of solution, since 49.04 is half of the molecular weight of H_2SO_4, 98.08. A one molar (M) solution contains 98.08 g H_2SO_4 per liter of solution.

(2) The *formality* of a solution is the number of gram-formula weights of the solute contained in one liter of solution. For solutes that have definite molecular weights, the molarity and the formality may be the same. Thus a 1 molar solution of H_2SO_4 is the same as a 1 formal (F) solution, since the molecular weight of H_2SO_4 is the same as the formula weight, 98.08. For solutes that do not have true molecular weights, like salts and hydroxides, it is not strictly correct to refer to moles, but use may still be made of the gram-formula weight. By usage, many chemists still use the term "molecular weight" for the formula weight of such substances, "mole" for gram-formula weight, and "molarity" for formality.

> **Example.**
> A one formal (F) solution of NaOH contains 40.00 g NaOH per liter of solution, since the formula weight of NaOH is 40.00. Chemists who refer to a one molar solution of NaOH are speaking of what is called a one formal solution here.

(3) The *normality* of a solution is the number of gram-equivalent weights of the solute contained in one liter of solution.

>   **Example.**
>
>   1 mole of HCl, 1/2 mole of H_2SO_4, and 1/6 gfw $K_2Cr_2O_7$ (as oxidizing agent), each in one liter of solution, give normal (N) solutions of these substances. A normal (N) solution of HCl is also a molar (M) solution. A normal (N) solution of H_2SO_4 is also a one-half molar (0.5 M) solution.

(4) The *molality* of a solution is the number of moles of the solute per kilogram of solvent contained in a solution. The molality (m), sometimes called the *weight molarity*, cannot be computed from the normality (N) or the molarity (M) unless the specific gravity of the solution is known.

>   **Example.**
>
>   A solution made up of 98.08 g of pure H_2SO_4 and 1000 g of water would be a 1 molal (m) solution of H_2SO_4.

(5) The *weight formality* of a solution is the number of gram-formula weights of the solute per kilogram of solvent contained in a solution. This term is not in very common use, and the word *molality* is more commonly used to express the same intent.

(6) *Mole fractions* or *mole percents* are used in theoretical work because many physical properties of solutions are expressed most clearly in terms of the relative numbers of solvent and solute molecules. (The number of moles of a substance is proportional to the number of molecules.)

The *mole fraction* of any component in a solution is defined as the number of moles of that component divided by the total number of moles of all components in the solution. The sum of the mole fractions of all components of a solution is 1. In a two-component solution, the mole fraction of the solute is equal to $\dfrac{\text{moles solute}}{\text{moles solute} + \text{moles solvent}}$. Similarly, the mole fraction of the solvent is equal to $\dfrac{\text{moles solvent}}{\text{moles solute} + \text{moles solvent}}$.

>   **Example.**
>
>   If a solution contains 2 moles of ethyl alcohol and 6 moles of water, the mole fraction of alcohol $= \dfrac{2}{2+6} = 0.25$, and the mole fraction of water $= \dfrac{6}{2+6} = 0.75$. The *mole percent* alcohol in this solution $= 100 \times 0.25 = 25$ mole percent alcohol, and the mole percent water $= 100 \times 0.75 = 75$ mole percent water.

The *molarity, formality,* and *normality* scales are useful for volumetric experiments, in which the amount of solute in a given portion of solution is related to the measured volume of solution. As will be seen in subsequent chapters, the normality scale is very convenient for comparing the relative volumes required for two solutions to react chemically with each other. A limitation of the normality scale is that a given solution may have more than one normality, depending on the reaction for which it is used. The molarity or formality of a solution, on the other hand, is a fixed number because the molecular weight or formula weight of a substance does not depend on the reaction for which the substance is used, as may the equivalent weight.

The *molality* scale is useful for experiments in which physical measurements are made (like freezing point, boiling point, vapor pressure, etc.) over a wide range of temperatures. The molality of a given solution, being a concentration unit that depends only on the weights of the solution components, is independent of temperature. The molarity, formality, or normality of a solution, on the other hand, being defined in terms of the volume, may vary appreciably as the temperature is changed because of the temperature dependence of the volume.

SUMMARY OF CONCENTRATION UNITS

(1) Molarity of a solution $= \dfrac{\text{number of moles of solute}}{\text{number of liters of solution}}$

(2) Formality of a solution $= \dfrac{\text{number of gram-formula weights of solute}}{\text{number of liters of solution}}$

(3) Normality of a solution $= \dfrac{\text{number of gram-equivalents of solute}}{\text{number of liters of solution}}$

$= \dfrac{\text{number of milli-equivalents of solute}}{\text{number of milliliters of solution}}$

A *milli-equivalent* (meq), or *milligram-equivalent* (mg-eq), is one one-thousandth of a gram-equivalent, or is that amount of the substance whose numerical part is the same as the equivalent weight when the unit is the milligram. For example, the equivalent weight of H_2SO_4 is 49.04. Then 1 gram-equivalent (1 g-eq) of H_2SO_4 is 49.04 g H_2SO_4, and 1 milli-equivalent (1 meq) is 49.04 mg H_2SO_4.

It should be clear that a normal solution of H_2SO_4 contains 49.04 g H_2SO_4 in 1 liter of solution, or 49.04 mg H_2SO_4 in 1 milliliter of solution.

(4) Molality of a solution $= \dfrac{\text{number of moles of solute}}{\text{number of kilograms of solvent}}$

(5) Weight formality of a solution $= \dfrac{\text{number of gram-formula weights of solute}}{\text{number of kilograms of solvent}}$

(6) Mole fraction of any component $= \dfrac{\text{number of moles of that component}}{\text{total number of moles of all components}}$

DILUTION PROBLEMS

The volumetric scales of concentration are those in which the concentration is expressed in terms of the amount of solute per fixed volume of solution. Examples are moles per liter (molarity), gram-formula weights per liter (formality), gram-equivalents per liter (normality), grams per liter, and grams per 100 ml. When the concentration is expressed on a volumetric scale, the amount of solute contained in a given volume of solution is equal to the product of the volume and the concentration. Thus,

Amount of dissolved solute = volume × concentration

When a solution is diluted, the volume is increased and the concentration is decreased, but the total amount of solute is constant. Hence, two solutions of different concentrations but containing the same amounts of solute will be related to each other as follows:

Amount of dissolved solute$_1$ = Amount of dissolved solute$_2$

or   volume$_1$ × concentration$_1$ = volume$_2$ × concentration$_2$

If any three terms in the above equation are known, the fourth can be calculated. The quantities on both sides of the equation must be expressed in the same units.

Solved Problems

CONCENTRATIONS EXPRESSED IN PHYSICAL UNITS

12.1. Explain how you would prepare 60 ml of an aqueous solution of $AgNO_3$ of strength 0.030 g $AgNO_3$ per ml.

Each ml of solution contains 0.030 g $AgNO_3$.

Therefore 60 ml of solution will contain   0.030 g/ml $\times$ 60 ml   =   1.8 g $AgNO_3$.

Dissolve 1.8 g $AgNO_3$ in about 50 ml of water. Then add sufficient water to make the volume exactly 60 ml. Stir thoroughly to insure uniformity.

12.2. How many grams of 5.0% NaCl solution are necessary to yield 3.2 g NaCl?

A 5.0% NaCl solution contains 5.0 g NaCl in 100 g of solution.

Then   5.0 g NaCl is contained in 100 g solution,

$$1 \text{ g NaCl is contained in } \frac{100}{5.0} \text{ g solution,}$$

and   3.2 g NaCl is contained in $3.2 \times \dfrac{100}{5.0}$ g solution = 64 g solution.

Or, by proportion, letting x = required number of grams of solution,

$$\frac{5.0 \text{ g NaCl}}{100 \text{ g solution}} = \frac{3.2 \text{ g NaCl}}{x} \qquad \text{from which} \qquad x = 64 \text{ g solution.}$$

Or, by the use of unit conversion factors,

$$x \text{ g solution} = 3.2 \text{ g NaCl} \times \frac{100 \text{ g solution}}{5.0 \text{ g NaCl}} = 64 \text{ g solution.}$$

12.3. How much $NaNO_3$ must be weighed out to make 50 ml of an aqueous solution containing 70 mg Na^+ per ml?

Weight of Na^+ in 50 ml of solution   =   50 ml $\times$ 70 mg/ml   =   3500 mg   =   3.5 g Na^+

Formula weight of $NaNO_3$ = 85;   atomic weight of Na = 23.

Then   23 g Na^+ is contained in 85 g $NaNO_3$,

$$1 \text{ g } Na^+ \text{ is contained in } \frac{85}{23} \text{ g } NaNO_3,$$

and   3.5 g Na^+ is contained in $3.5 \times \dfrac{85}{23}$ g = 12.9 g $NaNO_3$.

Or, by direct use of quantitative factors,

$$x \text{ g } NaNO_3 = 50 \text{ ml solution} \times \frac{70 \text{ mg } Na^+}{1 \text{ ml solution}} \times \frac{85 \text{ g } NaNO_3}{23 \text{ g } Na^+} \times \frac{1 \text{ g}}{1000 \text{ mg}} = 12.9 \text{ g } NaNO_3.$$

12.4. Calculate the weight of $Al_2(SO_4)_3 \cdot 18 H_2O$ needed to make up 50 ml of an aqueous solution of strength 40 mg Al^{+++} per ml.

Grams of Al^{+++} in 50 ml of solution   =   50 ml $\times$ 40 mg/ml   =   2000 mg   =   2.00 g Al^{+++}

Atomic weight of Al = 27;   formula weight of $Al_2(SO_4)_3 \cdot 18 H_2O$ = 666.

$$x \text{ g } Al_2(SO_4)_3 \cdot 18 H_2O = 2.00 \text{ g } Al^{+++} \times \frac{666 \text{ g } Al_2(SO_4)_3 \cdot 18 H_2O}{54 \text{ g } Al^{+++}}$$

$$= 24.7 \text{ g } Al_2(SO_4)_3 \cdot 18 H_2O$$

A solution of identical composition could be prepared from the appropriate amount of anhydrous $Al_2(SO_4)_3$. To compute the appropriate weight in this case, the formula weight of $Al_2(SO_4)_3$, 342, would be used instead of 666 above. In general, hydrated salts differ from the anhydrous salts only in the crystalline state. In solution, the water of hydration and the water of the solvent become indistinguishable from each other.

12.5. Describe how you would prepare 50 grams of a 12.0% by weight $BaCl_2$ solution, starting with $BaCl_2 \cdot 2H_2O$ and pure water.

A 12.0% $BaCl_2$ solution contains 12.0 g $BaCl_2$ per 100 g of solution, or 6.0 g $BaCl_2$ in 50 g of solution.

Formula weight of $BaCl_2$ = 208, of $BaCl_2 \cdot 2H_2O$ = 244.

Since 208 g $BaCl_2$ is contained in 244 g $BaCl_2 \cdot 2H_2O$, then

$$1 \text{ g } BaCl_2 \text{ is contained in } \frac{244}{208} \text{ g } BaCl_2 \cdot 2H_2O,$$

and   6.0 g $BaCl_2$ is contained in $6.0 \times \frac{244}{208}$ g = 7.0 g $BaCl_2 \cdot 2H_2O$.

Then 7.0 g $BaCl_2 \cdot 2H_2O$ contains 6.0 g $BaCl_2$ and 1.0 g H_2O. The solution is prepared by dissolving 7.0 g $BaCl_2 \cdot 2H_2O$ in 43 g or ml of water. This makes 50 g of solution.

12.6. Calculate the weight of anhydrous HCl in 5.00 ml of concentrated hydrochloric acid of specific gravity 1.19 and containing 37.23% HCl by weight.

One ml of solution weighs 1.19 g. Weight of 5.00 ml = 5.00 ml × 1.19 g/ml = 5.95 g.

The solution contains 37.23% HCl by weight.

Hence the weight of HCl in 5.95 g solution = 0.3723 × 5.95 g = 2.22 g anhydrous HCl.

12.7. Calculate the volume of concentrated sulfuric acid, of specific gravity 1.84 and containing 98% H_2SO_4 by weight, that would contain 40.0 g of pure H_2SO_4.

One ml of solution weighs 1.84 g and contains 0.98 × 1.84 g = 1.80 g pure H_2SO_4.

Then 40.0 g H_2SO_4 is contained in $\frac{40.0 \text{ g}}{1.80 \text{ g/ml}}$ = 22.2 ml of solution.

Or, by proportion, letting x = volume of solution that contains 40.0 g H_2SO_4,

$$\frac{1.80 \text{ g } H_2SO_4}{1.00 \text{ ml solution}} = \frac{40.0 \text{ g } H_2SO_4}{x} \qquad \text{which gives} \qquad x = 22.2 \text{ ml solution.}$$

12.8. Exactly 4.00 g of a solution of sulfuric acid was diluted with water and an excess of $BaCl_2$ was added. The washed and dried precipitate of $BaSO_4$ weighed 4.08 g. Find the percent H_2SO_4 in the original acid solution.

$$H_2SO_4 + BaCl_2 \longrightarrow 2HCl + BaSO_4$$

Formula weight of H_2SO_4 = 98.08, of $BaSO_4$ = 233.4.

(1) First determine the weight of H_2SO_4 required to precipitate 4.08 g $BaSO_4$. The equation shows that 1 gfw $BaSO_4$ (233.4 g) requires 1 mole H_2SO_4 (98.08 g).

$$n_{BaSO_4} = \frac{4.08 \text{ g } BaSO_4}{233.4 \text{ g/gfw}} = 0.0175 \text{ gfw } BaSO_4$$

Therefore 0.0175 mole H_2SO_4 is required.

Weight of H_2SO_4 required = 0.0175 mole H_2SO_4 × 98.08 g/mole = 1.72 g H_2SO_4.

(2) Fraction H_2SO_4 by weight = $\dfrac{\text{weight of } H_2SO_4}{\text{weight of solution}} = \dfrac{1.72 \text{ g}}{4.00 \text{ g}} = 0.430 = 43.0\%$.

CONCENTRATIONS EXPRESSED IN CHEMICAL UNITS

12.9. How many grams of solute are required to prepare 1 liter of 1 F $Pb(NO_3)_2$? What is the molarity of the solution with respect to each of the ions?

A formal solution contains 1 gfw of solute dissolved in 1 liter of solution. The formula weight of $Pb(NO_3)_2$ is 331.2; hence 331.2 g of $Pb(NO_3)_2$ is needed.

Although $Pb(NO_3)_2$ does not contain definite molecules, when dissolved in water it yields individual Pb^{++} and NO_3^- particles, which are ions (or charged molecules) of definite particle weight. It is therefore correct to refer to the formula weights of these ions as their molecular weights. A 1 F solution of $Pb(NO_3)_2$ is thus 1 M with respect to Pb^{++} and 2 M with respect to NO_3^-.

12.10. What is the molarity of a solution containing 16.0 g CH_3OH in 200 ml of solution?

Formula weight of CH_3OH = 32.0.

$$\text{Molarity} = \frac{\text{moles of solute}}{\text{liters of solution}} = \frac{\dfrac{16.0 \text{ g}}{32.0 \text{ g/mole}}}{0.200 \ l} = 2.50 \text{ mole}/l = 2.50 \text{ M}$$

12.11. Determine the formal concentration of each of the following solutions:
(a) 18.0 g $AgNO_3$ per liter of solution; (b) 12.00 g $AlCl_3 \cdot 6 H_2O$ per liter of solution.

Formula weight of $AgNO_3$ = 169.9, of $AlCl_3 \cdot 6 H_2O$ = 241.4.

(a) $\dfrac{18.0 \text{ g}/l}{169.9 \text{ g/gfw}}$ = 0.106 gfw/l = 0.106 F   (b) $\dfrac{12.00 \text{ g}/l}{241.4 \text{ g/gfw}}$ = 0.0497 gfw/l = 0.0497 F

12.12. What weight of $(NH_4)_2SO_4$ is required to prepare 400 ml of F/4 solution?

Formula weight of $(NH_4)_2SO_4$ = 132.1.

The notation F/4 is often used in place of $\frac{1}{4}$ F, M/2 for $\frac{1}{2}$ M, and so on.

One liter of F/4 solution contains $\frac{1}{4} \times 132.1$ g = 33.02 g $(NH_4)_2SO_4$.

Then 400 ml of F/4 solution requires   0.400 $l \times$ 33.02 g/l = 13.21 g $(NH_4)_2SO_4$.

Another Method.

Weight = formality $\times$ formula weight $\times$ volume

       = $\frac{1}{4}$ gfw/$l \times$ 132.1 g/gfw $\times$ 0.400 l = 13.21 g $(NH_4)_2SO_4$

12.13. What is the molality of a solution which contains 20.0 grams of cane sugar, $C_{12}H_{22}O_{11}$, dissolved in 125 grams of water?

Molecular weight of $C_{12}H_{22}O_{11}$ = 342.

$$\text{Molality} = \frac{\text{moles of solute}}{\text{kg of solvent}} = \frac{\dfrac{20.0 \text{ g}}{342 \text{ g/mole}}}{\dfrac{125 \text{ g solvent}}{1000 \text{ g/kg}}} = 0.468 \text{ moles/kg} = 0.468 \ m$$

12.14. A solution of ethyl alcohol, C_2H_5OH, in water is 1.54 molal. How many grams of alcohol are dissolved in 2500 grams of water?

Molecular weight of C_2H_5OH = 46.1.

Since the solution is 1.54 molal, 1000 g water dissolves 1.54 moles of alcohol. Then 2500 g water dissolves 2.500×1.54 = 3.85 moles of alcohol.

Weight of alcohol = 3.85 moles $\times$ 46.1 g/mole = 177 g alcohol.

12.15. Calculate the (a) molarity and (b) molality of a sulfuric acid solution of specific gravity 1.198, containing 27.0% H_2SO_4 by weight.

(a) Each ml of acid solution weighs 1.198 g and contains 0.270×1.198 g H_2SO_4.

The weight of H_2SO_4 in one liter (1000 ml) is $1000 \times 0.270 \times 1.198$ g = 324 g H_2SO_4.

Molecular wight of H_2SO_4 = 98.1.

In 1 liter of solution, $n_{H_2SO_4} = \dfrac{324 \text{ g}}{98.1 \text{ g/mole}}$ = 3.30 moles H_2SO_4.

One liter of solution contains 3.30 moles H_2SO_4; hence the acid is 3.30 M.

(b) The weight of water in one liter of solution is $(1198 - 324)$ g $= 874$ g H_2O.

$$\text{Molality} = \frac{\text{moles of solute}}{\text{kg of solvent}} = \frac{3.30 \text{ moles } H_2SO_4}{0.874 \text{ kg } H_2O} = 3.78 \; m$$

12.16. Determine the mole fraction of each substance in a solution containing 36.0 g of water and 46 g of glycerin, $C_3H_5(OH)_3$.

Molecular weight of $C_3H_5(OH)_3 = 92$, of $H_2O = 18.0$.

Moles glycerin $= \dfrac{46 \text{ g}}{92 \text{ g/mole}} = 0.50$ moles; moles water $= \dfrac{36.0 \text{ g}}{18.0 \text{ g/mole}} = 2.00$ moles.

Total moles $= 0.50 + 2.00 = 2.50$ moles.

$$\text{Mole fraction of glycerin} = \frac{\text{moles glycerin}}{\text{total moles}} = \frac{0.50}{2.50} = 0.20$$

$$\text{Mole fraction of water} = \frac{\text{moles water}}{\text{total moles}} = \frac{2.00}{2.50} = 0.80$$

Check: Sum of mole fractions $= 0.20 + 0.80 = 1.00$

12.17. How many gram-equivalents of solute are contained in:

(a) 1 liter of 2 N solution, (b) 1 liter of 0.5 N solution, (c) 0.5 liter of 0.2 N solution?

A normal solution contains 1 gram-equivalent of solute in 1 liter of solution.

(a) 1 liter of 2 N contains 2 gram-equivalents of solute.

(b) 1 liter of 0.5 N contains 0.5 gram-equivalent of solute.

(c) 0.5 liter of 0.2 N contains $0.5 \; l \times 0.2$ g-eq/l $= 0.1$ gram-equivalents of solute.

12.18. How many (a) gram-equivalents and (b) milli-equivalents of solute are present in 60 ml of 4.0 N solution?

(a) Number of gram-equivalents $=$ number of liters $\times$ normality
$$= 0.060 \; l \times 4.0 \text{ g-eq}/l = 0.24 \text{ g-eq}$$

(b) Since   1 g-eq $= 1000$ meq,
$$0.24 \text{ g-eq} = 0.24 \text{ g-eq} \times 1000 \text{ meq/g-eq} = 240 \text{ meq}$$

Another Method.   Number of meq $=$ number of ml $\times$ normality
$$= 60 \text{ ml} \times 4.0 \text{ meq/ml} = 240 \text{ meq}$$

12.19. How many grams of solute are required to prepare 1 liter of 1 N solution of each of the following: LiOH, Br_2 (as oxidizing agent), H_3PO_4 (for a reaction in which 3 H are replaceable)?

A normal solution contains 1 gram-equivalent of solute in 1 liter of solution.

Formula weight of LiOH $= 23.95$, of $Br_2 = 159.82$, of $H_3PO_4 = 97.99$.

$\quad\quad$ 1 liter of 1 N LiOH requires $23.95/1$ g $= 23.95$ g NaCl

$\quad\quad$ 1 liter of 1 N Br_2 requires $159.82/2$ g $= 79.91$ g Br_2

$\quad\quad$ 1 liter of 1 N H_3PO_4 requires $97.99/3$ g $= 32.66$ g H_3PO_4

12.20. Calculate the normality of each of the following solutions:

(a) 7.88 g of HNO_3 per liter of solution,

(b) 26.5 g of Na_2CO_3 per liter of solution (if acidified to form CO_2).

Formula weight $HNO_3 = 63.02$, $Na_2CO_3 = 106.0$.

(a) Equivalent weight of HNO_3 = formula weight = 63.02.

Then 1 N HNO_3 contains 63.02 g HNO_3 per liter of solution.

$$\text{Normality} = \frac{7.88 \text{ g/}l}{63.02 \text{ g/g-eq}} = 0.1251 \text{ g-eq/}l = 0.1251 \text{ N}$$

(b) Equivalent weight of Na_2CO_3 = 1/2 × formula weight = 1/2 × 106.0 = 53.0

$$\text{Normality} = \frac{26.5 \text{ g/}l}{53.0 \text{ g/g-eq}} = 0.500 \text{ g-eq/}l = 0.500 \text{ N}$$

12.21. How many ml of 2.00 F $Pb(NO_3)_2$ contains 600 mg Pb^{++}?

One liter of 1 F $Pb(NO_3)_2$ contains 1 g-at of Pb^{++}. Then 1 liter of 2 F contains 2 g-at of Pb^{++}, or 414 g of Pb^{++} per liter, or 414 mg of Pb^{++} per ml.

Hence 600 mg Pb^{++} is contained in $\dfrac{600 \text{ mg}}{414 \text{ mg/ml}} = 1.45 \text{ ml}$ of 2.00 F $Pb(NO_3)_2$.

12.22. How many kilograms of wet NaOH containing 12% water are required to prepare 60 liters of 0.50 N solution?

One liter of 0.50 N NaOH contains 0.50 × 40 g = 20 g = 0.020 kg NaOH.

Then 60 liters of 0.50 N NaOH contains 60 × 0.020 kg = 1.20 kg NaOH.

The wet NaOH contains 100% − 12% = 88% pure NaOH.

Then   88 kg pure NaOH is contained in 100 kg wet NaOH,

1 kg pure NaOH is contained in $\dfrac{100}{88}$ kg wet NaOH,

and   1.20 kg pure NaOH is contained in $1.20 \times \dfrac{100}{88}$ kg = 1.36 kg wet NaOH.

Or, by proportion, letting x = kg wet NaOH that gives 1.20 kg pure NaOH,

$$\frac{100 \text{ kg wet}}{88 \text{ kg pure}} = \frac{x}{1.20 \text{ kg pure}} \quad \text{from which} \quad x = 1.36 \text{ kg wet NaOH.}$$

12.23. $K^+MnO_4^- + K^+I^- + (H^+)_2SO_4^{--} \longrightarrow (K^+)_2SO_4^{--} + Mn^{++}SO_4^{--} + I_2 + H_2O$
$$\text{(unbalanced)}$$

(a) How many grams of $KMnO_4$ are needed to make 500 ml of 0.250 N solution?

(b) How many grams of KI are needed to make 25 ml of 0.36 N solution?

(a) In this oxidation-reduction reaction, the oxidation state of Mn changes from +VII in MnO_4^- to +II in Mn^{++}.

Hence the equivalent weight of $KMnO_4 = \dfrac{\text{formula weight}}{\text{oxidation state change}} = \dfrac{158}{5} = 31.6$.

Then 0.500 l of 0.250 N requires   0.500 l × 0.250 g-eq/l × 31.6 g/g-eq = 3.95 g $KMnO_4$.

(b) The oxidation state of I changes from −1 in I^- to 0 in I_2.

Hence the equivalent weight of KI = formula weight = 166.

Then 0.025 l of 0.36 N requires   0.025 l × 0.36 g-eq/l × 166 g/g-eq = 1.49 g KI.

12.24. $K^+MnO_4^- + Mn^{++}SO_4^{--} + H_2O \longrightarrow MnO_2 + (K^+)_2SO_4^{--} + (H^+)_2SO_4^{--}$ (unbalanced)
How many grams of $KMnO_4$ are needed to make 500 ml of 0.250 N solution?

In this oxidation-reduction reaction, the oxidation state of Mn changes from +VII in MnO_4^- to +IV in MnO_2.

Hence the equivalent weight of $KMnO_4 = \dfrac{\text{formula weight}}{\text{oxidation state change}} = \dfrac{158}{3} = 52.7$.

Then 0.500 l of 0.250 N requires   0.500 l × 0.250 g-eq/l × 52.7 g/g-eq = 6.59 g $KMnO_4$.

DILUTION PROBLEMS

12.25. To what extent must a given solution of concentration 40 mg $AgNO_3$ per ml be diluted to yield one of concentration 16 mg $AgNO_3$ per ml?

Let x = volume to which 1 ml of given solution must be diluted to yield one of concentration 16 mg $AgNo_3$ per ml.

The total amount of solute does not change with dilution. Therefore:

$$\text{Amount of dissolved solute}_1 = \text{Amount of dissolved solute}_2$$

or

$$\text{volume}_1 \times \text{concentration}_1 = \text{volume}_2 \times \text{concentration}_2$$

$$1 \text{ ml} \times 40 \text{ mg/ml} = x \times 16 \text{ mg/ml}$$

Solving, $x = 2.5$ ml. Each ml of the given solution must be diluted to a volume of 2.5 ml.

Another Method.

The amount of solute per ml of diluted solution will be 16/40 as much as in the original solution. Hence $40/16$ ml $= 2.5$ ml of the diluted solution will contain as much solute as 1 ml of the original solution.

Note that 2.5 is not the number of ml of water to be added, but the final volume of the solution after water has been added to 1 ml of the original solution. The dilution formula always gives answers in terms of the *total* volume of solution. If we can assume that there is no volume shrinkage or expansion on dilution, the amount of water to be added in this problem is 1.5 ml. Because this assumption cannot always be made, the answers in the subsequent problems will be left in terms of the *total* volumes of the solutions.

12.26. To what extent must a given 0.50 F $BaCl_2$ solution be diluted to yield one of concentration 20 mg Ba^{++} per ml?

(1) Convert the concentration 0.50 F $BaCl_2$ to mg Ba^{++} per ml.

0.50 F $BaCl_2$ contains 0.50 gfw of $BaCl_2$ or of Ba^{++} per liter.

Weight of Ba^{++} in 0.50 gfw $= 0.50$ g-at $\times 137.3$ g/g-at $= 68.6$ g Ba^{++}.

Thus 0.50 F $BaCl_2$ contains 68.6 g Ba^{++} per liter, or 68.6 mg Ba^{++} per ml.

(2) The problem now is to find the extent to which a given solution of strength 68.6 mg Ba^{++} per ml must be diluted to yield one of concentration 20 mg Ba^{++} per ml.

$$\text{Volume}_1 \times \text{concentration}_1 = \text{volume}_2 \times \text{concentration}_2$$

$$1 \text{ ml} \times 68.6 \text{ mg/ml} = x \times 20 \text{ mg/ml}$$

Solving, $x = 3.43$ ml. Each ml of 0.50 F $BaCl_2$ must be diluted with water to a volume of 3.43 ml.

12.27. A procedure calls for 100 ml of 20% H_2SO_4, specific gravity 1.14. How much of the concentrated acid, of specific gravity 1.84 containing 98% H_2SO_4 by weight, must be diluted with water to prepare 100 ml of acid of the required strength?

The concentrations must first be changed from a weight basis to a volumetric basis, so that the law of dilution will apply.

Weight of H_2SO_4 per ml of 20% acid $= 0.20 \times 1.14$ g/ml $= 0.228$ g/ml

Weight of H_2SO_4 per ml of 98% acid $= 0.98 \times 1.84$ g/ml $= 1.80$ g/ml

Let x = number of ml of 98% acid required for 100 ml of 20% acid.

$$\text{Volume}_1 \times \text{concentration}_1 = \text{volume}_2 \times \text{concentration}_2$$

$$100 \text{ ml} \times 0.228 \text{ g/ml} = x \times 1.80 \text{ g/ml}$$

Solving, $x = 12.7$ ml of the concentrated acid.

12.28. What volumes of N/2 and of N/10 HCl must be mixed to give 2 liters of N/5 HCl?

Let x = liters of N/2 required; then $(2-x)$ = liters of N/10 required.

$$\text{Gram-equivalents N/5} = \text{gram-equivalents N/2} + \text{gram-equivalents N/10}$$
$$2 \text{ liters} \times 1/5 \text{ N} = x \text{ liters} \times 1/2 \text{ N} + (2-x) \text{ liters} \times 1/10 \text{ N}$$

Solving, $x = 0.5$ liter. Then $0.5\ l$ of N/2 and $1.5\ l$ of N/10 are required.

12.29. How many milliliters of concentrated sulfuric acid, of specific gravity 1.84 containing 98% H_2SO_4 by weight, should be taken to make (a) one liter of normal solution, (b) one liter of 3.00 N solutions, (c) 200 cc of 0.500 N solution?

Equivalent weight of H_2SO_4 = $1/2 \times$ formula weight = $1/2 \times 98.1$ = 49.0.

The H_2SO_4 content of one liter of the concentrated acid is $.98 \times 1000 \text{ cc} \times 1.84 \text{ g/cc}$ = 1800 g H_2SO_4. The normality of the concentrated acid is $\dfrac{1800 \text{ g } H_2SO_4/l}{49.0 \text{ g } H_2SO_4/\text{g-eq}}$ = 36.7 g-eq/l. The dilution formula can now be applied to each case, $V_{conc.} \times N_{conc.} = V_{dil.} \times N_{dil.}$

(a) $V_{conc.} = \dfrac{1\ l \times 1.00 \text{ N}}{36.7 \text{ N}} = .0272\ l = 27.2$ cc of conc. acid.

(b) $V_{conc.} = \dfrac{1\ l \times 3.00 \text{ N}}{36.7 \text{ N}} = .0817\ l = 81.7$ cc of conc. acid.

(c) $V_{conc.} = \dfrac{200 \text{ cc} \times 0.500 \text{ N}}{36.7 \text{ N}} = 2.72$ cc.

Supplementary Problems

CONCENTRATIONS EXPRESSED IN PHYSICAL UNITS

12.30. What weight of NH_4Cl is required to prepare 100 ml of a solution of strength 70 mg NH_4Cl per ml? *Ans.* 7.0 g

12.31. How many grams of concentrated hydrochloric acid, containing 37.9% HCl by weight, will give 5.0 g HCl? *Ans.* 13.2 g

12.32. It is required to prepare 100 g of a 19.7% by weight solution of NaOH. How many grams each of NaOH and H_2O are required? *Ans.* 19.7 g NaOH, 80.3 g H_2O

12.33. What weight of $CrCl_3 \cdot 6\,H_2O$ is needed to prepare one liter of solution containing 20 mg Cr^{+++} per ml? *Ans.* 102 g

12.34. How many grams of Na_2CO_3 are needed to prepare 500 ml of a solution containing 10.0 mg CO_3^{--} per ml? *Ans.* 8.83 g

12.35. Calculate the volume occupied by 100 g of sodium hydroxide solution of specific gravity 1.20. *Ans.* 83.3 ml

12.36. What volume of dilute nitric acid, of specific gravity 1.11 and 19% HNO_3 by weight, contains 10 g HNO_3? *Ans.* 47 ml

12.37. How many ml of a solution containing 40 g $CaCl_2$ per liter are needed to react with 0.642 g of pure Na_2CO_3? $CaCO_3$ is formed in the reaction. *Ans.* 16.8 ml

12.38. Ammonia gas is passed into water, yielding a solution of specific gravity 0.93 and containing 18.6% by weight of NH_3. What is the weight of NH_3 per ml of solution? *Ans.* 173 mg/ml

12.39. Hydrogen chloride gas is passed into water, yielding a solution of specific gravity 1.12 and containing 30.5% HCl by weight. What is the weight of HCl per ml of solution?   *Ans.* 342 mg/ml

12.40. Given 100 ml of pure water at 4°C, what volume of a solution of hydrochloric acid, specific gravity 1.175 and containing 34.4% HCl by weight, could be prepared?   *Ans.* 130 ml

12.41. A volume of 105 ml of pure water at 4°C is saturated with NH_3 gas, yielding a solution of specific gravity 0.90 and containing 30% NH_3 by weight. Find the volume of the ammonia solution resulting, and the volume of the ammonia gas at 5°C and 775 mm which was used to saturate the water.
Ans. 167 ml, 59 l

12.42. An excellent solution for cleaning grease stains from cloth or leather consists of the following: Carbon tetrachloride 80% (by volume), ligroin 16%, amyl alcohol 4%. How many cc of each should be taken to make up 75 cc of solution?   *Ans.* 60 cc, 12 cc, 3 cc

12.43. A liter of milk weighs 1032 g. The butterfat which it contains to the extent of 4.0% by volume has specific gravity 0.865. What is the density of the fat-free "skimmed" milk?   *Ans.* 1.039 g/cc

12.44. To make a benzene soluble cement, melt 49 g of rosin in an iron pan and add 28 g each of shellac and beeswax. How many pounds of each component should be taken to make 75 lb of cement?
Ans. 35 lb rosin, 20 lb shellac, 20 lb beeswax

12.45. How much $CaCl_2 \cdot 6H_2O$ and water must be weighed out to make 100 grams of a solution that is 5.0% $CaCl_2$?   *Ans.* 9.9 g $CaCl_2 \cdot 6H_2O$, 90.1 g water

12.46. How much $BaCl_2$ would be needed to make 250 ml of a solution having the same concentration of Cl^- as one containing 3.78 g NaCl per 100 ml?   *Ans.* 16.8 g $BaCl_2$

CONCENTRATIONS EXPRESSED IN CHEMICAL UNITS

12.47. What is the formality of a solution containing 37.5 g $Ba(MnO_4)_2$ per liter, and what is the molarity with respect to each type of ion?   *Ans.* 0.100 F $Ba(MnO_4)_2$, 0.100 M Ba^{++}, 0.200 M MnO_4^-

12.48. How many grams of solute are required to prepare 1 liter of 1 F $CaCl_2 \cdot 6H_2O$?   *Ans.* 219.1 g

12.49. Exactly 100 g of NaCl is dissolved in sufficient water to give 1500 ml of solution. What is the formal concentration?   *Ans.* 1.14 F

12.50. Calculate the molality of a solution containing (*a*) 0.65 mole glucose, $C_6H_{12}O_6$, in 250 g water, (*b*) 45 g glucose in 1000 g water, (*c*) 18 g glucose in 200 g water.
Ans. (*a*) 2.6, (*b*) 0.25, (*c*) 0.50 molal

12.51. How many grams of $CaCl_2$ should be added to 300 ml of water to make up a 2.46 weight-formal (molal) solution?   *Ans.* 82 g

12.52. A solution contains 57.5 ml of ethyl alcohol (C_2H_5OH) and 600 ml of benzene (C_6H_6). How many grams of alcohol are in 1000 g benzene? What is the molality of the solution? Specific gravity of $C_2H_5OH = 0.80$, of $C_6H_6 = 0.90$.   *Ans.* 85 g, 1.85 molal

12.53. A solution contains 10.0 g acetic acid, CH_3COOH, in 125 g water. What is the concentration of the solution expressed as (*a*) mole fractions of CH_3COOH and H_2O, (*b*) molality?
Ans. (*a*) mole fraction acid = 0.024, water = 0.976;   (*b*) 1.33 molal

12.54. A solution contains 116 g acetone (CH_3COCH_3), 138 g ethyl alcohol (C_2H_5OH), and 126 g water. Determine the mole fraction of each.   *Ans.* acetone = 0.167, alcohol = 0.250, water = 0.583

12.55. What is the mole fraction of the solute in a 1.00 molal aqueous solution?   *Ans.* 0.0177

12.56. An aqueous solution labeled 35.0% $HClO_4$ had a density 1.251 g/cc. What are the molarity and molality of the solution?   *Ans.* 4.36 M, 5.36 m

12.57. A sucrose solution was prepared by dissolving 13.5 g $C_{12}H_{22}O_{11}$ in enough water to make exactly 100 ml of solution, which was then found to have a density of 1.050 g/cc. Compute the molarity and molality of the solution.   *Ans.* 0.395 M, 0.431 m

12.58. How many milli-equivalents of H_2SO_4 are in (a) 2.0 ml of 15 N, (b) 50 ml of N/4 H_2SO_4?
Ans. 30 meq, 12.5 meq

12.59. What volume of a 0.232 N solution contains (a) 3.17 meq of solute, (b) 6.5 g-eq of solute?
Ans. (a) 13.7 ml, (b) 28.0 l

12.60. Determine the formality of each of the following solutions:
    (a) 166 g KI per liter of solution.   *Ans.* 1.00 F
    (b) 33.0 g $(NH_4)_2SO_4$ in 200 ml of solution.   *Ans.* 1.25 F
    (c) 12.5 g $CuSO_4 \cdot 5 H_2O$ in 100 ml of solution.   *Ans.* 0.500 F
    (d) 10.0 mg Al^{+++} per ml of solution.   *Ans.* 0.371 F

12.61. What volume of 0.200 F $Ni(NO_3)_2 \cdot 6 H_2O$ contains 500 mg Ni^{++}?   *Ans.* 42.6 ml

12.62. Compute the volume of concentrated H_2SO_4 (specific gravity 1.835, 93.2% H_2SO_4 by weight) required to make up 500 ml of 3.00 N acid.   *Ans.* 43.0 ml

12.63. Compute the volume of concentrated HCl (specific gravity 1.19, 38% HCl by weight) required to make up 18 liters of 1/50 N acid.   *Ans.* 29 ml

12.64. Determine the weight of $KMnO_4$ required to make 80 ml of N/8 $KMnO_4$ when the latter acts as an oxidizing agent in acid solution and Mn^{++} is a product of the reaction.   *Ans.* 0.316 g

12.65. $Cr_2O_7^{--} + Fe^{++} + H^+ \longrightarrow Cr^{+++} + Fe^{+++} + H_2O$ (unbalanced)
    (a) What is the normality of a $K_2Cr_2O_7$ solution 35.0 ml of which contains 3.87 g of $K_2Cr_2O_7$?
    (b) What is the normality of a $FeSO_4$ solution 750 ml of which contains 96.3 g of $FeSO_4$?
    Ans. (a) 2.25 N, (b) 0.845 N

12.66. What weight of $Na_2S_2O_3 \cdot 5 H_2O$ is needed to make up 500 cc of 0.200 N solution for the following reaction?
$$2 S_2O_3^{2-} + I_2 \longrightarrow S_4O_6^{2-} + 2 I^-$$
    Ans. 24.8 g

DILUTION PROBLEMS

12.67. A solution contains 75 mg NaCl per ml. To what extent must it be diluted to give a solution of concentration 15 mg NaCl per ml of solution?
Ans. Each ml of given solution is diluted with water to a volume of 5 ml.

12.68. How many ml of a solution of concentration 100 mg Co^{++} per ml are needed to prepare 1.5 liters of solution of concentration 20 mg Co^{++} per ml?   *Ans.* 300 ml

12.69. Calculate the approximate volume of water that must be added to 250 ml of 1.25 N solution to make it 0.500 N (neglecting volume changes).   *Ans.* 375 ml

12.70. Determine the volume of dilute nitric acid (sp gr 1.11, 19.0% HNO_3 by weight) that can be prepared by diluting with water 50 ml of the concentrated acid (sp gr 1.42, 69.8% HNO_3 by weight). Calculate the molarities and molalities of the concentrated and dilute acid.
Ans. 235 ml; molarities, 15.7 and 3.35; molalities, 36.7 and 3.72

12.71. What volume of 95.0% alcohol by weight (sp gr 0.809) must be used to prepare 150 ml of 30.0% alcohol by weight (sp gr 0.957)?   *Ans.* 56.0 ml

12.72. What volumes of 12 N and 3 N HCl must be mixed to give 1 liter of 6 N HCl?
Ans. $\frac{1}{3}$ liter 12 N, $\frac{2}{3}$ liter 3 N

Chapter 13

Reactions Involving Standard Solutions

ADVANTAGES OF VOLUMETRIC STANDARD SOLUTIONS

Solutions of specified molarity, formality, or normality can be used conveniently for reactions involving quantitative procedures. Measured volumes of these solutions contain precisely determined amounts of solutes, and weighing of portions of the solutions can be avoided. Weight-weight problems can be extended to include volumes of such solutions by use of the following basic relationships in connection with balanced chemical equations.

Number of moles (or gfw)   =   number of liters $\times$ molarity (or formality)

or   Number of millimoles   =   number of milliliters $\times$ molarity

Stoichiometric calculations involving normal solutions are even simpler. By the definition of equivalent weights, two solutions containing the same number of gram-equivalents or milli-equivalents will react exactly with each other, since they contain the same number of equivalents. By definition,

$$\text{Normality} = \frac{\text{number of gram-equivalents of solute}}{\text{number of liters of solution}}$$

or

$$\text{Normality} = \frac{\text{number of milli-equivalents of solute}}{\text{number of milliliters of solution}}$$

Hence,   Number of gram-equivalents   =   number of liters $\times$ normality

and   Number of milli-equivalents   =   number of milliliters $\times$ normality

Then the following relation exists between two solutions which react exactly with each other:

Number of gram-equivalents$_1$ =   number of gram-equivalents$_2$

liters$_1$ $\times$ normality$_1$ =   liters$_2$ $\times$ normality$_2$

and   Number of milli-equivalents$_1$ =   number of milli-equivalents$_2$

milliliters$_1$ $\times$ normality$_1$ =   milliliters$_2$ $\times$ normality$_2$

Normal solutions are useful even when only one of the reactants is dissolved. In this case, the number of gram-equivalents or milli-equivalents of the non-dissolved reactant is computed in the usual way, by dividing the weight of the sample in grams or milligrams by the equivalent weight. The number of g-eq (or meq) of one reactant must still equal the number of g-eq (or meq) of the other.

Solved Problems

13.1. What volume of 1.40 M H_2SO_4 solution is needed to react exactly with 100 g Al?

The balanced molecular equation for the reaction is

$$2\,Al \;+\; 3\,H_2SO_4 \;\longrightarrow\; Al_2(SO_4)_3 \;+\; 3\,H_2$$

Mole Method.

Gram-atoms of Al in 100 g Al $= \dfrac{100 \text{ g}}{27.0 \text{ g/g-at}} = 3.70$ g-at

Moles of H_2SO_4 required for 3.70 g-at Al $= \frac{3}{2} \times 3.70 = 5.55$ moles

Volume of 1.40 M H_2SO_4 containing 5.55 moles $= \dfrac{5.55 \text{ moles}}{1.40 \text{ mole/}l} = 3.96 \ l$

Factor-Label Method.

x liters solution $= \dfrac{100 \text{ g Al}}{27.0 \text{ g Al/g-at Al}} \times \dfrac{3 \text{ moles } H_2SO_4}{2 \text{ g-at Al}} \times \dfrac{1 \ l \text{ solution}}{1.40 \text{ moles } H_2SO_4} = 3.96 \ l$

13.2. In standardizing a solution of $AgNO_3$ it was found that 40.0 ml was required to precipitate all the chloride ion contained in 36.0 ml of 0.520 F NaCl. How many grams of Ag could be obtained from 100 ml of the $AgNO_3$ solution?

In the precipitation of AgCl, equimolar amounts of Ag^+ and Cl^- are needed; thus equal numbers of gfw of $AgNO_3$ and NaCl must have been used.

Number of gfw of NaCl $= .0360 \ l \times 0.520$ gfw/$l = .01872$ gfw.

Then 40.0 ml of the $AgNO_3$ solution contains .01872 gfw $AgNO_3$, or

formality of the $AgNO_3$ solution $= \dfrac{.01872 \text{ gfw}}{.0400 \ l} = 0.468$ gfw/l

A liter of 0.468 F $AgNO_3$ contains 0.468 gfw Ag, or $0.468 \times 107.9 = 50.5$ g Ag. Hence 100 ml contains 5.05 g Ag.

13.3. Exactly 40.0 ml of 0.225 F $AgNO_3$ was required to react exactly with 25.0 ml of a solution of NaCN, according to the following equation:

$$Ag^+ + 2 \ CN^- \longrightarrow Ag(CN)_2^-$$

What is the formality of the NaCN solution?

The molarities of the two solutions with respect to Ag^+ and CN^- are equal to their respective formalities.

Number of gfw of $AgNO_3 = .0400 \ l \times 0.225$ gfw/$l = .00900$ gfw.

Number of gfw of NaCN required $= 2 \times$ gfw $AgNO_3 = .0180$ gfw.

Then 25.0 ml of the NaCN solution contains .0180 gfw NaCN, that is,

formality of the NaCN solution $= \dfrac{.0180 \text{ gfw}}{.025 \ l} = 0.72$ gfw/l

13.4. Determine the normality of a H_3PO_4 solution, 40 ml of which neutralized 120 ml of 0.531 N NaOH.

The solutions react exactly with each other. Therefore the following relation exists:

ml $H_3PO_4 \times$ normality $H_3PO_4 =$ ml NaOH $\times$ normality NaOH

40 ml $\times$ normality $H_3PO_4 =$ 120 ml $\times$ 0.531 N

Solving,   normality $H_3PO_4 = \dfrac{120 \text{ ml} \times 0.531 \text{ N}}{40 \text{ ml}} = 1.59$ N

Note. In this problem we do not have to know whether 1, 2, or 3 hydrogens of H_3PO_4 are replaceable. The normality was determined by reaction of the acid with a base of known concentration. Therefore the acid will have the same normality, 1.59 N, in reactions with any strong base under similar conditions. In order to know the molarity of the acid, however, it would be necessary to know the number of replaceable hydrogens in the reaction.

In a case like this, where a substance can have several equivalent weights, the normality determined by one type of reaction is not necessarily the normality in other types of reaction. If a weak base like NH_3 were used instead of a strong base for neutralizing the phosphoric acid, or if the method of detecting the point of neutralization (indicator) were changed, the equivalent weight of phosphoric acid and hence the normality of the above solution might be different. In order to predict the number of replaceable hydrogens of the acid in each case, detailed information about the chemistry of the acid must be known.

13.5. How many ml of 6.0 N NaOH are required to neutralize 30 ml of 4.0 N HCl?

$$\text{ml HCl} \times \text{normality HCl} = \text{ml NaOH} \times \text{normality NaOH}$$
$$30 \text{ ml} \times 4.0 \text{ N} = \text{ml NaOH} \times 6.0 \text{ N}$$

Solving, $\qquad \text{ml NaOH} = \dfrac{30 \text{ ml} \times 4.0 \text{ N}}{6.0 \text{ N}} = 20 \text{ ml of } 6.0 \text{ N NaOH}$

13.6. (a) What volume of 5.00 N H_2SO_4 is required to neutralize a solution containing 2.50 g NaOH? (b) How many grams of pure H_2SO_4 are required?

(a) One gram-equivalent H_2SO_4 reacts completely with one gram-equivalent NaOH.

Equivalent weight of NaOH = formula weight = 40.0.

Gram-equivalents in 2.50 g NaOH $= \dfrac{2.50 \text{ g}}{40.0 \text{ g/g-eq}} = 0.0625$ g-eq NaOH. Therefore 0.0625 gram-equivalent of H_2SO_4 is required.

0.0625 gram-equivalent makes up 0.0625 liter or 62.5 ml of 1.00 N solution, or 62.5/5.00 = 12.5 ml of 5.00 N solution. Then 12.5 ml of 5.00 N H_2SO_4 is required.

Another Method.   Gram-quivalents NaOH $=$ gram equivalents H_2SO_4

$\qquad\qquad\qquad\qquad\qquad\quad = $ liters $H_2SO_4 \times$ normality H_2SO_4

$$\dfrac{2.50 \text{ g}}{40.0 \text{ g/g-eq}} = \text{liters } H_2SO_4 \times 5.00 \text{ g-eq/liter}$$

Solving, liters $H_2SO_4 = \dfrac{2.50 \text{ g}}{40.0 \text{ g/g-eq} \times 5.00 \text{ g-eq/}l} = 0.0125 \; l = 12.5 \text{ ml.}$

(b) Equivalent weight of $H_2SO_4 = \frac{1}{2} \times$ formula weight $= \frac{1}{2} \times 98.08 = 49.04$

Weight of H_2SO_4 required $= 0.0625$ g-eq $\times 49.04$ g/g-eq $= 3.07$ g

13.7. A 0.250 g sample of a solid acid was dissolved in water and exactly neutralized by 40.0 ml of 0.125 N base. What is the equivalent weight of the acid?

Number of milli-equivalents of base $= 40.0 \text{ ml} \times 0.125 \text{ meq/ml} = 5.00 \text{ meq}$

Number of milli-equivalents of acid $=$ number of meq of base $= 5.00 \text{ meq}$

Equivalent weight of acid $= \dfrac{250 \text{ mg}}{5.00 \text{ meq}} = 50 \text{ mg/meq}$

13.8. A 48.4 ml sample of HCl solution requires 1.240 g of pure $CaCO_3$ for complete neutralization. Calculate the normality of the acid.

Each CO_3^{2-} ion requires two H^+ for neutralization: $CO_3^{2-} + 2\,H^+ \longrightarrow CO_2 + H_2O$. Thus, equivalent weight of $CaCO_3 = \frac{1}{2} \times$ formula weight $= 50.05$.

Gram-equivalents in 1.240 g $CaCO_3 = \dfrac{1.240 \text{ g}}{50.05 \text{ g/g-eq}} = 0.0248$ g-eq $CaCO_3$

Hence 48.4 ml of acid solution contains 0.0248 gram-equivalent HCl. Then 1000 ml contains 0.0248 g-eq $\times 1000/48.4 = 0.512$ g-eq HCl. The acid is 0.512 N.

Another Method.   Gram-equivalents $CaCO_3$ =   gram equivalents HCl

$$\text{=   liters HCl} \times \text{normality HCl}$$

$$\frac{1.240 \text{ g}}{50.05 \text{ g/g-eq}} \text{ =   } 0.0484 \text{ } l \times \text{normality HCl}$$

Solving,   normality HCl = 0.512 g-eq/l = 0.512 N.

13.9. Exactly 50.0 ml of Na_2CO_3 solution is equivalent to 56.3 ml of 0.102 N HCl in an acid-base neutralization. How many grams of $CaCO_3$ would be precipitated if an excess of $CaCl_2$ solution were added to 100 ml of this Na_2CO_3 solution?

56.3 ml acid contains   56.3 ml $\times$ 0.102 meq/ml = 5.74 meq HCl.   Hence 50.0 ml Na_2CO_3 solution contains 5.74 meq Na_2CO_3, and 100 ml contains 2×5.74 = 11.48 meq Na_2CO_3.

Since each CO_3^{2-} neutralizes 2 H^+, each meq of Na_2CO_3 has $\frac{1}{2}$ millimole of CO_3^{2-}.

Number of millimoles CO_3^{2-} = $\frac{1}{2} \times 11.48$ = 5.74 mmole.   The number of mmole of $CaCO_3$ in the precipitate equals the number of millimoles of CO_3^{2-} available.   The molecular (formula) weight of $CaCO_3$ is 100.1.

Grams in 5.74 mmole $CaCO_3$ = 5.74 mmole $\times$ 100.1 mg/mmole = 575 mg = 0.575 g

13.10. A 10.0 gram sample of "gas liquor" is boiled with an excess of NaOH and the resulting ammonia is passed into 60 ml of 0.90 N H_2SO_4. Exactly 10.0 ml of 0.40 N NaOH is required to neutralize the excess sulfuric acid (not neutralized by the NH_3). Determine the percent of ammonia in the "gas liquor" examined.

Number of milli-equivalents of NH_3 in 10.0 g of gas liquor

=   milli-equivalents H_2SO_4 − milli-equivalents NaOH

=   60 ml $\times$ 0.90 meq/ml − 10.0 ml $\times$ 0.40 meq/ml = 54 − 4 = 50 meq NH_3

In neutralization experiments, the equivalent weight of NH_3 is the same as the molecular weight in accord with the equation $NH_3 + H^+ \longrightarrow NH_4^+$.   Molecular wt of NH_3 = 17.0.

Amount of NH_3 in sample = 50 meq $\times$ 17.0 mg/meq = 850 mg = 0.85 g

Fraction of NH_3 in sample = $\frac{0.85 \text{ g}}{10.0 \text{ g}}$ = 0.085 = 8.5%

13.11. A 40.8 ml sample of an acid is equivalent to 50.0 ml of a Na_2CO_3 solution, 25.0 ml of which is equivalent to 23.8 ml of a 0.102 N HCl. What is the normality of the first acid?

1 ml of acid is equivalent to $\frac{50.0}{40.8}$ ml of Na_2CO_3 solution.

1 ml of Na_2CO_3 solution is equivalent to $\frac{23.8}{25.0}$ ml of 0.102 N HCl.

Therefore the normality of the Na_2CO_3 solution = $\frac{23.8}{25.0} \times 0.102$ N,

and the normality of the acid = $\frac{50.0}{40.8} \times \frac{23.8}{25.0} \times 0.102$ N = 0.119 N.

13.12. Calculate the number of grams of $FeSO_4$ that will be oxidized in a solution acidified with sulfuric acid by 24.0 ml of 0.250 N $KMnO_4$. The $KMnO_4$ is 0.250 N as an oxidizing agent for this reaction.

$$MnO_4^- + Fe^{++} + H^+ \longrightarrow Fe^{+++} + Mn^{++} + H_2O \quad \text{(unbalanced)}$$

It is not necessary to balance the complete equation. All that must be known from the equation is the equivalent weight of $FeSO_4$. The Fe changes in oxidation state from +II in Fe^{++} to +III in Fe^{+++}. The net change is 1.

$$\text{Equivalent weight of FeSO}_4 \ = \ \frac{\text{formula weight}}{\text{oxidation state change}} \ = \ \frac{152}{1} \ = \ 152$$

Another method of getting the same result is to write the balanced partial equation for the Fe^{++}.

$$Fe^{++} \ \longrightarrow \ Fe^{+++} + e$$

$$\text{Equivalent weight of FeSO}_4 \ = \ \frac{\text{formula weight}}{\text{electrons transferred}} \ = \ \frac{152}{1} \ = \ 152$$

Let x = required weight of $FeSO_4$.

$$\text{Milli-equivalents of KMnO}_4 \ = \ \text{milli-equivalents of FeSO}_4$$

or
$$\text{ml KMnO}_4 \times \text{normality KMnO}_4 \ = \ \frac{\text{mg}}{\text{mg/meq}} \text{ FeSO}_4$$

$$24.0 \text{ ml} \times 0.250 \text{ meq/ml} \ = \ \frac{x}{152 \text{ mg/meq}}$$

Solving, x = 912 mg = 0.912 g $FeSO_4$.

Note. The problem might have been solved directly in terms of gram-equivalents and grams, instead of meq and milligrams. Each student should select the procedure most convenient for him. If gram-equivalents are used, volumes must be expressed in *liters* instead of *milliliters*.

13.13. What volume of 0.1000 N $FeSO_4$ is required to reduce 4.000 g $KMnO_4$ in a solution acidified with sulfuric acid?

The normality of the $FeSO_4$ is given with respect to this oxidation-reduction reaction.

$$MnO_4^- + Fe^{++} + H^+ \ \longrightarrow \ Fe^{+++} + Mn^{++} + H_2O \qquad \text{(unbalanced)}$$

In this reaction the Mn changes in oxidation state from +VII in MnO_4^- to +II in Mn^{++}. The net change is 5. Or, from the balanced partial equation,

$$MnO_4^- + 8 H^+ + 5e \ \longrightarrow \ Mn^{++} + 4 H_2O$$

it can be seen that the electron transfer is 5 for each MnO_4^-. The equivalent weight of $KMnO_4$ in this reaction = 1/5 × formula weight = 1/5 × 158.0 = 31.6.

Let x = required volume of 0.1000 N $FeSO_4$.

$$\text{Milli-equivalents FeSO}_4 \ = \ \text{milli-equivalents KMnO}_4$$

or
$$\text{ml FeSO}_4 \times \text{normality FeSO}_4 \ = \ \frac{\text{mg}}{\text{mg/meq}} \text{ KMnO}_4$$

Substituting, $x \times 0.1000$ meq/ml $= \dfrac{4000 \text{ mg}}{31.6 \text{ mg/meq}}$ from which x = 1266 ml $FeSO_4$.

Supplementary Problems

13.14. How many ml of 0.25 F $BaCl_2$ are required to precipitate all the sulfate ion from 20 ml of a solution containing 100 grams/liter of Na_2SO_4?   *Ans.* 56 ml

13.15. A 50.0 ml sample of Na_2SO_4 solution is treated with an excess of $BaCl_2$. If the precipitated $BaSO_4$ weighs 1.756 g, what is the formality of the Na_2SO_4 solution?   *Ans.* 0.1505 F

13.16. What was the thorium content of a sample that required 35.0 ml of 0.0200 M $H_2C_2O_4$ to precipitate $Th(C_2O_4)_2$?   *Ans.* 81 mg

13.17. What molarity of $K_4Fe(CN)_6$ should be used so that 40.0 ml of the solution titrates 150.0 mg Zn (dissolved) by forming $K_2Zn_3[Fe(CN)_6]_2$?   *Ans.* 0.0382 M

13.18. A 50.0 ml sample of NaOH solution requires 27.8 ml of 0.100 N acid in titration. What is its normality? How many mg NaOH are in each ml?   *Ans.* 0.0556 N, 2.22 mg

13.19. In standardizing HCl, 22.5 ml was required to neutralize 25.0 ml of 0.100 N Na_2CO_3 solution. What is the normality of the HCl solution? How much water must be added to 200 ml of it to make it 0.100 N? Neglect volume changes.   *Ans.* 0.111 N, 22 ml

13.20. What volume of a 0.0224 N adipic acid solution would be used in the titration of 1.022 ml of 0.0317 N $Ba(OH)_2$?   *Ans.* 1.446 ml

13.21. Exactly 21 ml of 0.80 N acid was required to neutralize completely 1.12 g of an impure sample of calcium oxide. What is the purity of the CaO?   *Ans.* 42%

13.22. By the Kjeldahl method, the nitrogen contained in 5.0 g of a foodstuff is converted into ammonia. This is just sufficient to neutralize 20 ml of 0.100 N acid. Calculate the percentage of nitrogen in the foodstuff.   *Ans.* 0.56%

13.23. What is the purity of concentrated H_2SO_4 (specific gravity 1.800) if 5.00 ml is neutralized by 84.6 ml of 2.000 N NaOH?   *Ans.* 92.2%

13.24. A 10.0 ml portion of $(NH_4)_2SO_4$ solution was treated with an excess of NaOH. The NH_3 gas evolved was absorbed in 50.0 ml of 0.100 N HCl. To neutralize the remaining HCl, 21.5 ml of 0.098 N NaOH was required. What is the formality of the $(NH_4)_2SO_4$? How many grams of $(NH_4)_2SO_4$ are in a liter of solution?   *Ans.* 0.144 F, 19.0 g/l

13.25. Exactly 400 ml of an acid solution, when acted upon by an excess of zinc, evolved 2430 ml of H_2 gas measured over water at 21°C and 747.5 mm. What is the normality of the acid? Vapor pressure of water at 21°C = 18.6 mm.   *Ans.* 0.483 N

13.26. How many grams of copper will be replaced from two liters of 1.50 F $CuSO_4$ solution by 27 grams of aluminum?   *Ans.* 95 g

13.27. What volume of 3.00 N H_2SO_4 is needed to liberate 185 liters of hydrogen gas at S.T.P. when treated with an excess of zinc?   *Ans.* 5.51 liters

13.28. How many liters of hydrogen at S.T.P. would be replaced from 500 ml of 3.78 N HCl by 125 g of zinc?   *Ans.* 21.2 liters

13.29. Exactly 50.0 ml of a solution of Na_2CO_3 was titrated with 65.8 ml of 3.00 N HCl. If the specific gravity of the Na_2CO_3 solution is 1.25, what percent by weight of Na_2CO_3 does it contain? $CO_3^{--} + 2H^+ \longrightarrow CO_2 + H_2O$ *Ans.* 16.7%

13.30. What is the equivalent weight of an acid 1.243 g of which required 31.72 ml of 0.1923 N standard base for neutralization?   *Ans.* 203.7

13.31. The molecular weight of an organic acid was determined by the following study of its barium salt. 4.290 g of the salt was converted to the free acid by reaction with 21.64 ml of 0.954 N H_2SO_4. The barium salt was known to contain 2 moles of water of hydration per Ba^{++}, and the acid was known to be monoprotic (monobasic). What is the molecular weight of the anhydrous acid?
Ans. 122.1

13.32. A ferrous sulfate solution was standardized by titration. A 25.00 ml aliquot of the ferrous sulfate solution required 42.08 ml of 0.0800 N ceric sulfate for complete oxidation. What is the normality of the ferrous sulfate?   *Ans.* 0.1347 N

13.33. How many ml of 0.0257 N KIO_3 would be needed to reach the end point in the oxidation of 34.2 ml of 0.0416 N hydrazine in hydrochloric acid solution?   *Ans.* 55.4 ml

13.34. How many grams of $FeCl_2$ will be oxidized by 28 ml of 0.25 N $K_2Cr_2O_7$ in HCl solution?
$Fe^{++} + Cr_2O_7^{--} + H^+ \longrightarrow Fe^{+++} + Cr^{+++} + H_2O$ (unbalanced)   *Ans.* 0.89 g

13.35. What weight of MnO_2 is reduced by 35 ml of 0.16 N oxalic acid, $H_2C_2O_4$, in sulfuric acid solution?
$MnO_2 + H^+ + H_2C_2O_4 \longrightarrow CO_2 + H_2O + Mn^{++}$ (unbalanced)   *Ans.* 0.24 g

13.36. How many grams of $KMnO_4$ are required to oxidize 2.40 g of $FeSO_4$ in a solution acidified with sulfuric acid? What is the equivalent weight of $KMnO_4$ in this reaction?   *Ans.* 0.500 g, 31.6

13.37. $Mn^{++} + MnO_4^- + H_2O \longrightarrow MnO_2 + H^+$ (unbalanced). What is the equivalent weight of $KMnO_4$ in this reaction? How many grams of $MnSO_4$ are oxidized by 1.25 g $KMnO_4$?
Ans. 52.7, 1.79 g

13.38. $Cr_2O_7^{--} + Cl^- + H^+ \longrightarrow Cr^{+++} + Cl_2 + H_2O$ (unbalanced)
(*a*) What volume of 0.40 N $K_2Cr_2O_7$ is required to liberate the chlorine from 1.20 g of NaCl in a solution acidified with H_2SO_4? (*b*) How many grams of $K_2Cr_2O_7$ are required? (*c*) How many grams of chlorine are liberated?   *Ans.* 51 ml, 1.01 g, 0.73 g

13.39. If 25.0 ml of iodine solution is equivalent to 0.125 g of $K_2Cr_2O_7$, to what volume should 1000 ml be diluted to make the solution tenth normal?   *Ans.* 1020 ml

13.40. How many grams of $KMnO_4$ should be taken to make up 250 ml of a solution of such strength that one ml is equivalent to 5.00 mg of iron in $FeSO_4$?   *Ans.* 0.707 g

13.41. $S_2O_3^{--} + I_2 \longrightarrow S_4O_6^{--} + I^-$ (unbalanced). How many grams of iodine are present in a solution which requires 40 ml of 0.112 N $Na_2S_2O_3$ to react with it?   *Ans.* 0.57 g

13.42. To how many milligrams of iron (Fe^{++}) is one ml of 0.1055 N $K_2Cr_2O_7$ equivalent?
Ans. 5.89 mg

13.43. Reducing sugars are sometimes characterized by a number R_{Cu}, which is defined as the number of mg of copper reduced by one gram of the sugar, in which the half-reaction for the copper is
$Cu^{++} + OH^- \longrightarrow Cu_2O + H_2O$ (unbalanced).

It is sometimes more convenient to determine the reducing power of a carbohydrate by an indirect method. In this method 43.2 mg of the carbohydrate was oxidized by an excess of $K_3Fe(CN)_6$. The $Fe(CN)_6^{----}$ formed in this reaction required 5.29 ml of 0.0345 N $Ce(SO_4)_2$ for reoxidation to $Fe(CN)_6^{---}$. The normality of the ceric sulfate solution is given with respect to the reduction of Ce^{++++} to Ce^{+++}. Determine the R_{Cu} value for the sample.

(*Hint:* The number of milli-equivalents of Cu in a direct oxidation is the same as the number of milli-equivalents of Ce^{++++} in the indirect method.)   *Ans.* 268

13.44. A volume of 12.53 ml of 0.05093 M selenium dioxide, SeO_2, reacted exactly with 25.52 ml of 0.1000 M $CrSO_4$. In the reaction, Cr^{++} was oxidized to Cr^{+++}. To what oxidation state was the selenium converted by the reaction?   *Ans.* 0

13.45. An acid solution of a $KReO_4$ sample containing 0.02683 g of combined rhenium was reduced by passage through a column of granulated zinc. The effluent solution, including the washings from the column, was then titrated with 0.1000 N $KMnO_4$. 11.45 ml of the standard permanganate was required for the reoxidation of all the rhenium to the perrhenate ion, ReO_4^-. Assuming that rhenium was the only element reduced, what is the oxidation state to which rhenium was reduced by the zinc column?   *Ans.* −1

13.46. The iodide content of a solution was determined by titration with ceric sulfate in the presence of HCl, in which I^- is converted to ICl. A 250 ml sample of the solution required 20.0 ml of 0.050 N Ce^{4+} solution. What is the iodide concentration in the original solution, in mg/ml?
Ans. 0.25 mg/ml

13.47. A 0.518 gram sample of limestone is dissolved, and then the calcium is precipitated as calcium oxalate, CaC_2O_4. After filtering and washing the precipitate, it requires 40.0 ml of 0.250 N $KMnO_4$ solution acidified with sulfuric acid to titrate it. What is the percent CaO in the limestone? The unbalanced equation for the titration is:
$$MnO_4^- + CaC_2O_4 + (H^+)_2SO_4^{--} \longrightarrow CaSO_4 + Mn^{++} + CO_2 + H_2O \quad \text{(unbalanced)}$$
Ans. 54.2%

Chapter 14

Properties of Solutions

INTRODUCTION

Just as gases are characterized by more or less general adherence to a group of simple laws, so dilute solutions as a class have many properties that are determined by concentration alone without reference to the particular nature of the dissolved materials. All solutions obey the laws described below when the concentration is sufficiently low. These are the laws of the dilute solution. A few of the laws, specifically designated in the discussion below, are obeyed over the entire range of composition by certain pairs (or groups of more than two) of substances. Such pairs of substances are said to form *ideal solutions*.

VAPOR PRESSURE OF SOLUTIONS

The vapor pressures of all solutions of non-volatile solutes in a solvent are *less* than that of the pure solvent. If we prepare solutions of different solutes in a given solvent by adding *equal numbers of solute molecules* to a fixed amount of solvent, as we do by preparing solutions of equal molality, we find that the *depression* of the vapor pressure is the same in every case in dilute solutions of non-volatile non-electrolytes.

Raoult's Law states that in dilute solutions of non-volatile non-electrolytes *the depression is proportional to the mole fraction of the solute*, or *the solution vapor pressure is proportional to the mole fraction of the solvent*. This is explained by the interference that solute molecules at the liquid surface exert with the escape of solvent molecules into the vapor phase. Because of the vapor pressure lowering, the *boiling point of the solution is raised* and the *freezing point is lowered,* as compared with the pure solvent.

Depression of solvent vapor pressure

= ΔP = (vapor pressure of solvent) − (vapor pressure of solution)

= (vapor pressure of pure solvent) × (mole fraction of solute)

Vapor pressure of solvent over solution

= (vapor pressure of pure solvent) × (mole fraction of solvent)

In systems of liquids that mix with each other in all proportions to form ideal solutions, Raoult's law in the form of the second equation above applies to the partial pressure of each component separately.

Partial pressure of any component over solution

= (vapor pressure of that pure component) × (mole fraction of component)

Ideal solutions are those in which the chemical interactions between solvent and solute molecules are the same as between the molecules in the separate components. In the formation of an ideal solution from the separate components, there are no volume changes and no heat changes. Pairs of chemically similar substances, such as methanol (CH_3OH) and ethanol (C_2H_5OH), or benzene (C_6H_6) and toluene (C_7H_8), form ideal solutions. But dissimilar substances, such as C_2H_5OH and C_6H_6, form non-ideal solutions.

MOLALITY, m

$$\text{Molality of a solution} = \frac{\text{number of moles of solute in the solution}}{\text{number of kilograms of solvent in the solution}}.$$

A 1 molal (1 m) solution contains 1 mole of solute per kilogram of solvent.

FREEZING POINT LOWERING, ΔT_f

When most dilute solutions are cooled, pure solvent begins to crystallize before any solute crystallizes. The temperature at which the first crystals are in equilibrium with the solution is called the *freezing point of the solution*. The freezing point of such a solution is always *lower* than the freezing point of pure solvent. In dilute solutions, the lowering of the freezing point is directly proportional to the number of solute molecules (or moles) in a given weight of solvent. It is customary to relate the freezing point to solution concentrations on the molality scale, by the relationship

Lowering of freezing point

$$= \Delta T_f = (\text{freezing point of solvent}) - (\text{freezing point of solution})$$

$$= K_f \times m$$

If this equation were valid up to a concentration of 1 molal, the lowering of the freezing point of a 1 molal solution of any non-electrolyte dissolved in the solvent would be K_f, which is thus called the *molal freezing point constant* of the solvent. The lowering of the freezing point caused by solutes at some other concentration (within the dilute range) is equal to the product of this constant and the molality.

The numerical value of K_f is a property of the solvent alone and is independent of the nature of the solute.

The molal freezing point constant for water is $1.86°C \ m^{-1}$. Thus if 1 mole of cane sugar (342 g sugar) is dissolved in 1000 g of water, the solution will freeze at $-1.86°C$.

BOILING POINT ELEVATION, ΔT_b

The temperature at which a solution boils is *higher* than that of the pure solvent if the solute is comparatively non-volatile. In dilute solutions, the elevation of the boiling point is directly proportional to the number of solute molecules (or moles) in a given weight of solvent. Again, the molality scale is usually used, and the equation is

Elevation of boiling point

$$= \Delta T_b = (\text{boiling point of solution}) - (\text{boiling point of solvent})$$

$$= K_b \times m$$

K_b is called the *molal boiling point constant* of the solvent.

The numerical value of K_b is a property of the solvent alone and is independent of the nature of the solute, within the general requirements of non-volatility and non-dissociation into ions.

The molal boiling point constant for water is $0.513°C \ m^{-1}$. Thus if 1 mole of cane sugar (342 g sugar) is dissolved in 1000 g of water, the solution will boil at $100.513°C$ (at a pressure of one atmosphere). (If pure water boils at a temperature slightly different from $100°$ because the air pressure is not exactly one standard atmosphere, ΔT_b is still $0.513°$, and the boiling point of the solution is $0.513°$ higher than the actual boiling point of the water.) If a solution contains 1/2 mole of sugar (171 g) and 1000 g water, it will boil at $100.256°C$ at one atmosphere.

OSMOTIC PRESSURE

If a solution is separated from a sample of pure solvent by a porous sheet that allows solvent, but not solute molecules to pass through, solvent will move into the solution in an attempt to equalize the concentration on the two sides of the sheet. Such a dividing sheet is called a semi-permeable membrane. If the membrane is placed vertically and the vessel holding the solution can extend indefinitely in a horizontal direction to accommodate the incoming material, solvent will continue to flow until it is all used up or until the solution becomes so dilute that there is no more driving force due to the difference in solvent concentration on the two sides. If, however, the solution vessel is closed on all sides except for an extension tube on top, the incoming solvent will force some of the solution up the extension tube. The weight of this solution in the tube will exert a downward pressure that will tend to oppose the inward penetration of more solvent through the membrane. Eventually the two forces will just balance each other and no more solvent will enter. These two forces are the weight of the hydrostatic head of solution in the tube and the driving force tending to equalize the concentrations on the two sides of the membrane. The hydrostatic head at this balance point is called the osmotic pressure of the solution. It can be measured in usual units of pressure, such as atmospheres, centimeters of rise of the solution in the tube, or the calculated equivalent height of a mercury column $\left(\text{rise of solution} \ \times \ \dfrac{\text{density of solution}}{\text{density of mercury}} \right)$.

For dilute solutions the osmotic pressure, π, of solutions of non-electrolytes, if expressed in atmospheres, is given by the equation, $\pi = \dfrac{n}{V} RT$, where $\dfrac{n}{V}$ is the number of moles of solute per liter of solution and R is the usual gas law constant, $0.0821 \ l$-atm mole^{-1} deg^{-1}.

DEVIATIONS FROM LAWS OF DILUTE SOLUTIONS

The above laws are valid only in dilute solutions of non-electrolytes. For electrolyte solutions each ion contributes independently to the effective molality or molarity. On account of the electrical interactions between ions, however, none of the effects is as large as would be predicted on the basis of simple counting of ions. For example, a solution containing 0.100 gram-formula weight of KCl per kilogram of water freezes at $-0.345°$. The observed lowering is slightly less than $.2 \times 1.86 \ (.372°)$ which would be predicted if each K^+ and Cl^- were truly independently effective (.1 molal K^+ + .1 molal Cl^-). A solution containing 0.100 gram-formula weight of $BaCl_2$ per kilogram of water freezes at $-0.470°$, showing a lowering somewhat less than $.3 \times 1.86 \ (.558°)$ predicted from the simple additivity of molalities (.1 molal Ba^{++} + .2 molal Cl^-).

For any solution not too concentrated, whether electrolyte or non-electrolyte, the deviations from any one of the laws of the dilute solution are equal to the deviations from any of the others on a fractional or percentage basis, such that

$$\frac{\text{Lowering of freezing point}}{\text{L.f.p.}_0} = \frac{\text{Elevation of boiling point}}{\text{E.b.p.}_0}$$

$$= \frac{\text{Depression of vapor pressure}}{\text{D.v.p.}_0} = \frac{\pi}{\pi_0}$$

where the denominator in each of the four terms represents the magnitude of the effect that would be predicted by the laws of the dilute solution.

SOLUTIONS OF GASES IN LIQUIDS

At constant temperature, the concentration of a slightly soluble gas in a liquid (i.e., the weight or number of moles of a gas dissolved in a given volume of liquid) is directly proportional to the partial pressure of the gas. This is known as *Henry's Law.*

For example: One liter of water dissolves 0.1 g of a certain gas at 18°C and one atmosphere gas pressure. If the partial pressure is increased to two atmospheres and the temperature remains constant, then 0.2 g of the gas will dissolve per liter.

When a mixture of two gases is in contact with a solvent, the amount of each gas that is dissolved is the same as if it were present alone at a pressure equal to its own partial pressure in the gas mixture.

LAW OF DISTRIBUTION

A solute distributes itself between two immiscible solvents so that the ratio of its concentrations in dilute solutions in the two solvents is constant, regardless of the actual concentration in either solvent.

For the distribution of iodine between ether and water at room temperature, the value of this constant is about 200. Thus,

$$\frac{\text{Concentration of iodine in ether}}{\text{Concentration of iodine in water}} = K = 200$$

The value of this ratio of concentrations, called the distribution ratio or distribution coefficient, is equal to the ratio of the solubilities in the two solvents if saturated solutions in these solvents are dilute enough for the Law of Distribution to apply.

Solved Problems

FREEZING POINT LOWERING

14.1. The freezing point of pure camphor is 178.4°C, and its molal freezing point constant, K_f, is 40.0° m^{-1}. Find the freezing point of a solution containing 1.50 grams of a compound of molecular weight 125 in 35.0 grams of camphor.

The first step is to find the molality (m) of the solution.

$$m = \frac{\text{number of moles of solute}}{\text{number of kilograms of solvent}} = \frac{1.50/125 \text{ moles solute}}{35/1000 \text{ kg solvent}} = 0.343 \text{ molal}$$

Lowering of freezing point $= \Delta T_f = K_f \times m = 40.0 \times 0.343 = 13.7°$

Freezing point of solution $=$ (freezing point of pure solvent) $- \Delta T_f$
$$= 178.4° - 13.7° = 164.7°C$$

Or, the lowering could have been computed by the proportion method. Let $x =$ lowering of freezing point. The problem may be stated: A 1 molal solution has a lowering of 40.0°, and a

0.343 molal solution has a lowering of $x°$. Then

$$\frac{1.00\ m}{40.0°} = \frac{0.343\ m}{x} \quad \text{from which} \quad x = 13.7°$$

14.2. A solution containing 4.50 g of a non-electrolyte dissolved in 125 g of water freezes at $-0.372°C$. Calculate the approximate molecular weight of the solute.

First Method.

First compute the molality from the freezing point equation.

Lowering of freezing point $= \Delta T_f = K_f \times m$

$$0.372 = 1.86 \times m$$

$$m = 0.372/1.86 = 0.200 \text{ moles solute/kg solvent}$$

From the definition of molality compute the number of moles of solute in the sample, corresponding to 0.125 kg (125 g) water.

$$n_{\text{solute}} = 0.200 \text{ moles solute/kg solvent} \times 0.125 \text{ kg solvent}$$

$$= 0.025 \text{ moles solute}$$

$$\text{Molecular weight} = \frac{4.50 \text{ grams solute}}{0.025 \text{ moles solute}} = 180 \text{ g/mole}$$

Second Method.

One mole of solute lowers the freezing point of 1000 g of water by 1.86°C.

First calculate the weight of solute that would be dissolved in 1000 g water to give a solution of the same concentration (as 4.50 g solute in 125 g water).

In 1 g H_2O there is $\frac{4.50}{125}$ g solute; in 1000 g H_2O, $1000 \times \frac{4.50}{125} = 36.0$ g solute.

Since   1.86°C lowering is produced by 1 mole solute in 1000 g H_2O,

$$0.372°C \text{ lowering is produced by } \frac{0.372}{1.86} = 0.200 \text{ moles in 1000 g } H_2O.$$

Then   0.200 moles solute is contained in 36.0 g solute,   and

$$1 \text{ mole solute is contained in } \frac{36.0 \text{ g}}{0.200} = 180 \text{ g.} \qquad \text{M. W.} = 180.$$

Third Method.

We know, from the beginning of the second method solution, that (1) 36.0 g solute per kg of solvent produces a lowering of 0.372°C and that (2) 1 mole solute produces a lowering of 1.86°C. Then

$$\frac{36.0 \text{ g}}{0.372°C} = \frac{1 \text{ mole}}{1.86°C} \quad \text{from which} \quad 1 \text{ mole} = 180 \text{ g}$$

BOILING POINT ELEVATION

14.3. The molecular weight of an organic compound is 58.0. Compute the boiling point of a solution containing 24.0 g of the solute and 600 g of water, when the barometric pressure is such that pure water boils at 99.725°C.

First Method.

$$\text{Molality, } m = \frac{n_{\text{solute}}}{\text{number of kg solvent}} = \frac{24.0/58.0 \text{ moles solute}}{0.600 \text{ kg solvent}} = 0.690 \text{ molal}$$

$$\text{Elevation of boiling point} = \Delta T_b = K_b \times m = 0.513 \times 0.690 = 0.354°$$

Boiling point of solution $=$ (boiling point of water) $+ \Delta T_b$

$$= 99.725° + 0.354° = 100.079°C$$

Second Method.

Number of moles of solute in 24.0 g $= \frac{24.0 \text{ g}}{58.0 \text{ g/mole}} = 0.414$ moles in 600 g water.

Number of moles of solute in 1000 g water $= 1000 \times 0.414/600$ moles $= 0.690$ moles.

The problem now reads: 1 mole of solute in 1000 g water produces an elevation of 0.513°C; hence 0.690 mole produces an elevation of $x°C$. Then

$$\frac{1 \text{ mole}}{0.513°C} = \frac{0.690 \text{ mole}}{x} \quad \text{and} \quad x = 0.354°C \text{ elevation}$$

14.4. A solution was made up by dissolving 3.75 grams of a pure hydrocarbon in 95 grams of acetone. The boiling point of pure acetone was observed to be 55.95°C, and of the solution, 56.50°C. If the molal boiling point constant of acetone is $1.71° \ m^{-1}$, what is the approximate molecular weight of the hydrocarbon?

First Method.

Compute the molality (m) from the boiling point equation.

$$\text{Elevation of boiling point} \ = \ \Delta T_b \ = \ K_b \times m$$
$$56.50 - 55.95 \ = \ 1.71 \times m$$

Solving, $m \ = \ 0.322 \text{ molal} \ = \ 0.322 \text{ moles solute/kg solvent.}$

Find the number of moles of solute in the weighed sample.

$$n_{\text{solute}} \ = \ 0.322 \ \frac{\text{moles solute}}{\text{kg solvent}} \ \times \ 0.095 \text{ kg solvent} \ = \ 0.0306 \text{ moles solute}$$

$$\text{Molecular weight} \ = \ \frac{3.75 \text{ grams solute}}{0.0306 \text{ mole solute}} \ = \ 123 \text{ g/mole}$$

Second Method.

1 g acetone contains $\frac{3.75}{95}$ g solute; 1000 g acetone would contain $1000 \times \frac{3.75}{95} = 39.5$ g solute.

Since $1.71°$ elevation is produced by 1 mole solute per 1000 g acetone,

$0.55°$ elevation is produced by $\frac{0.55}{1.71} = 0.322$ mole per kg acetone.

Then   0.322 mole solute is contained in 39.5 g solute, and

$$1 \text{ mole solute is contained in } \frac{39.5 \text{ g}}{0.322} = 123 \text{ g.} \qquad \text{M. W.} = 123.$$

VAPOR PRESSURE OF SOLUTIONS

14.5. The vapor pressure of water at 28°C is 28.35 mm. Compute the vapor pressure at 28°C of a solution containing 68 g of cane sugar, $C_{12}H_{22}O_{11}$, in 1000 g of water.

Molecular weight of H_2O = 18.02, of $C_{12}H_{22}O_{11}$ = 342.

$$\text{Moles of } C_{12}H_{22}O_{11} \text{ in 68 g} \ = \ \frac{68 \text{ g}}{342 \text{ g/mole}} \ = \ 0.20 \text{ mole } C_{12}H_{22}O_{11}$$

$$\text{Moles of } H_2O \text{ in 1000 g} \ = \ \frac{1000 \text{ g}}{18.02 \text{ g/mole}} \ = \ 55.49 \text{ moles } H_2O$$

$$\text{Total moles} \ = \ 0.20 + 55.49 \ = \ 55.69 \text{ moles}$$

$$\text{Mole fraction } C_{12}H_{22}O_{11} = \frac{0.20}{55.69} = 0.0036 \qquad \text{Mole fraction } H_2O = \frac{55.49}{55.69} = 0.9964$$

First Method.

$$\text{Vapor pressure of solution} \ = \ \text{v.p. of pure solvent} \times \text{mole fraction of solvent}$$
$$= \ 28.35 \text{ mm} \times 0.9964 \ = \ 28.25 \text{ mm}$$

Second Method.

$$\text{Vapor pressure depression} \ = \ \Delta P \ = \ \text{v.p. of pure solvent} \times \text{mole fraction of solute}$$
$$= \ 28.35 \text{ mm} \times 0.0036 \ = \ 0.10 \text{ mm}$$
$$\text{Vapor pressure of solution} \ = \ (28.35 - 0.10) \text{ mm} \ = \ 28.25 \text{ mm}$$

Comparison of the two methods.

The first method gives the vapor pressure of the solution directly. The second method gives the depression directly, which must then be subtracted from the vapor pressure of the pure solvent. The advantage of the second method is that its calculations may be made on the slide rule. To achieve the same accuracy by the first method, the number 0.9964 must be inserted in the calculations and logarithms must be used.

14.6. At 30°C, pure benzene (molecular weight 78.1) has a vapor pressure of 121.8 mm. Dissolving 15.0 g of a non-volatile solute in 250 g of benzene produced a solution having a vapor pressure of 120.2 mm. Determine the approximate molecular weight of the solute.

$$\text{Moles of benzene in 250 g} = \frac{250 \text{ g}}{78.1 \text{ g/mole}} = 3.20 \text{ moles benzene.}$$

$$\text{Moles of solute in 15.0 g} = \frac{15.0}{x} \text{ mole, where } x = \text{molecular weight of the solute.}$$

$$\text{Vapor pressure of solution} = \text{v.p. of pure solvent} \times \frac{\text{moles solvent}}{\text{moles solute + moles solvent}}$$

$$\text{Substituting,} \quad 120.2 \text{ mm} = 121.8 \text{ mm} \times \frac{3.20 \text{ moles}}{15.0/x \text{ moles} + 3.20 \text{ moles}}$$

$$\text{Simplifying,} \quad 120.2 \text{ mm} = 121.8 \text{ mm} \times \frac{3.20x}{15.0 + 3.20x} \qquad \text{Solving, } x = 350$$

Note that the accuracy of the calculation is limited by the term $(121.8 - 120.2)$ that appears in the expansion. The answer is significant only to 1 part in 16.

14.7. At 20°C the vapor pressure of methyl alcohol (CH_3OH) is 94 mm, and the vapor pressure of ethyl alcohol (C_2H_5OH) is 44 mm. Being closely related, these compounds form a two-component system which adheres quite closely to Raoult's Law throughout the entire range of concentrations. If 20 g of CH_3OH is mixed with 100 g of C_2H_5OH, determine the partial pressure exerted by each and the total pressure of the solution. Calculate the composition of the vapor above the solution by applying Dalton's Law.

In an ideal solution of two liquids, there is no distinction between solute and solvent, and Raoult's Law holds for each component of such solutions. Hence when two liquids are mixed to give an ideal solution, the partial pressure of each liquid is equal to its vapor pressure multiplied by its mole fraction in the solution.

Molecular weight of $CH_3OH = 32$, of $C_2H_5OH = 46$.

$$\text{Partial pressure of component} = \text{v.p. of pure component} \times \text{its mole fraction}$$

$$\text{Partial pressure of } CH_3OH = 94 \text{ mm} \times \frac{20/32 \text{ mole } CH_3OH}{20/32 \text{ mole } CH_3OH + 100/46 \text{ mole } C_2H_5OH}$$

$$= 94 \text{ mm} \times 0.22 = 21 \text{ mm}$$

$$\text{Partial pressure of } C_2H_5OH = 44 \text{ mm} \times \frac{100/46 \text{ mole } C_2H_5OH}{20/32 \text{ mole } CH_3OH + 100/46 \text{ mole } C_2H_5OH}$$

$$= 44 \text{ mm} \times 0.78 = 34 \text{ mm}$$

The total pressure of the gaseous mixture is the sum of the partial pressures of all the components (Dalton's Law). Total pressure of solution = 21 mm + 34 mm = 55 mm.

Dalton's Law also indicates that the mole fraction of any component of a gaseous mixture is equal to its pressure fraction, i.e. its partial pressure divided by the total pressure.

$$\text{Mole fraction of } CH_3OH \text{ in vapor} = \frac{\text{partial pressure of } CH_3OH}{\text{total pressure}} = \frac{21 \text{ mm}}{55 \text{ mm}} = 0.38$$

$$\text{Mole fraction of } C_2H_5OH \text{ in vapor} = \frac{\text{partial pressure of } C_2H_5OH}{\text{total pressure}} = \frac{34 \text{ mm}}{55 \text{ mm}} = 0.62$$

Since the mole fraction for gases is the same as the volume fraction, we may also say that the vapor consists of 38% CH_3OH by volume. Note that the vapor is relatively richer in the more volatile component, methyl alcohol (mole fraction 0.38), than is the liquid (mole fraction of CH_3OH, 0.22).

OSMOTIC PRESSURE OF SOLUTIONS

14.8. What would be the osmotic pressure at 17°C of an aqueous solution containing 1.75 g of sucrose ($C_{12}H_{22}O_{11}$) per 150 ml of solution?

Molecular weight of $C_{12}H_{22}O_{11}$ = 342.

$$\frac{n}{V} = \frac{\text{number of moles of solute}}{\text{number of liters of solution}} = \frac{\dfrac{1.75 \text{ g}}{342 \text{ g/mole}}}{0.150 \text{ } l} = 0.0341 \text{ mole}/l$$

Osmotic pressure, $\pi = \dfrac{n}{V}RT$ = 0.0341 mole/l × 0.0821 l atm mole^{-1} °K^{-1} × 290°K

$$= 0.81 \text{ atmospheres}$$

14.9. The osmotic pressure of a solution of a synthetic polyisobutylene in benzene was determined at 25°C. A sample containing 0.20 g of solute per 100 ml of solution developed a rise of 2.4 mm at osmotic equilibrium. The density of the solution was 0.88 g/cc. What is the molecular weight of the polyisobutylene?

The osmotic pressure must be converted to atmospheres. First, the equivalent length of a mercury column may be computed from the formula,

$$\text{Height of mercury column} = \text{height of solution} \times \frac{\text{density of solution}}{\text{density of mercury}}$$

$$= 2.4 \text{ mm} \times \frac{0.88 \text{ g/cc}}{13.6 \text{ g/cc}} = 0.155 \text{ mm mercury}$$

This equation is based on the fact that the pressure at the base of a standing column of liquid is equal to the height of the column times the density of the liquid. A 0.155 mm column of mercury constitutes $0.155/760 = 2.0 \times 10^{-4}$ atmospheres.

The molarity can now be determined from the osmotic pressure equation.

$$\frac{n}{V} = \frac{\pi}{RT} = \frac{2.0 \times 10^{-4} \text{ atm}}{0.0821 \text{ } l \text{ atm mole}^{-1} \text{ deg}^{-1} \times 298 \text{ deg}} = 8.2 \times 10^{-6} \text{ mole}/l$$

The solution contained 0.20 g solute per 100 ml solution, or 2.0 g per liter, and is known to contain 8.2×10^{-6} mole/l. Then

$$\text{Molecular weight} = \frac{2.0 \text{ g}}{8.2 \times 10^{-6} \text{ mole}} = 2.4 \times 10^5 \text{ g/mole}$$

14.10. An aqueous solution of urea had a freezing point of −0.52°C. Predict the osmotic pressure of the same solution at 37°C. Assume that the molarity and the molality are the same.

The concentration of the solution is not specified, but the effective concentration may be inferred from the freezing point lowering, where

$$\text{Molality } (m) = \frac{\text{freezing point lowering}}{K_f} = \frac{0.52}{1.86} = 0.280 \text{ molal}$$

The assumption that the molality and molarity are the same is not very bad for dilute aqueous solutions. Then 0.280 may be used for the molarity in the osmotic pressure equation,

Osmotic pressure, $\pi = \dfrac{n}{V}RT$ = 0.280 mole/l × 0.082 l atm mole^{-1} deg^{-1} × 310 deg

$$= 7.1 \text{ atm}$$

SOLUTIONS OF GASES IN LIQUIDS

14.11. At 20°C and a total pressure of 760 mm, one liter of water dissolves 0.043 g of pure oxygen. Under corresponding conditions, one liter of water dissolves 0.019 g of pure nitrogen. Assuming dry air is composed of 20% oxygen and 80% nitrogen (by volume), determine the weight of oxygen and of nitrogen dissolved by one liter of water at 20°C exposed to air at a total pressure of 760 mm.

The quantities of various gases (oxygen and nitrogen) dissolved from a gaseous mixture (air) by a given volume of liquid are directly proportional to the solubilities and partial pressures of the gases.

When water is exposed to pure oxygen at a total pressure of 760 mm, the partial pressure of oxygen is (760 − v.p.), where v.p. is the vapor pressure of water at 20°C. When the water is exposed to air at a total pressure of 760 mm, the partial pressure of oxygen is $1/5 \times (760 - \text{v.p.})$, and of nitrogen $4/5 \times (760 - \text{v.p.})$.

$$\text{Solubility of O}_2 \text{ from air} = \frac{1/5 \times (760 - \text{v.p.})}{(760 - \text{v.p.})} \times 0.043 \text{ g/l}$$

$$= 1/5 \times 0.043 \text{ g/l} = 0.0086 \text{ g/l}$$

Similarly, the solubility of N_2 from air $= 4/5 \times 0.019$ g/l $= 0.015$ g/l.

14.12. A gaseous mixture of hydrogen and oxygen contains 70% hydrogen and 30% oxygen by volume. If the gas mixture at a pressure of 2.5 atmospheres (excluding the vapor pressure of water) is allowed to saturate water at 20°C, the water is found to contain 31.5 ml of hydrogen (S.T.P.) per liter. Find the solubility of hydrogen (reduced to S.T.P.) at 20°C and one atmosphere partial pressure of hydrogen.

Since the volume of a gas at S.T.P. depends only on the mass, the volume of the dissolved gas (reduced to S.T.P.) is proportional to the partial pressure of the gas.

Partial pressure of hydrogen $= 0.70 \times 2.5$ atm $= 1.75$ atm

Solubility of H_2 at 20°C and 1 atm $= \dfrac{1.00 \text{ atm}}{1.75 \text{ atm}} \times 31.5$ ml/l $= 18.0$ ml (S.T.P.)/l

LAW OF DISTRIBUTION — EXTRACTION

14.13. (*a*) A volume of 25 ml of an aqueous solution of iodine containing 2 mg of iodine is shaken with 5 ml of CCl_4, and the CCl_4 is allowed to separate. Given that iodine is 85 times more soluble in CCl_4 than in water and that both saturated solutions may be considered to be "dilute." Calculate the quantity of iodine remaining in the water layer.

(*b*) Suppose a second extraction is made of the water layer using another 5 ml of CCl_4, calculate the quantity of iodine remaining after the second extraction.

(*a*) Let $x =$ milligrams of iodine in H_2O layer at equilibrium. Then

$2 - x =$ milligrams of iodine in CCl_4 layer at equilibrium.

The concentration of iodine in the water layer will be $x/25$ (mg/ml of water), and the concentration of iodine in the CCl_4 layer will be $(2 - x)/5$ (mg/ml of CCl_4). Hence

$$\frac{\text{Conc. I}_2 \text{ in CCl}_4}{\text{Conc. I}_2 \text{ in H}_2\text{O}} = \frac{85}{1} \quad \text{or} \quad \frac{(2-x)/5}{x/25} = \frac{85}{1} \quad \text{or} \quad \frac{2-x}{x} = 17$$

Solving, $x = 0.11$ mg of iodine left in the water layer after first extraction.

(*b*) Let $y =$ milligrams of iodine in H_2O layer after second extraction. Then

$0.11 - y =$ milligrams of iodine in CCl_4 layer after second extraction.

The concentration of iodine in the water layer will be $y/25$, and the concentration in the CCl_4 layer will be $(0.11 - y)/5$. Hence

$$\frac{\text{Conc. I}_2 \text{ in CCl}_4}{\text{Conc. I}_2 \text{ in H}_2\text{O}} = \frac{85}{1} \quad \text{or} \quad \frac{(0.11-y)/5}{y/25} = \frac{85}{1} \quad \text{or} \quad \frac{0.11-y}{y} = 17$$

Solving, $y = 0.0061$ mg of iodine left in water layer after second extraction.

Note. Any concentration units could have been used in this problem, so long as the same units were used in both numerator and denominator. The choice of mg/ml was the most convenient for this particular case.

Supplementary Problems

14.14. A solution containing 6.35 g of a non-electrolyte dissolved in 500 g of water freezes at $-0.465°C$. Determine the molecular weight of the solute.   *Ans.* 50.8

14.15. A solution containing 3.24 g of a non-volatile non-electrolyte and 200 g of water boils at $100.130°C$ at one atmosphere. What is the molecular weight of the solute?   *Ans.* 64

14.16. Calculate the freezing point and the boiling point at one atmosphere of a solution containing 30.0 g cane sugar (molecular weight 342) and 150 g water.   *Ans.* $-1.09°C$, $100.300°C$

14.17. If glycerin, $C_3H_5(OH)_3$, and methyl alcohol, CH_3OH, sold at the same price per pound, which would be cheaper for preparing an anti-freeze solution for the radiator of an automobile?

Ans. methyl alcohol

14.18. What weight of ethyl alcohol, C_2H_5OH, must be added to one liter of water so that the solution will not freeze at $-4°$ Fahrenheit?   *Ans.* 495 g

14.19. If the radiator of an automobile contains 12 liters water, how much would the freezing point be lowered by the addition of 5 kg of Prestone (glycol, $C_2H_4(OH)_2$)? How many kilograms of Zerone (methyl alcohol, CH_3OH) would be required to produce the same result? Assume 100% purity.

Ans. $12°C$, 2.6 kg

14.20. What is the freezing point of a 10% (by weight) solution of CH_3OH in water?   *Ans.* $-6.5°C$

14.21. When 10.6 g of a non-volatile substance is dissolved in 740 g of ether, its boiling point is raised $0.284°C$. What is the molecular weight of the substance? Molal boiling point elevation for ether is $2.11°C$.   *Ans.* 106

14.22. The freezing point of a sample of naphthalene was found to be $80.6°C$. When 0.512 g of a substance is dissolved in 7.03 g naphthalene, the solution has a freezing point of $75.2°C$. What is the molecular weight of the solute? The molal freezing point lowering of naphthalene is $6.80°C$.

Ans. 92

14.23. Pure benzene freezes at $5.45°C$. A solution containing 7.24 g of $C_2Cl_4H_2$ in 115.3 g of benzene was observed to freeze at $3.55°C$. What is the molal freezing point constant of benzene?

Ans. $5.08°C \, m^{-1}$

14.24. What is the osmotic pressure at $0°C$ of an aqueous solution containing 46.0 grams of glycerin $(C_3H_8O_3)$ per liter?   *Ans.* 11.2 atm

14.25. A solution of crab hemocyanin, a pigmented protein extracted from crabs, was prepared by dissolving 0.750 g in 125 ml of an aqueous medium. At $4°C$ an osmotic pressure rise of 2.6 mm of the solution was observed. The solution had a density of 1.00 g/cc. Determine the molecular weight of the protein.   *Ans.* 5.5×10^5

14.26. The osmotic pressure of blood is 7.65 atmospheres at $37°C$. How much glucose should be used per liter for an intravenous injection that is to have the same osmotic pressure as blood?

Ans. 54.3 g

14.27. The vapor pressure of pure water at $26°C$ is 25.21 mm. What is the vapor pressure of a solution which contains 20.0 g glucose, $C_6H_{12}O_6$, in 70 g water?   *Ans.* 24.51 mm

14.28. The vapor pressure of pure water at $25°C$ is 23.76 mm. The vapor pressure of a solution containing 5.40 g of a non-volatile substance in 90 g water is 23.32 mm. Compute the molecular weight of the solute.   *Ans.* 57

14.29. Ethylene bromide, $C_2H_4Br_2$, and 1,2-dibromopropane, $C_3H_6Br_2$, form a series of ideal solutions over the whole range of composition. At $85°C$ the vapor pressures of these two pure liquids are 173 mm and 127 mm respectively.

   (a) If 10.0 g of ethylene bromide is dissolved in 80.0 g of 1,2-dibromopropane, calculate the partial pressure of each component and the total pressure of the solution at $85°C$.

   (b) Calculate the composition of the vapor in equilibrium with the above solution and express as mole fraction of ethylene bromide.

   (c) What would be the mole fraction of ethylene bromide in a solution at $85°C$ equilibrated with a 50 : 50 mole mixture in the vapor?

   Ans. (a) ethylene bromide, 20.5 mm; 1, 2-dibromopropane, 112 mm; total, 132 mm

        (b) 0.155, (c) 0.42

14.30. At 40°C the vapor pressure of methyl alcohol—ethyl alcohol solutions in mm is represented by the equation $P = 119x_{CH_3OH} + 135$, where x_{CH_3OH} is the mole fraction of methyl alcohol. What are the vapor pressures of the pure components at this temperature?

Ans. methyl alcohol, 254 mm; ethyl alcohol, 135 mm

14.31. A 0.100 weight formal solution of $NaClO_3$ freezes at $-0.3433°C$. What would you predict for the boiling point of this aqueous solution at 1 atmosphere pressure? At 0.001 weight formal concentration of this same salt, the electrical interferences between the ions no longer exist because the ions, on the average, are too far apart from each other. Predict the freezing point of this more dilute solution.   *Ans.* 100.095°C, $-0.0037°C$

14.32. The molecular weight of a newly synthesized organic compound was determined by the method of isothermal distillation. In this procedure two solutions, each in an open calibrated vial, are placed side by side in a closed chamber. One of the solutions contained 9.3 mg of the new compound, the other 13.2 mg of azobenzene (molecular weight = 182). Both were dissolved in portions of the same solvent. During a period of three days of equilibration, solvent distilled from one vial into the other until the same partial pressure of solvent was reached in the two vials. At this point the distillation of solvent stopped. Neither of the solutes distilled at all. The volumes of the two solutions at equilibrium were then read on the calibration marks of the vials. The solution containing the new compound occupied 1.72 cc and the azobenzene solution occupied 1.02 cc. What is the molecular weight of the new compound? The weight of solvent in solution may be assumed to be proportional to the volume of the solution.   *Ans.* 76

14.33. If 29 mg of N_2 dissolves in one liter of water at 0°C and 760 mm N_2 pressure, what weight of N_2 will dissolve in one liter of water at 0°C and 5.00 atmospheres pressure?   *Ans.* 145 mg

14.34. At 20°C and 1.00 atmosphere partial pressure of hydrogen, 18 ml of hydrogen measured at S.T.P. dissolves in one liter of water. If water at 20°C is exposed to a gaseous mixture having a total pressure of 1400 mm (excluding the vapor pressure of water) and containing 68.5% H_2 by volume, find the volume of H_2, measured at S.T.P., which will dissolve in one liter of water.   *Ans.* 23 ml

14.35. A liter of CO_2 gas at 15°C and 1.00 atmosphere dissolves in a liter of water at the same temperature when the pressure of CO_2 is 1.00 atmosphere. Compute the molal concentration of CO_2 in a solution over which the partial pressure of CO_2 is 150 mm at this temperature.

Ans. 0.0083 molal

14.36. (a) Iodine is 200 times more soluble in ether than in water. If 30 ml of an aqueous solution of iodine containing 2.0 mg of iodine is shaken with 30 ml of ether and the ether is allowed to separate, what quantity of iodine remains in the water layer?

(b) What quantity of iodine remains in the water layer if only 3 ml of ether is used?

(c) How much iodine is left in the water layer after a second extraction, again using 3 ml of ether?

(d) Which method is more efficient – a single large washing, or repeated small washings?

Ans. 0.010 mg, 0.095 mg, 0.0045 mg

14.37. The ratio of the solubility of stearic acid in *n*-heptane to that in 97.5% acetic acid is 4.95. How many extractions of a 10 ml solution of stearic acid in 97.5% acetic acid with successive 10 ml portions of *n*-heptane are needed to reduce the residual stearic acid content of the acetic acid layer to less than 0.5% of its original value?   *Ans.* 3

14.38. One method of purifying penicillin is by extraction. The distribution coefficient for penicillin G between isopropyl ether and an aqueous phosphate medium is 0.34 (lower solubility in the ether). The corresponding ratio for penicillin F is 0.68. A preparation of penicillin G has penicillin F as a 10.0% impurity.

(a) If an aqueous phosphate solution of this preparation is extracted with an equal volume of isopropyl ether, what will be the % recovery of the initial G in the residual aqueous phase product after one extraction, and what will be the % impurity in this product?

(b) Compute the same two quantities for the product remaining in the aqueous phase after a second extraction with an equal volume of ether.

Ans. (a) 75% recovery, 8.1% impurity   (b) 56% recovery, 6.5% impurity

Chapter 15

Heat

HEAT

Heat is a form of energy. Other forms of energy, such as mechanical, chemical, electrical, etc., tend in natural processes to be transformed into heat energy. When any other kinds of energy are transformed into heat energy, or vice versa, the heat energy is exactly equivalent to the amount of transformed energy.

All substances require heat when their temperature is raised (other variables remaining constant), and yield the same amount of heat when they are cooled to the initial temperature. Heat is always absorbed when a solid melts or when a liquid evaporates.

HEAT UNITS

The three units most commonly used in measuring the quantity of heat are defined as follows:

1 calorie (cal) = quantity of heat required to raise the temperature of 1 gram of water through 1 centigrade degree.

Note. Since the calorie was originally defined as stated above, it has been recognized that the energy requirement for raising the temperature of 1 gram of water by 1 degree depends slightly on the temperature, with a variation of about half a percent over the interval from $0°$ to $100°C$. For work requiring an accuracy no greater than one percent, the above definition is satisfactory. For the most precise work, it has been agreed to define the calorie in terms of electrical units of energy, so that 1 calorie = 4.1840 joules. This is very close to the amount of energy required to raise the temperature of 1 gram of water from $16.5°$ to $17.5°C$.

1 kilocalorie or *kilogram-calorie* (kcal or kg-cal) = 1000 calories. (The nutritional *Calorie*, spelled with an upper case C, is one *kilocalorie*.)

1 British thermal unit (Btu) = quantity of heat required to raise the temperature of 1 pound of water through 1 Fahrenheit degree.

Since 1 lb = 453.6 g and $1°F = 5/9 °C$, 1 Btu $= 453.6 × 5/9$ cal = 252 cal.

SPECIFIC HEAT

Specific heat of a substance (is numerically equal to the number of)

= calories required to raise temperature of 1 gram of substance $1°C$.

= Btu required to raise temperature of 1 pound of substance $1°F$.

Specific heat is expressed in calories per gram per degree centigrade (cal/g-$°C$) in the metric system, and in Btu per pound per degree Fahrenheit (Btu/lb-$°F$) in the British system.

The heat capacity of a substance is the amount of heat required to raise the temperature of that substance 1°. The heat capacity is equal to the specific heat times the mass of the substance. Heat capacity may be expressed in such units as calories per degree centigrade or in Btu per degree Fahrenheit. Frequent use is made of such quantities as the *molar heat capacity* (heat capacity per mole) and the *atomic heat capacity* (heat capacity per gram-atom).

From the definitions of the calorie and the Btu, it follows that the specific heat of water is numerically equal to 1 in either of the above two systems of units (1 cal/g-°C or 1 Btu/lb-°F), neglecting the variations with temperature. Also, the specific heat of any substance is numerically the same in these two systems of units. Thus, the specific heat of pyrex glass is 0.20 cal/g-°C or 0.20 Btu/lb-°F.

LAW OF DULONG AND PETIT

This is a useful approximate rule, first discovered empirically a century and a half ago and later given some theoretical justification. For many elementary solids, the atomic heat capacity is approximately 6.4 calories per gram-atom per degree Celsius. This approximation is valid near room temperature to within 6% for most of the elements beyond the third period.

CALORIMETRY

An object undergoing a temperature change without a chemical reaction or change of state absorbs or discharges an amount of heat equal to its heat capacity times the temperature change.

Heat (in calories) = mass (grams) × sp ht (cal/g-°C) × temp. change (°C)

Heat (in Btu)   = mass (pounds) × sp ht (Btu/lb-°F) × temp. change (°F)

The heating or cooling of a body of matter of known heat capacity can be used in *calorimetry*, the measurement of quantities of heat.

HEAT OF FUSION

Heat of fusion of a solid is the amount of heat required to melt 1 unit mass of the solid without changing its temperature.

Heat of fusion of ice = 80 cal per gram (at 0°C and 1 atmosphere)

= 80 cal/g × 18.0 g/mole = 1.44×10^3 cal/mole

= 144 Btu per pound (at 32°F and 1 atmosphere)

HEAT OF VAPORIZATION

Heat of vaporization of a liquid is the amount of heat required to vaporize 1 unit mass of the liquid without changing its temperature.

Heat of vaporization of water = 540 cal per gram (at 100°C and 1 atm)

= 972 Btu per pound (at 212°F and 1 atm)

HEAT OF SUBLIMATION

Heat of sublimation of a substance is the amount of heat required to convert 1 unit mass of it from the solid to the gaseous state at a given temperature. The heat of sublimation of a substance at a given temperature is equal to the sum of its heats of fusion and vaporization for that same temperature.

ENTHALPY

When a system absorbs heat part of the absorbed energy may be used for doing work, such as lifting a weight, expanding against the atmosphere, or operating a battery, and part is stored within the system itself as the energy of the internal motions of the atoms and molecules themselves and as the energy of interaction among the atoms and molecules. This stored portion is known as the *internal energy*. The *enthalpy*, H, is a quantity closely related to the internal energy and so defined that the *increase in enthalpy* of any system undergoing a change at constant pressure and temperature exactly equals the heat absorbed in the process, if the only kind of work done in the process is expansion against the atmosphere. Although these defining restrictions may seem severe, they are the approximate conditions under which most chemical processes are carried out in open vessels. As a consequence of the law of conservation of energy, the amount of enthalpy contained in a system is a fixed property of the system, and the enthalpy changes for two processes which are the opposite of each other are equal in magnitude but opposite in sign. For example, the enthalpy change, written as ΔH, for the fusion of ice is 1440 cal/mole, since 1440 cal is the amount of heat absorbed in the melting of one mole at constant temperature, $0°C$, and constant pressure, 1 atmosphere. The enthalpy gained by water when it melts is lost if it refreezes, and ΔH of freezing at $0°$ is -1440 cal/mole; this amount of heat is liberated to the environment on freezing.

An *endothermic* process is one for which ΔH is positive, i.e., heat is absorbed. An *exothermic* process is one for which ΔH is negative, i.e., heat is liberated.

Most chemical reactions are accompanied by enthalpy changes, arising mainly from the difference in the chemical bond strengths of the products and reactants. Calculations of enthalpy changes of reactions are important not only for predicting heat effects, such as for fuels, but also, in conjunction with probability considerations, for predicting the direction in which chemical reactions will naturally proceed. Although the exact calculation of the probability factor cannot be done in an elementary course, a rough qualitative rule may be used, based on enthalpy changes alone. At normal temperatures, most reactions having a large negative ΔH can in principle proceed spontaneously, although nothing can be said in general about the rate of such reactions. Conversely, most reactions having a large positive ΔH cannot proceed very far at normal temperatures.

THERMOCHEMICAL EQUATIONS

A ΔH value assigned to a chemical equation defines the enthalpy change occurring when the number of moles of each reactant consumed is equal to its coefficient in the balanced equation.

$$C(graphite) + O_2(gas) \longrightarrow CO_2(gas) \qquad \Delta H = -94.1 \text{ kcal}$$

$$H_2(gas) + I_2(crystal) \longrightarrow 2\,HI(gas) \qquad \Delta H = +12.4 \text{ kcal}$$

In the above reactions, 94.1 kcal is *liberated* when 1 mole of $CO_2(gas)$ is formed from graphite and oxygen, and 12.4 kcal is *absorbed* in the formation of 2 moles of HI from hydrogen and crystalline iodine. ΔH is an extensive quantity, proportional to the amount of material undergoing reaction. Thus, ΔH for the formation of 2 moles of CO_2 from C and O_2 is $2(-94.1) = -188.2$ kcal, and for the formation of 1 gram ($\frac{1}{44}$ mole) CO_2, $\frac{1}{44}(-94.1) = -2.14$ kcal.

(In some books, thermochemical equations are written as if heat is either a product or reactant. In such a notation the above equations are written:

$$C + O_2 \longrightarrow CO_2 + 94.1 \text{ kcal}$$

$$H_2 + I_2 + 12.4 \text{ kcal} \longrightarrow 2 \text{ HI}$$

The magnitude of the heat term is often called the *heat of reaction*. Because of the ambiguity of the sign, the term *heat of reaction* will not generally be used. The enthalpy convention will be used in the remainder of the book, a positive ΔH representing an endothermic process and a negative ΔH an exothermic process.)

The *enthalpy of formation* of a substance, ΔH_f, is the enthalpy change accompanying the formation of *one mole* (or gfw) of the substance from its elements. For this purpose, the elements are defined in their *standard,* or normal stable states at 25°C. For example, the *standard state* of H_2, O_2, Cl_2, or N_2 is the pure gas at 1 atmosphere pressure; the *standard state* of bromine or mercury is the liquid; the *standard state* of iron, sodium, or iodine is the solid; the *standard state* of carbon is graphite, which is more stable than diamond. With reference to the examples in the preceding paragraph, the enthalpy of formation of $CO_2(gas)$ is -94.1 kcal and of HI(gas) is $+6.2$ kcal (half of the $+12.4$ kcal enthalpy increase on forming two moles of HI).

LAW OF CONSTANT HEAT SUMMATION

The enthalpy change in a chemical process is the same whether it takes place in one or in several steps, since the overall change depends only on the properties of the initial and final substances. This law enables us to determine indirectly the enthalpy change in a reaction, even though it may be impossible to measure the heat of the reaction experimentally.

For example: It is impossible to measure accurately the heat liberated when C burns to CO, because the oxidation cannot be stopped exactly at the CO stage. We can, however, measure accurately the heat liberated when C burns to CO_2 (94.1 kcal per mole), and also the heat liberated when CO burns to CO_2 (67.7 kcal per mole of CO). The enthalpy change for the burning of C to CO is determined by treating algebraically these last two experimentally determined thermochemical equations. If two chemical equations are added or subtracted, their corresponding enthalpy changes are to be added or subtracted. Thus:

$$2 \text{ C}(graphite) + 2 \text{ O}_2(gas) \longrightarrow 2 \text{ CO}_2(gas) \qquad \Delta H = 2(-94.1) = -188.2 \text{ kcal}$$

$$2 \text{ CO}(gas) + \text{ O}_2(gas) \longrightarrow 2 \text{ CO}_2(gas) \qquad \Delta H = 2(-67.7) = -135.4 \text{ kcal}$$

Subtracting the second equations (both chemical and enthalpic) from the first, and transposing -2 CO to the right, we have

$$2 \text{ C}(graphite) + \text{ O}_2(gas) \longrightarrow 2 \text{ CO}(gas) \qquad \Delta H = -52.8 \text{ kcal}$$

ΔH_f, the enthalpy of formation of CO(gas), is thus evaluated as -26.4 kcal (half of -52.8).

USE OF THERMOCHEMICAL TABLES

The law of constant heat summation greatly simplifies the tabulation of thermochemical data. It is not necessary to list the enthalpy change for every chemical reaction in order to convey the available thermochemical information. It is sufficient to list only the minimum number of ΔH values from which others can be evaluated by the principle of additivity. It has been agreed by chemists throughout the world to prepare and use tables of *enthalpies of formation* for as many pure chemical substances as possible. A portion of such a listing appears in Table 15-1, where enthalpies of formation at 25°C are tabulated. Problems and data in this chapter refer to 25°C unless a specific statement to the contrary is made. The following abbreviations are used: (g), gas; (l), liquid; (s), solid; (aq), dissolved in water.

The enthalpy change of any reaction is equal to the sum of the enthalpies of formation of all products minus the sum of the enthalpies of formation of all reactants, each ΔH_f being multiplied by the number of moles of the substance in the balanced chemical equation. The validity of this rule can be checked for the following example.

			ΔH
(1)	$P(white) + \tfrac{3}{2} Cl_2(g) + \tfrac{1}{2} O_2(g)$	$\longrightarrow \quad POCl_3(g)$	-141.5 kcal
(2)	$H_2(g) \quad + \quad Cl_2(g)$	$\longrightarrow \quad 2\,HCl(g)$	-44.2 kcal
(3)	$PCl_5(g)$	$\longrightarrow \quad P(white) + \tfrac{5}{2} Cl_2(g)$	$+95.4$ kcal
(4)	$H_2O(g)$	$\longrightarrow \quad H_2(g) + \tfrac{1}{2} O_2(g)$	$+57.8$ kcal
Sum:	$PCl_5(g) + H_2O(g)$	$\longrightarrow \quad POCl_3(g) + 2\,HCl(g)$	-32.5 kcal

Table 15-1

Standard Enthalpies of Formation at 25°C			
Substance	ΔH_f, in kcal	Substance	ΔH_f, in kcal
$Al_2O_3(s)$	-399.1	$H_2O(g)$	-57.80
$B_2O_3(s)$	-302	$H_2O(l)$	-68.32
C(diamond)	$+0.5$	$H_2O_2(l)$	-46
$CF_4(g)$	-163	$H_2S(g)$	-4.8
$CH_3OH(g)$	-48.1	$H_2S(aq)$	-9.4
$C_9H_{20}(l)$	-65.8	$I_2(g)$	$+14.9$
$(CH_3)_2N_2H_2(l)$	$+54$	$KCl(s)$	-104.2
$C(NO_2)_4(l)$	-8.8	$KClO_3(s)$	-93.5
$CO(g)$	-26.4	$KClO_4(s)$	-104
$CO_2(g)$	-94.1	$LiAlH_4(s)$	-24
$CaC_2(s)$	-15.0	$LiBH_4(s)$	-45
$CaO(s)$	-151.9	$Li_2O(s)$	-142
$Ca(OH)_2(s)$	-235.8	$NH_3(l)$	-16.0
$CaCO_3(s)$	-288.5	$N_2H_4(l)$	$+12.0$
$ClF_3(l)$	-42.7	$NO(g)$	$+21.6$
$Cl^-(aq)$	-40.0	$NO_2(g)$	$+12.4$
$Cu^{2+}(aq)$	$+15.4$	$N_2O_4(g)$	$+2.3$
$CuSO_4(s)$	-184.0	$N_2O_4(l)$	-6.8
$F_2O(l)$	$+3.0$	$O_3(g)$	$+34.0$
$Fe^{2+}(aq)$	-21.0	$OH^-(aq)$	-54.96
$Fe_2O_3(s)$	-196.5	$PCl_3(g)$	-73.2
$FeS(s)$	-22.7	$PCl_5(g)$	-95.4
$H^+(aq)$	0.0	$POCl_3(g)$	-141.5
$HCl(g)$	-22.1	$SO_2(g)$	-71.0
$HF(g)$	-64.2	$SO_3(g)$	-94.4
$HI(g)$	$+6.2$	$Zn^{2+}(aq)$	-36.4
$HNO_3(l)$	-41		

The ΔH listing for (1) is ΔH_f for $POCl_3$, for (2) twice ΔH_f of HCl, for (3) and (4) the negative of ΔH_f for PCl_5 and H_2O respectively. Note the use of fractional coefficients in the formation equations of some of the compounds; this is not uncommon in thermochemistry.

After the student has verified the procedure for the above case, he can prove the validity of the general rule as a consequence of the laws of balancing chemical equations.

Solved Problems

SPECIFIC HEAT AND CALORIMETRY

15.1. (a) How many calories are required to heat 100 grams of copper from 10°C to 100°C?

(b) The same quantity of heat is added to 100 grams of aluminum at 10°C. Which gets hotter, the copper or the aluminum? Specific heat of Cu = 0.093, of Al = 0.217 cal/g-°C.

(a) Heat required = mass × specific heat × temperature change

$$= \ 100 \text{ g} \ \times \ 0.093 \text{ cal/g-°C} \ \times \ (100-10)\text{°C} \ = \ 840 \text{ calories}$$

(b) Since the specific heat of copper is less than that of aluminum, less heat is required to raise the temperature of a mass of copper by 1° than is required for an equal mass of aluminum. Hence the copper is hotter.

15.2. One gram of anthracite coal when burned evolves about 7300 cal. What weight of coal is required to heat 4.0 liters of water from room temperature (20°C) to the boiling point (at 1 atmosphere pressure), assuming that all the heat is available?

Heat required to raise the temperature of 4000 g of water from 20°C to 100°C

$$= \ \text{mass} \ \times \ \text{specific heat} \ \times \ \text{temperature change}$$

$$= \ 4000 \text{ g} \ \times \ 1 \text{ cal/g-°C} \ \times \ (100-20)\text{°C} \ = \ 3.2 \times 10^5 \text{ cal}$$

$$\text{Weight of coal required} \ = \ \frac{3.2 \times 10^5 \text{ cal}}{7300 \text{ cal/g}} \ = \ 44 \text{ g coal}$$

15.3. A steam boiler is made of steel and weighs 900 lb. The boiler contains 400 lb of water. Assuming that 70% of the heat is delivered to boiler and water, how many Btu are required to raise the temperature of the whole from 42°F to 212°F (the boiling point of water at 1 atmosphere pressure)? Specific heat of steel is 0.11 cal/g-°C or 0.11 Btu/lb-°F.

Heat needed = mass × specific heat × temperature change

Heat needed for boiler = 900 lb × 0.11 Btu/lb-°F × (212−42)°F = 1.7×10^4 Btu

Heat needed for water = 400 lb × 1 Btu/lb-°F × (212−42)°F = 6.8×10^4 Btu

$$\text{Total Btu required} \ = \ \frac{(1.7+6.8) \times 10^4 \text{ Btu}}{0.70} \ = \ 1.21 \times 10^5 \text{ Btu}$$

15.4. Exactly 3 g of carbon was burned to CO_2 in a copper calorimeter. The mass of the calorimeter is 1500 g, and the mass of water in the calorimeter is 2000 g. The initial temperature was 20°C and the final temperature 31°C. Calculate the heat value of carbon in calories per gram. Specific heat of copper is 0.093.

Heat gained = mass × specific heat × temperature change

Heat gained by calorimeter = 1500 g × 0.093 cal/g-°C × (31−20)°C = 1530 cal

Heat gained by water = 2000 g × 1 cal/g-°C × (31−20)°C = 22,000 cal

$$\text{Heat value of carbon} \ = \ \frac{1530 \text{ cal} + 22,000 \text{ cal}}{3 \text{ g}} \ = \ 7.8 \times 10^3 \text{ cal/g}$$

15.5. A 1.250 g sample of benzoic acid, $C_7H_6O_2$, was placed in a combustion bomb. The bomb was filled with an excess of oxygen at high pressure, sealed, and immersed in a pail of water which served as a calorimeter. The heat capacity of the entire apparatus was found to be 2422 cal/°C, including the bomb, the pail, a thermometer, and the water. The oxidation of the benzoic acid was triggered by passing an electric spark through the sample. After complete combustion of the sample, the thermometer immersed in the water registered a temperature 3.256 deg greater than before the combustion. How much heat is liberated per mole of benzoic acid combusted in a bomb-type calorimeter? Assume that no correction must be made for the sparking process.

$$\text{Heat evolved by reaction} = \text{Heat gained by calorimeter and contents}$$

$$= 2422 \text{ cal/}^\circ\text{C} \times 3.256^\circ\text{C} = 7.89 \times 10^3 \text{ cal}$$

$$\text{Heat evolved per mole} = \frac{7.89 \times 10^3 \text{ cal}}{1.250 \text{ g acid}} \times 122.1 \text{ g acid/mole acid}$$

$$= 771 \times 10^3 \text{ cal/mole benzoic acid}$$

15.6. The equivalent weight of a metal in its reaction with oxygen is 69.67 and its specific heat is 0.0305 (calories per gram per degree centigrade). Compute the oxidation state and exact atomic weight of the metal.

Law of Dulong and Petit: Specific heat $\times$ atomic weight = 6.4 approximately.

Therefore the approximate atomic weight of the metal = 6.4/0.0305 = 210.

The law of Dulong and Petit gives only the approximate atomic weight. The equivalent weight is determined by chemical analysis and may be known very exactly.

The exact atomic weight must be a whole number multiple of the equivalent weight. This number, which represents the number of equivalents contained in the atomic weight of the element, is the oxidation state. In this case, 3 is the obvious factor (3×69.67 is approximately 210); hence the oxidation state of the metal is 3.

Exact atomic weight = oxidation state $\times$ equivalent weight = 3×69.67 = 209.01

15.7. A 25.0 g sample of an alloy was heated to 100.0°C and dropped into a beaker containing 90 grams of water at 25.32°C. The temperature of the water rose to a final value of 27.18°. Neglecting heat losses to the room and the heat capacity of the beaker itself, what is the specific heat of the alloy?

Let x = specific heat of the alloy.

The heat lost by the alloy = heat absorbed by water

$$25.0 \text{ g} \times x \text{ cal/g-}^\circ\text{C} \times (100.0 - 27.2)^\circ\text{C} = 90 \text{ g} \times 1 \text{ cal/g-}^\circ\text{C} \times (27.18 - 25.32)^\circ\text{C}$$

from which $x = 0.092$ cal/g-°C.

HEATS OF FUSION AND VAPORIZATION

15.8. Determine the resulting temperature, t, when 150 g of ice at 0°C is mixed with 300 g of water at 50°C.

Heat to melt ice = mass $\times$ heat of fusion = 150 g $\times$ 80 cal/g = 1.20×10^4 cal

Heat to raise temperature of 150 g of water at 0°C to final temperature

= mass $\times$ specific heat $\times$ temperature change = $150 \times 1 \times (t - 0)$ cal

Heat lost by 300 g water = mass $\times$ sp ht $\times$ temp. change = $300 \times 1 \times (50 - t)$ cal

Heat lost = heat gained

$$300(50 - t) = 1.20 \times 10^4 + 150t \qquad \text{from which} \qquad t = 6.7^\circ\text{C}$$

15.9. How much heat is given up when 20 g of steam at 100°C is condensed and cooled to 20°C?

Heat given up in condensing 20 g of steam at 100°C to water at 100°C
= mass × heat of vaporization = 20 g × 540 cal/g = 1.08×10^4 cal

Heat given up in cooling 20 g of water at 100°C to 20°C
= mass × sp ht × temperature change = $20 \times 1 \times (100 - 20)$ = 0.16×10^4 cal

Total heat given up = $(1.08 + 0.16) \times 10^4$ cal = 1.24×10^4 cal

15.10. How much heat is required to convert 40 g of ice at −10°C to steam at 120°C? Specific heat of ice = 0.5, of steam = 0.5 cal/g-°C.

Heat to raise temperature of ice at −10°C to ice at 0°C
= mass × sp ht × temperature change = $40 \times 0.5 \times 10$ = 200 cal = 0.2 kcal

Heat to melt ice at 0°C = mass × heat of fusion = 40 g × 80 cal/g = 3.2 kcal

Heat to raise temperature of 40 g of water at 0°C to 100°C
= mass × sp ht × temperature change = $40 \times 1 \times 100$ = 4.0 kcal

Heat to change water at 100°C to steam at 100°C
= mass × heat of vaporization = 40 g × 540 cal/g = 21.6 kcal

Heat to raise temperature of steam at 100°C to steam at 120°C
= mass × sp ht × temperature change = $40 \times 0.5 \times 20$ = 0.4 kcal

Total heat required = (0.2 + 3.2 + 4.0 + 21.6 + 0.4) kcal = 29.4 kcal

15.11. (a) Convert the heat of fusion of ice, 80 cal/g, to its equivalent in Btu/lb.

(b) How many Btu are absorbed by an electric refrigerator in changing 5.0 lb water at 60°F to ice at 32°F?

(a) $80 \dfrac{\text{cal}}{\text{g}}$ = $80 \dfrac{\text{cal}}{\text{g}} \times \dfrac{1 \text{ Btu}}{252 \text{ cal}} \times \dfrac{454 \text{ g}}{1 \text{ lb}}$ = $80 \times \dfrac{454 \text{ Btu}}{252 \text{ lb}}$ = 144 Btu/lb

(b) Heat absorbed in changing water at 60°F to water at 32°F
= mass × specific heat of water × temperature change
= 5.0 lb × 1 Btu/lb-°F × (60 − 32)°F = 140 Btu

Heat absorbed in changing water at 32°F to ice at 32°F
= mass × heat of fusion = 5.0 lb × 144 Btu/lb = 720 Btu

Total heat absorbed = 140 Btu + 720 Btu = 860 Btu

15.12. What is the heat of vaporization of water per gram at 25°C?

We can write the thermochemical equation for the process: $H_2O(l) \longrightarrow H_2O(g)$.

ΔH can be evaluated by subtracting ΔH_f of reactants from ΔH_f of products, as tabulated in Table 15-1.

ΔH = ΔH_f (products) − ΔH_f (reactants) = −57.80 − (−68.32) = 10.52 kcal

The enthalpy of vaporization per gram is $\dfrac{10.52 \times 10^3 \text{ cal/mole}}{18.02 \text{ g/mole}}$ = 584 cal/g.

Note that the heat of vaporization at 25°C is greater than the value (540 cal/g) at 100°C.

THERMOCHEMICAL EQUATIONS

15.13. The thermochemical equation for the combustion of ethylene gas, C_2H_4, is

$$C_2H_4(g) \;+\; 3\,O_2(g) \;\longrightarrow\; 2\,CO_2(g) \;+\; 2\,H_2O(l) \qquad \Delta H = -337 \text{ kcal}$$

Assuming 70% efficiency, how many kilograms of water at 20°C can be converted into steam at 100°C by burning 1000 liters of C_2H_4 gas measured at S.T.P.?

Since 22.4 l of C_2H_4 at S.T.P. (1 mole) produces 337 kcal,   then

$$\text{1000 } l \text{ at S.T.P. produces} \quad \frac{1000\ l}{22.4\ l/\text{mole}} \times 337\ \text{kcal/mole} \;=\; 15.0 \times 10^3\ \text{kcal}$$

Useful heat $=$ efficiency $\times$ total heat $=$ $0.70(15.0 \times 10^3\ \text{kcal})$ $=$ 10.5×10^3 kcal.

To change 1 kg of water at 20°C to water at 100°C, 80 kcal is required.

To convert 1 kg of water at 100°C to steam at 100°C, 540 kcal is required.

Heat required for each kg of water $=$ 80 kcal $+$ 540 kcal $=$ 620 kcal.

$$\text{Kilograms of water converted into steam at 100°C} \;=\; \frac{10.5 \times 10^3\ \text{kcal}}{620\ \text{kcal/kg}} \;=\; 16.9\ \text{kg.}$$

15.14. Calculate the enthalpy of reduction of ferric oxide by aluminum (thermite reaction) at 25°C.

We must start with a balanced equation for the reaction. We may then write in parentheses under each formula the enthalpy of formation, taken from Table 15-1, and outside the parentheses the number of moles, gfw, or g-at in the balanced equation. Remember that ΔH_f for any element in its standard state is zero, by definition.

$$\begin{array}{cccccc}
2\,Al & + & Fe_2O_3 & \longrightarrow & 2\,Fe & + & Al_2O_3 \\
2(0) & & 1(-196.5) & & 2(0) & & 1(-399.1)
\end{array}$$

We then combine all terms on the left by multiplying each ΔH_f by the number of moles and adding the product, and then we do the same on the right. ΔH of the reaction is the difference of the sum for the products minus the sum for the reactants.

$$\Delta H = (\text{sum of } \Delta H_f \text{ of products}) - (\text{sum of } \Delta H_f \text{ of reactants})$$
$$= -399.1 - (-196.5) = -202.6\ \text{kcal}$$

This is ΔH for the reduction of 1 gfw Fe_2O_3.

15.15. Calculate the enthalpy of decomposition of $CaCO_3$ into CaO and CO_2.

$$CaCO_3(s) \;\longrightarrow\; CaO(s) \;+\; CO_2(g)$$

Enthalpies of formation: -288.5 -151.9 -94.1

ΔH of reaction $=$ enthalpies of form. of products $-$ enthalpies of form. of reactants

$$= -151.9 - 94.1 + 288.5 = +42.5\ \text{kcal per mole of } CaCO_3$$

A positive value signifies an endothermic reaction.

15.16. Calculate the enthalpy of neutralization of a strong acid by a strong base in water. The heat liberated on neutralization of HCN (weak acid) by NaOH is 2.5 kcal per mole. How many kcal are absorbed in ionizing one mole of HCN in water?

The basic equation for neutralization is as follows:

$$H^+(aq) \;+\; OH^-(aq) \;\longrightarrow\; H_2O(l)$$

Enthalpies of formation: 0 -54.96 -68.32

Thus $\Delta H = -68.32 - (-54.96) = -13.36$ kcal.

The neutralization of HCN(aq) by NaOH(aq) may be thought of as the sum of two processes, ionization of HCN(aq) and neutralization of $H^+(aq)$ with $OH^-(aq)$. (Since NaOH is a strong base, NaOH(aq) implies complete ionization, and a separate thermochemical equation for the ionization need not be written.) We may therefore construct the following thermochemical cycle.

$$HCN(aq) \longrightarrow H^+(aq) + CN^-(aq) \qquad \Delta H_{ionization} = x$$

$$\underline{H^+(aq) + OH^-(aq) \longrightarrow H_2O(l)} \qquad \underline{\Delta H \qquad\qquad = -13.4}$$

Sum: $HCN(aq) + OH^-(aq) \longrightarrow H_2O(l) + CN^-(aq) \qquad \Delta H_{experimental} = -2.5$

The sign of $\Delta H_{experimental}$ is negative because heat is liberated on neutralization.

From the principal of additivity,
$$x + (-13.4) = -2.5 \qquad \text{and} \qquad x = 10.9 \text{ kcal} = \Delta H_{ionization}$$
The ionization process is endothermic to the extent of 10.9 kcal per mole.

15.17. The heat evolved on combustion of acetylene, C_2H_2, at 25° is 310.7 kcal/mole. Determine the enthalpy of formation of acetylene gas.

The complete combustion of an organic compound involves the formation of CO_2 and H_2O.

$$C_2H_2(g) + \tfrac{5}{2} O_2(g) \longrightarrow 2\, CO_2(g) + H_2O(l)$$
Enthalpies of formation: $\quad x \qquad\qquad$ zero $\qquad\quad$ $2(-94.1) \qquad -68.3$

ΔH of reaction $\quad=\quad$ enthalpies of form. of products $-$ enthalpies of form. of reactants
$$-310.7 \quad = \quad [2(-94.1) + (-68.3)] - x$$

Solving, $x = \Delta H_f$ for $C_2H_2 = +54.2$ kcal/mole.

15.18. How much heat will be required to make 1 kg of CaC_2 according to the reaction given below?

$$CaO(s) + 3\, C(s) \longrightarrow CaC_2(s) + CO(g)$$
Enthalpies of formation: $-151.9 \qquad$ zero $\qquad\qquad -15.0 \qquad -26.4$ kcal/mole

ΔH of reaction $\quad=\quad$ enthalpies of form. of products $-$ enthalpies of form. of reactants
$$= \quad -(15.0 + 26.4) - (-151.9) \quad = \quad +110.5 \text{ kcal per mole } CaC_2.$$

Number of moles in 1 kg of $CaC_2 = \dfrac{1000 \text{ g}}{64.10 \text{ g/mole}} = 15.60$ moles CaC_2

Heat required for the endothermic formation of 15.60 moles CaC_2
$$= \quad 110.5 \text{ kcal/mole} \times 15.60 \text{ moles} \quad = \quad 1724 \text{ kcal.}$$

15.19. How much heat will be evolved in making 22.4 liters at S.T.P. (1 mole) of H_2S from FeS and dilute hydrochloric acid?

$$FeS(s) + 2\, H^+(aq) \longrightarrow Fe^{2+}(aq) + H_2S(g)$$
$$-22.7 \qquad\quad \text{zero} \qquad\qquad -21.0 \qquad -4.8$$

Since HCl and $FeCl_2$ are strong electrolytes, only their essential ions need be written.

$\Delta H \quad=\quad$ enthalpies of form. of products $-$ enthalpies of form. of reactants
$$= \quad -(21.0 + 4.8) + 22.7 \quad = \quad -3.1 \text{ kcal per mole } H_2S.$$

Supplementary Problems

HEAT CAPACITY AND CALORIMETRY

15.20. How many calories are required to heat each of the following from 15°C to 65°C: (a) 1 g water, (b) 5 g pyrex glass, (c) 20 g platinum? Specific heat of pyrex glass = 0.20, of platinum = 0.032 cal/g-°C.   *Ans.* 50 cal, 50 cal, 32 cal

15.21. How many Btu are removed in cooling each of the following from 212°F to 68°F: (a) 1 lb water, (b) 2 lb leather, (c) 3 lb asbestos? Specific heat of leather = 0.36, of asbestos = 0.20 Btu/lb-°F. *Ans.* 144 Btu, 104 Btu, 86 Btu

15.22. The combustion of 5.00 g of coke raised the temperature of one liter of water from 10°C to 47°C. Calculate the heat value of coke in kcal/g. *Ans.* 7.4 kcal/g

15.23. Furnace oil has a heat of combustion of 19,000 Btu/lb. Assuming that 70% of the heat is useful, how many pounds of oil are required to heat 1000 lb of water from 50°F to 190°F? *Ans.* 10.5 lb

15.24. Assuming that 50% of the heat is useful, how many kg of water at 15°C can be heated to 95°C by burning 200 liters of methane, CH_4, measured at S.T.P.? The heat of combustion of methane is 213 kcal/mole. *Ans.* 11.9 kg

15.25. The heat of combustion of ethane gas, C_2H_6, is 373 kcal/mole. Assuming that 60% of the heat is useful, how many liters of ethane measured at S.T.P. must be burned to supply enough heat to raise the temperature of 50 kg of water at 10°C to steam at 100°C? *Ans.* 3150 liters

15.26. A 50 gallon tank full of water is heated from 40°F to 160°F using coal having a heat of combustion of 14,000 Btu/lb. How many pounds of coal are required if 55% of the heat is useful? One gallon of water weighs 8.34 lb. *Ans.* 6.5 lb

15.27. A 45.0 g sample of a metal was heated to 90.0°C and then dropped into a beaker containing 82 grams of water at 23.50°C. The temperature of the water rose to a final temperature of 26.25°C. What is the specific heat of the alloy? *Ans.* 0.079 cal/g-°C

15.28. Two solutions, initially at 25.08°C, were mixed in an insulated bottle. One contained 400 ml of a 0.200 M weak monoprotic acid. The other contained 100 ml of 0.800 M NaOH. After mixing, the temperature rose to 26.25°C. How much heat is evolved in the neutralization of one mole of the acid? Assume that the densities of all solutions are 1.00 g/cc and that their specific heats are all 1.00 cal/g-°C. (These assumptions are in error by several percent, but the consequent errors in the final result partly cancel each other.) *Ans.* 7.3 kcal/mole

15.29. An element was found to have a heat capacity of 0.0276 calories per degree per gram. 114.79 grams of a chloride of this element contained 79.34 grams of the metallic element. What is the exact atomic weight of this element? *Ans.* 238.0

15.30. The heat capacity of a solid element is 0.0442 calories per degree per gram. A sulfate of this element was purified and found to contain 42.2% of the element by weight. (a) What is the exact atomic weight from this determination? (b) What is the formula of the sulfate? *Ans.* (a) 140.3 (b) $Ce(SO_4)_2$

HEATS OF PHASE CHANGES

15.31. Determine the resulting temperature when 1 kg of ice at 0°C is mixed with 9 kg of water at 50°C. Heat of fusion of ice is 80 cal/g. *Ans.* 37°C

15.32. How much heat is required to change 10 g of ice at 0°C to steam at 100°C? Heat of vaporization of water at 100°C is 540 cal/g. *Ans.* 7.2 kcal

15.33. Ten pounds of steam at 212°F is passed into 500 lb of water at 40°F. What is the resulting temperature? *Ans.* 62.4°F

15.34. What is the heat of sublimation of solid iodine at 25°C? *Ans.* 14.9 kcal/mole of I_2

15.35. Is the process of dissolving H_2S gas in water endothermic or exothermic? To what extent? *Ans.* Exothermic, 4.6 kcal per mole

15.36. How much heat is released on dissolving a mole of HCl(g) in a large amount of water? (*Hint.* HCl is completely ionized in dilute solution.) *Ans.* 17.9 kcal

15.37. In an ice calorimeter, a chemical reaction is allowed to occur in thermal contact with an ice-water mixture at 0°C. Any heat liberated by the reaction is used to melt some ice; an observation of the extent of volume change of the ice-water mixture can be used to indicate the amount of melting. When 10 ml of 0.10 F $AgNO_3$ was mixed with 10 ml of 0.10 F NaCl in such a calorimeter, both solutions having been pre-cooled to 0°C, 0.20 grams of ice melted. Assuming complete reaction in this experiment, what is ΔH for the reaction: $Ag^+ + Cl^- \longrightarrow AgCl$? *Ans.* −16 kcal

THERMOCHEMICAL EQUATIONS

15.38. Given: $N_2(g) + 3 H_2(g) \longrightarrow 2 NH_3(g)$; $\Delta H = -22.0$ kcal. What is the enthalpy of formation of NH_3 gas? *Ans.* −11.0 kcal/mole

15.39. Determine ΔH of decomposition of 1 mole of solid $KClO_3$ into solid KCl and gaseous oxygen.
Ans. -10.7 kcal

15.40. The heat released on neutralization of $CsOH$ with all strong acids is 13.4 kcal/mole. The heat released on neutralization of $CsOH$ with HF (weak acid) is 16.4 kcal/mole. Calculate ΔH of ionization of HF in water.   *Ans.* -3.0 kcal/mole

15.41. Calculate ΔH for the reaction: $CuSO_4(aq) + Zn(s) \longrightarrow ZnSO_4(aq) + Cu(s)$
Ans. -51.8 kcal/mole

15.42. How much heat will be evolved in slaking 1 kg of quicklime (CaO) according to the reaction:
$CaO(s) + H_2O(l) \longrightarrow Ca(OH)_2(s)$?   *Ans.* 278 kcal

15.43. The heat liberated on complete combustion of 1 mole of CH_4 gas to $CO_2(g)$ and $H_2O(l)$ is 212.8 kcal. Determine the enthalpy of formation of 1 mole of CH_4 gas.   *Ans.* -17.9 kcal/mole

15.44. The heat evolved on combustion of 1 g of starch, $(C_6H_{10}O_5)_x$, into CO_2 gas and H_2O liquid is 4.18 kcal/g. Compute the enthalpy of formation of 1 gram of starch.   *Ans.* -1.42 kcal/g

15.45. The amount of heat evolved in dissolving $CuSO_4$ is 17.5 kcal/gfw. What is ΔH_f for $SO_4^{2-}(aq)$?
Ans. -216.9 kcal

15.46. The heat of solution of $CuSO_4 \cdot 5\,H_2O$ in a large amount of water is 1.3 kcal/mole (endothermic). Calculate the heat of reaction for

$$CuSO_4(s) + 5\,H_2O(l) \longrightarrow CuSO_4 \cdot 5\,H_2O(s)$$

Use data from the preceding problem.   *Ans.* 18.8 kcal/mole (exothermic)

15.47. The heat evolved on combustion into $CO_2(g)$ and $H_2O(l)$ of one mole of C_2H_6 is 372.9 kcal, and of C_2H_4 is 337.3 kcal. Calculate ΔH of the following reaction: $C_2H_4 + H_2(g) \longrightarrow C_2H_6$.
Ans. -32.7 kcal/mole

15.48. The solution of $CaCl_2 \cdot 6\,H_2O$ in a large volume of water is endothermic to the extent of 3.5 kcal/mole. For the reaction $CaCl_2(s) + 6\,H_2O(l) \longrightarrow CaCl_2 \cdot 6\,H_2O(s)$, $\Delta H = -23.2$ kcal. What is the heat of solution of $CaCl_2$ (anhydrous) in a large volume of water?
Ans. 19.7 kcal/mole exothermic

15.49. The thermochemical equation for the dissociation of hydrogen gas into atoms may be written: $H_2 \longrightarrow 2\,H$; $\Delta H = 104.2$ kcal. What is the ratio of the energy yield on combustion of hydrogen atoms to steam to the yield on combustion of an equal weight of hydrogen molecules to steam?
Ans. 2.8

15.50. The commercial production of water gas utilizes the reaction: $C + H_2O(g) \longrightarrow H_2 + CO$. The required heat for this endothermic reaction may be supplied by adding a limited amount of air and burning some carbon to carbon dioxide. How many grams of carbon must be burned to CO_2 to provide enough heat for the water gas conversion of 100 grams of carbon? Neglect all heat losses to the environment.   *Ans.* 33.4 g

15.51. The reversible reaction, $Na_2SO_4 \cdot 10\,H_2O \longrightarrow Na_2SO_4 + 10\,H_2O$, for which $\Delta H = +18.8$ kcal, goes completely to the right at temperatures above 32.4°C, and remains completely on the left below this temperature. This system has been used in some solar houses for heating at night with the energy absorbed from the sun's radiation during the day. How many cubic feet of fuel gas could be saved per night by the reversal of the dehydration of a fixed charge of 100 lb $Na_2SO_4 \cdot 10\,H_2O$? Assume that the fuel value of the gas is 2000 Btu per cubic foot.   *Ans.* 5.3 cu ft

15.52. An important criterion for the desirability of fuel reactions for rockets is the fuel value per gram of reactant or per cc of reactant. Compute each of these quantities for each of the following reactions.

    (a) $N_2H_4(l) + 2\,H_2O_2(l) \longrightarrow N_2(g) + 4\,H_2O(g)$

    (b) $2\,LiBH_4(s) + KClO_4(s) \longrightarrow Li_2O(s) + B_2O_3(s) + KCl(s) + 4\,H_2(g)$

    (c) $6\,LiAlH_4(s) + 2\,C(NO_2)_4(l) \longrightarrow 3\,Al_2O_3(s) + 3\,Li_2O(s) + 2\,CO_2(g) + 4\,N_2(g) + 12\,H_2(g)$

    (d) $4\,HNO_3(l) + 5\,N_2H_4(l) \longrightarrow 7\,N_2(g) + 12\,H_2O(g)$

    (e) $7\,N_2O_4(l) + C_9H_{20}(l) \longrightarrow 9\,CO_2(g) + 10\,H_2O(g) + 7\,N_2(g)$

    (f) $4\,ClF_3(l) + (CH_3)_2N_2H_2(l) \longrightarrow 2\,CF_4(g) + N_2(g) + 4\,HCl(g) + 4\,HF(g)$

    (g) $3\,F_2O(l) + 4\,NH_3(l) \longrightarrow 2\,N_2(g) + 3\,H_2O(g) + 6\,HF(g)$

Use the following density values: $N_2H_4(l)$, 1.01 g/cc; $H_2O_2(l)$, 1.46 g/cc; $LiBH_4(s)$, 0.66 g/cc; $KClO_4(s)$, 2.52 g/cc; $LiAlH_4(s)$, 0.92 g/cc; $C(NO_2)_4(l)$, 1.65 g/cc; $HNO_3(l)$, 1.50 g/cc; $N_2O_4(l)$, 1.45 g/cc; $C_9H_{20}(l)$, 0.72 g/cc; $ClF_3(l)$, 1.77 g/cc; $(CH_3)_2N_2H_2(l)$, 0.78 g/cc; $F_2O(l)$, 1.90 g/cc; $NH_3(l)$, 0.80 g/cc. In computing the volume of each reaction mixture, assume that the reactants are present in stoichiometric proportions.

Ans.   (a) 1.51 kcal/g; 1.93 kcal/cc   (d) 1.44 kcal/g; 1.82 kcal/cc   (g) 2.19 kcal/g; 2.96 kcal/cc

          (b) 1.95 kcal/g; 2.93 kcal/cc   (e) 1.70 kcal/g; 2.11 kcal/cc

          (c) 2.66 kcal/g; 3.40 kcal/cc   (f) 1.28 kcal/g; 1.93 kcal/cc

Chapter 16

Chemical Equilibrium

INTRODUCTION

Theoretically, all chemical reactions are reversible, that is, there is a possibility that the atoms in the product molecules rearrange to form the reactant molecules. Often the *driving force* of a reaction favors one direction so greatly that the extent of the reverse reaction is infinitesimal and impossible to measure. The driving force of a chemical reaction, or the *change in chemical potential* accompanying the reaction, is an exact measure of the tendency of a reaction to go to completion. When the magnitude of this quantity is very large (and the sign negative), the reaction may go practically to completion in the forward direction; if it is only slightly negative, the reaction may proceed to a small extent but could go in the opposite direction with a slight change in conditions. In the latter case, the reaction is said to be reversible. Since many organic and metallurgical reactions are of this reversible type, it is necessary to learn how conditions should be altered to obtain economical yields, hasten desirable reactions, and minimize undesirable reactions.

Reversible reactions involve products which themselves react to form the initial substances to such an extent that there is a measurable amount of the reactants as well as the products. Both reactions proceed continually until a state of dynamic equilibrium is reached, in which the opposing forward and backward reactions proceed with equal speed. The rate or "speed" of a chemical reaction is defined as the quantity reacting per unit time, usually per unit volume.

Chemical equilibrium is said to exist when the reactants are forming as rapidly as the products, so that the composition of the mixture remains constant and does not change with time. It must be understood that this condition is a *dynamic* one, constantly reactive in both directions, in contrast to the lifeless *static* equilibrium of pulleys, levers and springs.

The most important characteristic of the state of chemical equilibrium is that the forward and backward reactions are proceeding at equal speeds, thus balancing each other and preventing any change in the composition of the reacting mixture. The exact equilibrium composition in a given reversible reaction depends upon the nature of the reacting substances, the temperature, and usually the pressure and concentration, but is unaffected by catalysts.

THE EQUILIBRIUM CONSTANT

Consider the following reversible reaction in a homogenous solution.

$$A + B \rightleftharpoons C + D$$

A reaction mixture could be made up by starting with A and B, with C and D, or with a combination of substances appearing on both sides of the equation. In each case a net reaction would occur in one direction or the other until the system comes to an eventual state of no further net change. At this point the situation could be described by specifying the concentrations of the four substances at equilibrium. Because of the variety of ways of making up the initial mixture, differing in the relative amounts of the various substances used, there is an infinite number of equilibrium states, each describable by a set of the concentrations of the four participating chemicals. A unifying relationship exists, however, which systematizes this infinity of solutions to the equilibrium problem. It has been found, both by theoretical argument and by experimental verification in a large number of cases, that the concentrations of these materials co-existing at equilibrium must satisfy the following equation

$$\frac{[C] \times [D]}{[A] \times [B]} = K$$

where the bracketed symbol, e.g. [C], refers to the concentration of the particular species, usually in moles per liter. At any one temperature K has a fixed numerical value characteristic of the particular chemical equation. K is called the *equilibrium constant*. It must be emphasized that the above equation must be satisfied at equilibrium, regardless of the choice of initial amounts of substances for the reaction mixture. The value of K is not affected by a catalyst.

For a reaction in which more than two substances react, such as

$$A + B + C \rightleftharpoons D + E$$

the equilibrium equation is $\quad \dfrac{[D] \times [E]}{[A] \times [B] \times [C]} = K$

When two or more molecules of a substance appear in the equation, as in

$$A + 3B \rightleftharpoons 2C$$

the equilibrium equation is $\quad \dfrac{[C]^2}{[A] \times [B]^3} = K$

Here the concentration of B is cubed and the concentration of C is squared.

In general, for the reaction

$$mA + nB \rightleftharpoons yC + zD$$

the equilibrium equation is $\quad \dfrac{[C]^y \times [D]^z}{[A]^m \times [B]^n} = K$

The numerator always contains the product of the concentrations of all species appearing on the right side of the chemical equation and the denominator contains the product of the concentrations of the species on the left. Each concentration is raised to a power equal to the coefficient in the balanced chemical equation.

Experimental measurements show that molecules in highly compressed gases or highly concentrated solutions, especially if electrically charged, abnormally affect each other. In such cases the true *activity* or *effective concentration* may be greater or less than the measured concentration. Hence when the molecules involved in equilibrium are relatively close together, the concentration should be multiplied by an *activity coefficient* (determined experimentally). At moderate pressures and dilutions, the activity coefficient for non-ionic compounds is close to unity. In any event, the activity coefficient correction will not be made in the problems of this chapter.

The magnitude of the equilibrium constant for any given reaction depends upon the temperature and upon the units in which activity or effective concentration is expressed. It will be assumed that concentrations of dissolved substances are expressed in moles per liter, unless a statement is made to the contrary. The value of K is independent of the units only in cases where the sum of the concentration exponents in the numerator equals the sum of the exponents in the denominator.

Terms representing the concentrations of undissolved *solid* reactants or products are conventionally omitted from the K equation, because their concentrations cannot be varied. For example, one may have a saturated solution of iodine in equilibrium with 1 milligram or with 1 pound of solid, undissolved iodine. The concentration of the solid iodine crystals, in moles per liter *of crystals,* depends on the density but not on the number of crystals. The numerical value of K for a reaction involving solids is so chosen as to take into account the constant concentration of the solids.

Another special case occurs when the solvent is one of the products or reactants, such as the hydrolysis of urea in aqueous solution.

$$CO(NH_2)_2 + H_2O \; \rightleftharpoons \; CO_2 + 2\,NH_3 \qquad K = \frac{[CO_2] \times [NH_3]^2}{[CO(NH_2)_2]}$$

As long as the solution is moderately dilute, the concentration of water in moles per liter cannot be much less than in pure water itself. It is therefore customary not to write the concentration of water in the equilibrium constant equation. The numerical value of the equilibrium constant is so chosen as to take this into account.

All of these equilibrium relationships are connected with the fact that at equilibrium the rates of forward and backward reactions must be equal. The rate of any chemical reaction depends on the concentrations of the reactants in a manner that depends on the particular reaction. In simple cases where the reactants interact directly in a *single event* to form the products without the intervention of intermediate substances, the rate of that particular reaction is proportional to the products of the concentrations of the reactants, each concentration raised to a power equal to the number of molecules (or ions or atoms) of that substance that must participate in the *single event*. In the first example cited

$$A + B \; \rightleftharpoons \; C + D$$

above, if A reacts with B in a single event to give the forward reaction, and C and D react in a single event to give the reverse reaction, then

$$\text{rate of forward reaction} \;=\; k_1 \times [A] \times [B]$$
$$\text{rate of reverse reaction} \;=\; k_2 \times [C] \times [D]$$

where k_1 and k_2 are proportionality constants. At equilibrium the two rates are equal, so that

$$k_1 \times [A] \times [B] \;=\; k_2 \times [C] \times [D]$$

and
$$\frac{[C] \times [D]}{[A] \times [B]} \;=\; \frac{k_1}{k_2}$$

The ratio of the two k's must itself be a constant and is what we have called the equilibrium constant, K. In most cases, the reactions expressed in the usual chemical equations occur in a sequence of two or more events. Although the arguments based on the equality of forward and reverse reaction rates become complex in such cases, the general formulation of the equilibrium constant as given by the rules in this section are independent of the complexity of the actual mechanism by which the reactions proceed.

THE PARTIAL PRESSURE EQUILIBRIUM CONSTANT (K_p)

In many gaseous systems it is more convenient to express the concentrations of gases in terms of partial pressures. The equilibrium constant is then usually designated as K_p.

Partial pressure of any gas in a gas mixture

$$= \frac{\text{number of moles of that gas in mixture}}{\text{total number of moles in gas mixture}} \times \text{total pressure of gas mixture.}$$

When all substances in the reversible reaction

$$m\text{A} + n\text{B} \; \rightleftharpoons \; y\text{C} + z\text{D}$$

are gases, the equilibrium expression may be written as follows:

$$\frac{p_\text{C}^y \times p_\text{D}^z}{p_\text{A}^m \times p_\text{B}^n} \;=\; K_p$$

where p_A is the partial pressure of A, etc.

When the equation shows no change in total number of moles of gas as the reaction proceeds (e.g., $\text{N}_2 + \text{O}_2 \rightleftharpoons 2\,\text{NO}$), K_p will be identical with the K expressed in molar concentrations.

Partial pressures will be expressed in atmospheres for calculations of K_p in this chapter, unless otherwise stated.

EFFECT OF VARYING THE CONCENTRATION

Once the equilibrium constant for a given reaction is determined by investigating one equilibrium mixture, we can calculate what will happen to any other mixture of the substances concerned *at the same temperature*. For example, if a system at equilibrium with respect to the reaction $\text{A} + \text{B} \rightleftharpoons \text{C} + \text{D}$ is disturbed by an increase in the concentration of A (by inserting an additional quantity of A), then the composition of the mixture will adjust itself until, when equilibrium is restored, the ratio $\dfrac{[C] \times [D]}{[A] \times [B]}$ has again attained the value of K.

In this case the denominator [A] of the fraction has been increased by the addition of excess A. To restore equilibrium, further reaction takes place between A and B to form more of C and D, thus increasing the concentrations of C and D (the numerator) while reducing the resulting concentration of B (the denominator) until the fraction attains the value of K. At this point equilibrium is again established. The resulting concentration of A is greater than before the supplementary addition of A but less than it would have been had there been no net chemical reaction of adjustment following the addition.

LE CHATELIER'S PRINCIPLE

If some stress (such as a change in temperature, pressure or concentration) is brought to bear upon a system in equilibrium, a reaction occurs which displaces the equilibrium in the direction which tends to undo the effect of the stress. This generalization is extremely useful in predicting the effects of changes in temperature, pressure, and concentration upon a system at equilibrium.

(1) **Effect of Changes in Temperature.** When the temperature of a system at equilibrium is raised, the equilibrium is displaced in the direction which absorbs heat (Van't Hoff's Law). For example, in the thermochemical equation for the synthetic methanol process (all substances are in the gaseous state)

$$\text{CO} + 2\,\text{H}_2 \; \rightleftharpoons \; \text{CH}_3\text{OH} \qquad \Delta H = -22\,\text{kcal}$$

the forward reaction liberates heat (exothermic) while the reverse reaction absorbs heat (endothermic). If the temperature of the system is raised, the equilibrium is displaced in the direction which absorbs heat — to the left. Conversely, the equilibrium yield of methanol is increased by lowering the temperature of this system, and the value of K is correspondingly increased.

(2) **Effect of Changes in Pressure.** When the pressure of a system in equilibrium is increased, the equilibrium is displaced so as to lower the volume as much as possible. For example, in the synthetic methanol process (all substances are in the gaseous state)

$$CO + 2 H_2 \quad \rightleftharpoons \quad CH_3OH$$

| 3 molecules of gas | 1 molecule of gas |
| 3 volumes of gas | 1 volume of gas |

the forward reaction is accompanied by a decrease in volume. Then an increase in pressure will increase the equilibrium yield of CH_3OH. (This increase in yield occurs even though the value of K, depending only on temperature, does not change.)

A pressure change will *not* affect the relative amounts of the substances at equilibrium in any gaseous system where the number of molecules reacted equals the number produced. For example, $H_2 + CO_2 \rightleftharpoons CO + H_2O$ (steam) and $H_2 + I_2 \rightleftharpoons 2 HI$ (all substances in the gaseous state).

The effect of pressure on equilibrium systems involving gases and liquids or solids is usually due to a change in the number of gaseous molecules, since molar volumes of gases are so much larger than of liquids or solids. In the case of equilibrium reactions of solids or liquids, with no gases, the pressure effect is usually small, and it can be computed in terms of the general rule of volume change if the densities of products and reactants are known.

(3) **Effect of Changes in Amount of Solvent.** Reactions that take place in solution are affected by changes in the amount of solvent in a way analogous to the effect of pressure on gas reactions. Increasing the amount of solvent (dilution) will displace the equilibrium in the direction of the larger number of dissolved particles. This is analogous to decreasing the pressure in a gas reaction. This can be illustrated by the dimerization of acetic acid in benzene solutions.

$$2 HC_2H_3O_2 \text{ (in solution)} \quad \rightleftharpoons \quad (HC_2H_3O_2)_2 \text{ (in solution)}$$

| 2 dissolved particles | 1 dissolved particle |

$$K = \frac{[(HC_2H_3O_2)_2]}{[HC_2H_3O_2]^2}$$

Let us imagine an equilibrium benzene solution of acetic acid. If the solution is suddenly diluted to twice its original volume, and if there were no change in the relative amounts of the two forms of acetic acid, the concentration of each form would be just one half of what it had been before dilution. In the equilibrium constant equation, the numerator would be $\frac{1}{2}$ its former value and the denominator would be $\frac{1}{4}$ its former value ($\frac{1}{2}$ squared). The ratio of numerator to denominator would be 2 times its original value ($\frac{1}{2}$ divided by $\frac{1}{4}$). But this ratio must return to its original value of K. It can do this if the numerator becomes smaller and the denominator larger. In other words, some of the dimer, $(HC_2H_3O_2)_2$, must react backwards to form $2 HC_2H_3O_2$.

Changes in the amount of solvent will not affect the equilibrium in any system where the number of dissolved particles reacted equals the number produced. For example, the esterification of methyl alcohol with formic acid in an inert solvent:

$$CH_3OH + HCO_2H \quad \rightleftharpoons \quad HCO_2CH_3 + H_2O$$

(4) **Effect of Varying the Concentration.** Increasing the concentration of any component of a system in equilibrium will promote the action which tends to consume some of the added substance. For example, in the reaction $H_2 + I_2 \rightleftharpoons 2 HI$ the consumption of iodine is improved by adding excess hydrogen.

(5) **Effect of Catalysts.** Catalysts accelerate both forward and backward reaction rates equally. They speed up the approach to equilibrium, but do not alter the equilibrium concentrations.

Solved Problems

16.1. $PCl_5(g) \rightleftharpoons PCl_3(g) + Cl_2(g)$

Explain the effect upon the material distribution in this equilibrium of (*a*) increased temperature, (*b*) increased pressure, (*c*) higher concentration of Cl_2, (*d*) higher concentration of PCl_5, (*e*) presence of a catalyst.

(*a*) When the temperature of a system in equilibrium is raised, the equilibrium point is displaced in the direction which absorbs heat. Table 15-1, Page 141, can be used to determine that the *forward* reaction as written is endothermic.

$$\Delta H = \Delta H_f(PCl_3) + \Delta H_f(Cl_2) - \Delta H_f(PCl_5) = -73.2 + 0 - (-95.4) = +22.2 \text{ kcal}$$

Hence increasing the temperature will cause more PCl_5 to dissociate.

(*b*) When the pressure of a system in equilibrium is increased, the equilibrium point is displaced in the direction of the smaller volume.

One volume each of PCl_3 and Cl_2, a total of 2 gas volumes, form 1 volume of PCl_5. Hence a pressure increase will promote the reaction between PCl_3 and Cl_2 to form more PCl_5.

(*c*) Increasing the concentration of any component will displace the equilibrium in the direction which tends to lower the concentration of the component added.

Increasing the concentration of Cl_2 will result in the consumption of more PCl_3 and the formation of more PCl_5, and this action will tend to offset the increased concentration of Cl_2.

(*d*) Increasing the concentration of PCl_5 will result in the formation of more PCl_3 and Cl_2.

(*e*) A catalyst accelerates both forward and backward reactions equally. It speeds up the approach to equilibrium, but does not favor reaction in either direction over the other.

16.2. $N_2(g) + 3 H_2(g) \rightleftharpoons 2 NH_3(g)$ $\Delta H = -22 \text{ kcal}$

What conditions would you suggest for the manufacture of ammonia by the Haber process?

From the sign of ΔH, we see that the forward reaction gives off heat (exothermic). Therefore the equilibrium formation of NH_3 is favored by as low a temperature as practicable. In a reaction like this, the choice of temperature requires a compromise between equilibrium considerations and rate considerations. The equilibrium yield of NH_3 is higher, the lower the temperature. On the other hand, the rate at which the system will reach equilibrium is lower, the lower the temperature.

The forward reaction is accompanied by a decrease in volume, since 4 volumes of initial gases yield 2 volumes of NH_3. Therefore increased pressure will give a larger proportion of NH_3 in the equilibrium mixture.

A catalyst should be employed to speed up the approach to equilibrium.

16.3. $PCl_5(g) \rightleftharpoons PCl_3(g) + Cl_2(g)$

A quantity of PCl_5 was heated in a 12 liter vessel at 250°C. At equilibrium the vessel contains 0.21 mole PCl_5, 0.32 mole PCl_3, and 0.32 mole Cl_2. Compute the equilibrium constant K for the dissociation of PCl_5 at 250°C when concentrations are expressed in moles per liter.

$$K = \frac{[PCl_3] \times [Cl_2]}{[PCl_5]} = \frac{\frac{0.32 \text{ mole}}{12\,l} \times \frac{0.32 \text{ mole}}{12\,l}}{\frac{0.21 \text{ mole}}{12\,l}} = \frac{0.32 \times 0.32}{0.21 \times 12} \text{ mole}/l = 0.041 \text{ mole}/l$$

16.4. When 1 mole of pure ethyl alcohol is mixed with 1 mole of acetic acid at room temperature, the equilibrium mixture contains 2/3 of a mole each of ester and water. (a) What is the equilibrium constant? (b) How many moles of ester are formed at equilibrium when 3 moles of alcohol is mixed with 1 mole of acid? All substances are liquids (l) at the reaction temperature.

(a)

	alcohol $C_2H_5OH(l)$	acid $CH_3COOH(l)$ $\rightleftharpoons$	ester $CH_3COOC_2H_5(l)$	water $H_2O(l)$
(1) Moles at start:	1	1	none	none
(2) Change by reaction:	$-\frac{2}{3}$	$-\frac{2}{3}$	$+\frac{2}{3}$	$+\frac{2}{3}$
(3) Moles at equilibrium:	$1-\frac{2}{3}=\frac{1}{3}$	$1-\frac{2}{3}=\frac{1}{3}$	$\frac{2}{3}$	$\frac{2}{3}$

This tabular representation is a convenient way of doing the bookkeeping for equilibrium problems. Under each substance in the balanced equation an entry is made on three lines: (1) the amount of starting material, (2) the change in the number of moles (plus or minus) due to the attainment of equilibrium, and (3) the equilibrium amount, which is the algebraic sum of entries (1) and (2). The entries in line (2) must be in the same ratio to each other as the coefficients in the balanced chemical equation. The equilibrium constant equation can be written in terms of the entries in line (3).

Let V = volume of mixture, in liters.

$$K = \frac{[\text{ester}] \times [\text{water}]}{[\text{alcohol}] \times [\text{acid}]} = \frac{\frac{2/3}{V} \text{ mole}/l \times \frac{2/3}{V} \text{ mole}/l}{\frac{1/3}{V} \text{ mole}/l \times \frac{1/3}{V} \text{ mole}/l} = \frac{2/3 \times 2/3}{1/3 \times 1/3} = 4$$

Note that the volume of the mixture, V, cancels and does not appear in the answer. In this case, the numerical value of K is independent of the concentration units. Note also that water is not the solvent in this case; its concentration must therefore appear in the equilibrium constant equation along with those of the reactants and of the other product.

(b) Let x = number of moles of alcohol reacting.

	C_2H_5OH	CH_3COOH $\rightleftharpoons$	$CH_3COOC_2H_5$	H_2O
Moles at start:	3	1	none	none
Change by reaction:	$-x$	$-x$	$+x$	$+x$
Moles at equilibrium:	$3-x$	$1-x$	x	x

$$K = 4 = \frac{\frac{x}{V} \times \frac{x}{V}}{\frac{3-x}{V} \times \frac{1-x}{V}} = \frac{x^2}{(3-x)(1-x)} = \frac{x^2}{3-4x+x^2}$$

Then $x^2 = 4(3-4x+x^2)$ or $3x^2 - 16x + 12 = 0$. This is the familiar form, $ax^2 + bx + c = 0$, in which a (the coefficient of x^2) = 3, b (the coefficient of x) = -16, and c (the constant term) = 12. The general solution of this type of equation follows:

$$x = \frac{-b \pm \sqrt{b^2 - 4ac}}{2a} = \frac{+16 \pm \sqrt{(16)^2 - 4(3)(12)}}{2(3)} = \frac{16 \pm 10.6}{6} = 4.4 \text{ or } 0.9 \text{ moles}$$

There are two mathematical roots to the quadratic equation. Only one of the two has physical meaning. The applicable root can generally be selected very easily. In this problem, we started with 3 moles of alcohol and 1 mole of acid. We can see from the reaction equation that we cannot form more than 1 mole of ester, even if we use up all the acid. Therefore the correct root of the quadratic is 0.9. The other root, 4.4, has no physical meaning; it would lead to a negative value for the equilibrium concentrations of both the alcohol and the acid. Such a root must be rejected.

When the algebraic equation is complicated, it is always advisable for the student to check the result by substituting the answer into the original equation.

$$\text{Check:} \quad \frac{x^2}{(3 - x)(1 - x)} = \frac{(0.9)^2}{(3 - 0.9)(1 - 0.9)} = \frac{0.81}{2.1 \times 0.1} = 4$$

Note that the number of moles of ester formed in this problem is greater than the number of moles of ester in equilibrium in the previous problem (0.9 compared with 2/3). This answer was to be expected because of the increased concentration of alcohol, one of the reactants. A further addition of alcohol would lead to an even greater yield of ester, but in no case could the amount of ester formed exceed 1 mole, since 1 mole would represent 100% conversion of the acid. In practice, the actual selection of the excess concentration to be used may depend on economic factors. If alcohol is cheap compared to acid and ester, a large excess of alcohol might be used to assure a high percentage conversion of the acid. If, on the other hand, alcohol costs more per mole than acid, it would be more sensible to use an excess of acid and aim for a high percentage conversion of the alcohol.

16.5. $H_2(g) + I_2(g) \rightleftharpoons 2 HI(g)$

In a 10 liter evacuated chamber 0.5 mole H_2 and 0.5 mole I_2 are reacted at 448°C, at which temperature $K = 50$ for this reaction, for concentrations in moles per liter.

(a) What is the value of K_p?

(b) What is the total pressure in the chamber?

(c) How many moles of the iodine remain unreacted at equilibrium?

(d) What is the partial pressure of each component in the equilibrium mixture?

(a) The same number of gas moles (or gas volumes) occurs on both sides of the equation. Hence this reaction is not affected by a volume change or a pressure change, and $K_p = K = 50$. This can be proved in detail as follows.

Let V = volume of chamber in liters, and P = total pressure in atmospheres.

$$K = \frac{[HI]^2}{[H_2] \times [I_2]} = \frac{\left(\dfrac{\text{moles HI}}{V}\right)^2}{\left(\dfrac{\text{moles } H_2}{V}\right) \times \left(\dfrac{\text{moles } I_2}{V}\right)} = \frac{(\text{moles HI})^2}{(\text{moles } H_2)(\text{moles } I_2)}$$

$$K_p = \frac{(p_{HI})^2}{p_{H_2} \times p_{I_2}} = \frac{\left(\dfrac{\text{moles HI}}{\text{total moles}} \times P\right)^2}{\left(\dfrac{\text{moles } H_2}{\text{total moles}} \times P\right)\left(\dfrac{\text{moles } I_2}{\text{total moles}} \times P\right)} = \frac{(\text{moles HI})^2}{(\text{moles } H_2)(\text{moles } I_2)}$$

Note that the volume (V) and pressure (P) terms cancel out, as does the term in total moles. Thus, $K_p = K = 50$.

(b) Before the reaction sets in, the total number of gas moles $= 0.5 + 0.5 = 1$. During the reaction, there is no change in the total number of moles. (For every mole of H_2 that reacts, one mole of I_2 will react and 2 moles of HI will be formed.) Similarly, the total pressure will remain the same because it depends only on the total number of moles of gas. The total pressure can be computed from the ideal gas law, where n is the total number of moles.

$$P = \frac{nRT}{V} = \frac{1 \text{ mole} \times 0.0821 \ l \text{ atm mole}^{-1} \text{ °K}^{-1} \times 721 \text{°K}}{10 \ l} = 5.9 \text{ atm}$$

(c)	H_2	+	I_2	$\rightleftharpoons$	2 HI
Moles at start:	0.5		0.5		0.0
Change by reaction:	$-x$		$-x$		$2x$
Moles at equilibrium:	$0.5 - x$		$0.5 - x$		$2x$

Note the 2 : 1 ratio of moles of HI formed to moles of H_2 reacted. This ratio is demanded by the coefficients in the balanced chemical equation. Regardless of how complete or incomplete the reaction may be, 2 moles of HI are always formed for every 1 mole of H_2 that reacts.

It was shown in (b) that the volume of the container cancels out.

$$K = 50 = \frac{(2x)^2}{(0.5-x)(0.5-x)} \quad \text{or} \quad \sqrt{50} = 7.1 = \frac{2x}{0.5-x}$$

Then $2x = 7.1(0.5 - x)$. Solving, $x = 0.39$ mole I_2 reacted.

Moles of I_2 remaining unreacted at equilibrium $= 0.5 - 0.39 = 0.11$ mole

Check: $\dfrac{(2x)^2}{(0.5-x)^2} = \dfrac{(2 \times 0.39)^2}{(0.5-0.39)^2} = \dfrac{0.61}{0.012} = 51$ (close enough to the 50 in original equation)

(d) $p_{I_2} = \dfrac{\text{moles } I_2}{\text{total moles}} \times \text{total pressure} = \dfrac{0.11}{1} \times 5.9 \text{ atm} = 0.65 \text{ atm}$

$p_{H_2} = p_{I_2} = 0.65 \text{ atm}$

$p_{HI} = \text{total pressure} - (p_{H_2} + p_{I_2}) = 5.9 - 1.3 = 4.6 \text{ atm}$

Or, $p_{HI} = \dfrac{\text{moles } HI}{\text{total moles}} \times \text{total pressure} = \dfrac{0.78}{1} \times 5.9 \text{ atm} = 4.6 \text{ atm}$

16.6. Sulfide ion in alkaline solution reacts with solid sulfur to form polysulfide ions having formulas S_2^{--}, S_3^{--}, S_4^{--}, and so on. The equilibrium constant for the formation of S_2^{--} is 1.7 and for the formation of S_3^{--} is 5.3, both from S and S^{--}. What is the equilibrium constant for the formation of S_3^{--} from S_2^{--} and S?

To avoid confusion, let us designate the equilibrium constants for the various reactions by subscripts. Also we note that only the ion concentrations appear in the equilibrium constant equations because sulfur, S, is a solid.

$$S + S^{--} \rightleftharpoons S_2^{--} \qquad 2 S + S^{--} \rightleftharpoons S_3^{--} \qquad S + S_2^{--} \rightleftharpoons S_3^{--}$$

$$K_1 = \frac{[S_2^{--}]}{[S^{--}]} = 1.7 \qquad K_2 = \frac{[S_3^{--}]}{[S^{--}]} = 5.3 \qquad K_3 = \frac{[S_3^{--}]}{[S_2^{--}]} = x$$

The desired constant, K_3, expresses the equilibrium ratio of S_3^{--} to S_2^{--} concentrations in a solution equilibrated with solid sulfur. Such a solution must also contain sulfide ion, S^{--}, resulting from the dissociation of S_2^{--} by the reverse of the first equation. Since all four species, S, S^{--}, S_2^{--}, and S_3^{--}, are present, all of the equilibria represented above must be satisfied. The three equilibrium ratios are not all independent, because

$$\frac{[S_3^{--}]}{[S_2^{--}]} = \frac{\dfrac{[S_3^{--}]}{[S^{--}]}}{\dfrac{[S_2^{--}]}{[S^{--}]}}. \qquad \text{Or, } K_3 = \frac{K_2}{K_1} = \frac{5.3}{1.7} = 3.1$$

16.7. At 27°C and 1 atmosphere, N_2O_4 is 20% dissociated into NO_2. Find (a) K_p and (b) the percent dissociation at 27°C and a total pressure of 0.10 atmosphere. (c) What is the extent of dissociation in a 69 gram sample of N_2O_4 confined in a 20 liter vessel at 27°C?

(a) When 1 mole of N_2O_4 gas dissociates completely, 2 moles of NO_2 gas are formed. Since this problem does not specify a particular size of reaction vessel or a particular weight of sample, we are free to choose any convenient amount. Let us choose a sample that would be 1 mole of pure N_2O_4 if dissociation did not occur, i.e. 92 grams (molecular weight of N_2O_4). In the tabular organization of the data in a gaseous equilibrium for which K_p is used, an additional column should be added for the total gas in the mixture. The symbol P will be used for total gas pressure and p for partial pressure.

	$N_2O_4(g)$	$\rightleftharpoons$	$2\ NO_2(g)$	Total gas
Moles, assuming no dissociation:	1		zero	1
Change by reaction:	-0.2		$+0.4$	
Moles at equilibrium:	$1 - 0.2 = 0.8$		0.4	1.2
Mole fraction:	$0.8/1.2 = .67$		$0.4/1.2 = .33$	1.0
Part. press. (atm) = mole fract. $\times P$:	.67		.33	1.0

The number of moles of NO_2 formed equals twice the number of moles of N_2O_4 dissociated, because of the relative coefficients $2:1$ in the balanced chemical equation.

$$K_p = \frac{(p_{NO_2})^2}{p_{N_2O_4}} = \frac{(.33\ \text{atm})^2}{.67\ \text{atm}} = 0.17\ \text{atm}$$

(b) Let α = degree of dissociation at equilibrium at 0.1 atm pressure.

	N_2O_4	$\rightleftharpoons$	$2\ NO_2$	Total gas
Moles, assuming no dissociation:	1		zero	1
Change by reaction:	$-\alpha$		2α	
Moles at equilibrium:	$1 - \alpha$		2α	$1 + \alpha$
Mole fraction:	$\dfrac{1-\alpha}{1+\alpha}$		$\dfrac{2\alpha}{1+\alpha}$	1
p = mole fraction $\times P$:	$\dfrac{1-\alpha}{1+\alpha} \times 0.1$		$\dfrac{2\alpha}{1+\alpha} \times 0.1$	0.1

$$K_p = \frac{(p_{NO_2})^2}{p_{N_2O_4}} = \frac{\left(\dfrac{2\alpha}{1+\alpha} \times 0.1\right)^2 \text{atm}}{\left(\dfrac{1-\alpha}{1+\alpha} \times 0.1\right)} = \frac{4\alpha^2 \times 0.1\ \text{atm}}{(1+\alpha)(1-\alpha)} = \frac{0.4\alpha^2\ \text{atm}}{1-\alpha^2} = 0.17\ \text{atm}$$

Then $0.4\alpha^2 = 0.17(1 - \alpha^2)$. Solving, $\alpha = 0.55 = 55\%$ dissociated at $27°C$ and 0.1 atm.

Note that a larger fraction of the N_2O_4 is dissociated at 0.1 atm than at 1 atm. This is in agreement with the qualitative considerations of Le Chatelier's Principle. Decreasing the pressure should favor the side with the greater volume (NO_2).

(c) If the sample were all N_2O_4 it would contain 0.75 mole $\left(\dfrac{69\ g}{92\ g/mole}\right)$. Let α be the fractional dissociation. The tabular analysis follows:

	N_2O_4	$\rightleftharpoons$	$2\ NO_2$	Total gas
Moles, assuming no dissociation:	0.75		0	0.75
Change by reaction:	-0.75α		$+2 \times 0.75\alpha$	
Moles at equilibrium:	$0.75(1 - \alpha)$		1.50α	$0.75(1 + \alpha)$
Mole fraction:	$\dfrac{1-\alpha}{1+\alpha}$		$\dfrac{2\alpha}{1+\alpha}$	1
p:	$\dfrac{1-\alpha}{1+\alpha}P$		$\dfrac{2\alpha}{1+\alpha}P$	P

The total pressure cannot be evaluated immediately but it is related by the perfect gas law to the total number of moles, n.

$$P = \frac{nRT}{V} = \frac{0.75(1+\alpha)\ \text{mole} \times 0.082\ l\ \text{atm mole}^{-1}\ \text{deg}^{-1} \times 300\ \text{deg}}{20\ l}$$

$$= 0.92(1+\alpha)\ \text{atm}$$

$$K_p = \frac{(p_{NO_2})^2}{p_{N_2O_4}} = \frac{\left(\dfrac{2\alpha}{1+\alpha}\right)^2 P^2}{\dfrac{1-\alpha}{1+\alpha}P} = \frac{4\alpha^2 P}{(1+\alpha)(1-\alpha)}$$

Substituting for P from the perfect gas equation,

$$K_p = \frac{4\alpha^2 \times 0.92(1+\alpha)}{(1+\alpha)(1-\alpha)} = \frac{4 \times 0.92\alpha^2}{1-\alpha} = 0.17, \qquad 3.68\alpha^2 + 0.17\alpha - 0.17 = 0,$$

and $\quad \alpha = \dfrac{-0.17 \pm \sqrt{0.17^2 + 4(0.17)(3.68)}}{2(3.68)} = \dfrac{-0.17 \pm 1.59}{7.36} = -0.24 \text{ or } +0.19$

The negative root must obviously be discarded. The extent of dissociation is 19%.

16.8. At 817°C, K_p for the reaction between pure CO_2 and excess hot graphite is 10 atm.
(a) What is the analysis of the gases at equilibrium at 817°C and a total pressure of 4 atmospheres? What is the partial pressure of CO_2 at equilibrium?
(b) At what total pressure will the gases analyze 6% CO_2 by volume?

(a) Consider 1 mole of CO_2 at the start. Let α = the fraction of CO_2 that converts.

	$CO_2(gas)$ + C(solid) $\rightleftharpoons$	2 CO(gas)	Total gas
Moles at start:	1	0	1
Change by reaction:	$-\alpha$	2α	
Moles at equilibrium:	$1-\alpha$	2α	$1+\alpha$
p:	$\dfrac{1-\alpha}{1+\alpha} \times 4$	$\dfrac{2\alpha}{1+\alpha} \times 4$	4

Since the effective concentration of all solids is a constant, graphite is not included in the equilibrium constant, so long as it is present in excess.

$$K_p = \frac{(p_{CO})^2}{p_{CO_2}} \quad \text{or} \quad 10 = \frac{\left(\dfrac{2\alpha}{1+\alpha} \times 4\right)^2}{\left(\dfrac{1-\alpha}{1+\alpha} \times 4\right)} = \frac{16\alpha^2}{1-\alpha^2}$$

Solving, $16\alpha^2 = 10 - 10\alpha^2$, $26\alpha^2 = 10$, and $\alpha = \sqrt{10/26} = 0.62$.

Hence, at equilibrium: Moles $CO_2 = 1 - \alpha = 1 - 0.62 = 0.38$

$$\text{Moles CO} = 2\alpha = 2(0.62) = \underline{1.24}$$

$$\text{Total moles} = 1.62$$

Mole fraction CO_2 at equilibrium $= \dfrac{1-\alpha}{1+\alpha} = \dfrac{0.38}{1.62} = 23\% \ CO_2 \text{ (by volume)}$

Mole fraction CO at equilibrium $= \dfrac{2\alpha}{1+\alpha} = 77\% \text{ (by volume)}$

$p_{CO_2} = 0.23 \times 4 \text{ atm} = 0.92 \text{ atm}$

(b) % $CO_2 = 6\%$ by volume; hence, % CO $= 100\% - 6\% = 94\%$ by volume; then $p_{CO_2} = 0.06P$, and $p_{CO} = 0.94P$.

$$K_p = 10 \text{ atm} = \frac{(0.94P)^2}{0.06P} \qquad \text{from which} \qquad P = 0.68 \text{ atm}$$

We could have predicted, without solving the problem numerically, that the pressure for 6% CO_2 would be less than 4 atmospheres, since decreasing the pressure favors an increase in the number of gas molecules. The equilibrium gases in (b) have more CO (side with the larger number of gas molecules), 94%, than the equilibrium gases in (a), 77% CO.

16.9. Would 1.00% CO_2 in the air be sufficient to prevent any loss in weight while drying Ag_2CO_3 at 110°C? How low would the partial pressure of CO_2 have to be to promote this reaction at 110°C?

$$Ag_2CO_3(solid) \;\rightleftharpoons\; Ag_2O(solid) + CO_2(gas) \qquad K_p = 0.0095 \text{ atm at } 110°C$$

Since the concentrations of the solids are constant, they do not appear in the equilibrium equation. Hence K_p = partial pressure of CO_2 = .0095 atm. This is the partial pressure of CO_2 in equilibrium with the two solids, Ag_2CO_3 and Ag_2O. If some Ag_2CO_3 were placed in a closed vessel, a small amount would decompose into CO_2 and Ag_2O until the partial pressure of CO_2 reached .0095 atm. The decomposition would then stop, as the system would have reached equilibrium. If a partial pressure of CO_2 (e.g. 1%, or .01 atm) greater than this value is applied to the system, the above reaction will be reversed, all of the Ag_2O will be converted to Ag_2CO_3, and there will no longer be an equilibrium of the three components. In other words, there would be no loss in weight on heating Ag_2CO_3.

If on the other hand the partial pressure of CO_2 near the system is less than .0095 atm, some Ag_2CO_3 will dissociate according to the above equation. The amount of decomposition will depend on how much the surrounding air is stirred. If the air remains quiet, a small amount of decomposition will suffice to build up the partial pressure to the equilibrium value. If the air is changed rapidly, the CO_2 is removed from contact with the Ag_2CO_3 and more Ag_2CO_3 will have to decompose to again make up the equilibrium partial pressure of CO_2. Regardless of the amount of stirring, Ag_2CO_3 would eventually decompose completely in an open vessel at a temperature at which the equilibrium partial pressure of CO_2 is 1 atm. At this temperature the CO_2 from the decomposition would push the air away from the sample, and the falling CO_2 pressure near the sample resulting from the outward movement of gas would have to be compensated by progressive decomposition.

16.10. Under what conditions will $CuSO_4 \cdot 5\,H_2O$ be efflorescent at 25°C? How good a drying agent is $CuSO_4 \cdot 3\,H_2O$ at the same temperature? Given, K_p at 25°C is 1.086×10^{-4} atm^2 for the following reaction: $CuSO_4 \cdot 5\,H_2O$ (solid) $\rightleftharpoons CuSO_4 \cdot 3\,H_2O$ (solid) $+ 2\,H_2O$ (gas). The vapor pressure of water at 25°C is 23.8 mm.

An efflorescent salt is one that loses water to the atmosphere. This will occur if the water vapor pressure in equilibrium with the salt is greater than the water vapor pressure in the atmosphere. The mechanism by which $CuSO_4 \cdot 5\,H_2O$ could be efflorescent is that the salt would lose 2 molecules of H_2O and simultaneously form 1 gfw of $CuSO_4 \cdot 3\,H_2O$ for each gfw of the original salt that dissociates. Then the above equilibrium equation would apply, since all three components would be present.

Since $CuSO_4 \cdot 5\,H_2O$ and $CuSO_4 \cdot 3\,H_2O$ are both solids, $K_p = (p_{H_2O})^2 = 1.086 \times 10^{-4}$ atm^2, where p_{H_2O} is the partial pressure of water vapor in equilibrium with the two solids.

Solving, $p_{H_2O} = 1.042 \times 10^{-2}$ atm $= 1.042 \times 10^{-2}$ atm $\times$ 760 mm/atm = 7.92 mm.

Since p_{H_2O} is less than the vapor pressure of water at the same temperature, $CuSO_4 \cdot 5\,H_2O$ will not always effloresce. It will effloresce only on a dry day, when the partial pressure of moisture in the air is less than 7.92 mm. This will occur when the relative humidity is less than $\frac{7.92 \text{ mm}}{23.8 \text{ mm}} = 0.333 = 33.3\%$.

The mechanism by which $CuSO_4 \cdot 3\,H_2O$ acts as a drying agent is that it can react with 2 molecules of H_2O to form $CuSO_4 \cdot 5\,H_2O$. The same equilibrium would be set up as above, and the vapor pressure of water would be fixed at 7.92 mm. In other words, $CuSO_4 \cdot 3\,H_2O$ can reduce the moisture content of any confined volume of gas to 7.92 mm. It cannot reduce the moisture content below this value.

If we wanted to know the conditions under which $CuSO_4 \cdot 3\,H_2O$ would be efflorescent, we would have to know the equilibrium constant for another reaction, which shows the product of dehydration of $CuSO_4 \cdot 3\,H_2O$. This reaction is

$$CuSO_4 \cdot 3\,H_2O\ (s) \;\rightleftharpoons\; CuSO_4 \cdot H_2O\ (s) + 2\,H_2O\ (g)$$

Supplementary Problems

16.11. $N_2 + O_2 \rightleftharpoons 2\,NO$; $\Delta H = 43$ kcal. Explain the effect upon the reaction equilibrium of (a) increased temperature, (b) decreased pressure, (c) higher concentration of O_2, (d) lower concentration of N_2, (e) higher concentration of NO, (f) presence of a catalyst.

   Ans. Favors (a) forward reaction, (b) neither reaction, (c) forward reaction, (d) backward reaction, (e) backward reaction, (f) neither reaction

16.12. Predict the effect upon the following reaction equilibria of: (a) increased temperature, (b) increased pressure. Use thermal data from Table 15-1, Page 141.

   1. $CO(g) + H_2O(g) \rightleftharpoons CO_2(g) + H_2(g)$
   2. $2\,SO_2(g) + O_2(g) \rightleftharpoons 2\,SO_3(g)$
   3. $N_2O_4(g) \rightleftharpoons 2\,NO_2(g)$
   4. $H_2O(g) \rightleftharpoons H_2(g) + \frac{1}{2}\,O_2(g)$
   5. $2\,O_3(g) \rightleftharpoons 3\,O_2(g)$
   6. $CO(g) + 2\,H_2(g) \rightleftharpoons CH_3OH(g)$
   7. $CaCO_3(s) \rightleftharpoons CaO(s) + CO_2(g)$
   8. $C(s) + H_2O(g) \rightleftharpoons H_2(g) + CO(g)$
   9. $4\,HCl(g) + O_2(g) \rightleftharpoons 2\,H_2O(g) + 2\,Cl_2(g)$
   10. $C(s,\ diamond) \rightleftharpoons C(s,\ graphite)$ (This equilibrium can exist only under very special conditions.) Specific gravity of diamond $= 3.5$, of graphite $= 2.3$.

   Ans. F = favors forward reaction, B = favors backward reaction.

1. (a) B, (b) neither	4. (a) F, (b) B	7. (a) F, (b) B	10. (a) B, (b) B
2. (a) B, (b) F	5. (a) B, (b) B	8. (a) F, (b) B	
3. (a) F, (b) B	6. (a) B, (b) F	9. (a) B, (b) F	

16.13. When α-d-glucose is dissolved in water it undergoes a partial conversion to β-d-glucose, a sugar of the same molecular weight but of slightly different physical properties. This conversion, called mutarotation, stops when 63.6% of the glucose is in the β-form. Assuming that equilibrium has been attained, calculate K for the reaction: α-d-glucose $\rightleftharpoons$ β-d-glucose.   *Ans.* 1.75

16.14. The equilibrium between fumaric acid, $H_2C_4H_2O_4$, and malic acid, $H_2C_4H_4O_5$, can be made to occur in water solution in the presence of the enzyme fumarase. K for the reaction

$$H_2C_4H_2O_4 + H_2O \rightleftharpoons H_2C_4H_4O_5$$

in aqueous solution is 3.5 at a specified acidity. How much pure fumaric acid should be weighed out to make one liter of a solution at the standard acidity which is to contain 0.20 mole malic acid after enzymatic equilibration?   *Ans.* 0.26 mole

16.15. The equilibrium constant for the reaction $H_3BO_3 + glycerin \rightleftharpoons (H_3BO_3\text{-}glycerin)$ is 0.9 l/mole. How much glycerin should be added per liter of 0.10 molar H_3BO_3 solution so that 60% of the H_3BO_3 is converted to the boric acid-glycerin complex?   *Ans.* 1.7 moles

16.16. The equilibrium

$$p\text{-xyloquinone} + \text{methylene white} \rightleftharpoons p\text{-xylohydroquinone} + \text{methylene blue}$$

may be studied conveniently by observing the difference in color between methylene blue and methylene white. One millimole of methylene blue was added to a liter of solution that was 0.24 M in p-xylohydroquinone and 0.0120 M in p-xyloquinone. It was then found that 4.0% of the added methylene blue was reduced to methylene white. What is the equilibrium constant for the above reaction? The equation is balanced with one molecule of each of the four substances.

   Ans. 4.8×10^2

16.17. A saturated solution of iodine in water contains 0.33 g I_2 per liter. More than this can dissolve in a KI solution because of the following equilibrium:

$$I_2 + I^- \rightleftharpoons I_3^-$$

A 0.10 F KI solution (0.10 M I^-) actually dissolves 12.5 g of iodine per liter, most of which is converted to I_3^-. Assuming that the concentration of I_2 in all saturated solutions is the same, calculate the equilibrium constant for the above reaction. What is the effect of adding water to a clear saturated solution of I_2 in the KI solution?

   Ans. 7.1×10^2 l/mole; backward reaction is favored.

16.18. The following equilibria of amino acids have been studied in dilute aqueous solution.

    (1) Glutamic acid + Pyruvic acid ⇌ α-Ketoglutaric acid + Alanine

    (2) Glutamic acid + Oxalacetic acid ⇌ α-Ketoglutaric acid + Aspartic acid

    (3) Alanine + Oxalacetic acid ⇌ Pyruvic acid + Aspartic acid

One molecule of each substance appears in each of these balanced equations. K for reaction (1) is 2.07, and for reaction (2) is 5.44. What is K for reaction (3)?

Hint. The equilibrium of reaction (3) in the presence of a small amount of glutamic acid is the same as in the absence of glutamic acid.   *Ans.* 2.63

16.19. $H_2(g) + I_2(g) \rightleftharpoons 2\,HI(g)$. When 46 g of I_2 and 1.0 g of H_2 are heated to equilibrium at 450°C, the equilibrium mixture contains 1.9 g I_2. (a) How many moles of each gas are present in the equilibrium mixture? (b) Compute the equilibrium constant.

Ans. 0.0075 mole I_2,   0.32 mole H_2,   0.35 mole HI, $K = 50$

16.20. Exactly 1 mole each of H_2 and I_2 are heated in a 30 liter evacuated chamber to 448°C, at which temperature $K = 50$ for this reaction.

    (a) How many moles of I_2 remain unreacted when equilibrium is established?

    (b) What is the total pressure in the chamber?

    (c) What is the partial pressure of I_2 and of HI in the equilibrium mixture?

    (d) Now if one additional mole of H_2 is introduced into this equilibrium system, how many moles of the original iodine will remain unreacted?

Ans. (a) 0.22 mole, (b) 3.9 atm, (c) $p_{H_2} = p_{I_2} = 0.43$ atm, $p_{HI} = 3.0$ atm, (d) 0.065 mole

16.21. $PCl_5(g) \rightleftharpoons PCl_3(g) + Cl_2(g)$. Calculate the number of moles of Cl_2 produced at equilibrium when one mole of PCl_5 is heated at 250°C in a vessel having a capacity of 10 liters. At 250°C, $K = 0.041$ mole/l for this dissociation.   *Ans.* 0.47 mole

16.22. Pure PCl_5 is introduced into an evacuated chamber and comes to equilibrium at 250°C and 2.00 atm. The equilibrium gas contains 40.7% chlorine by volume.

    (a) 1. What are the partial pressures of PCl_3 and PCl_5 at equilibrium?

        2. Calculate K_p at 250°C.

    (b) If the gas mixture is expanded to 0.200 atm at 250°C, calculate:

        1. The percent of PCl_5 that would be dissociated at equilibrium.

        2. The percent by volume of Cl_2 in the equilibrium gas mixture.

        3. The partial pressure of Cl_2 in the equilibrium mixture.

Ans. (a) 1. $p_{Cl_2} = p_{PCl_3} = 0.814$ atm, $p_{PCl_5} = 0.372$ atm   2. 1.78 atm

      (b) 1. 94.8%   2. % Cl_2 = % PCl_3 = 48.7%, % PCl_5 = 2.6%

         3. $p_{Cl_2} = p_{PCl_3} = 0.0974$ atm, $p_{PCl_5} = 0.0052$ atm

16.23. One molecule of sucrose phosphate is synthesized biologically by the condensation of one molecule each of fructose and glucose phosphate. K for the reaction is 0.050 l/mole.

$$\text{Fructose + Glucose phosphate} \rightleftharpoons \text{Sucrose phosphate}$$

To what volume should a solution containing 0.050 mole each of fructose and glucose phosphate be diluted so that there will be a 3.0% conversion to sucrose phosphate at enzymatic equilibrium?

Ans. 78 ml

16.24. A theoretically computed equilibrium constant for the polymerization of formaldehyde, HCHO, to glucose, $C_6H_{12}O_6$, in aqueous solution is 6×10^{22} (mole/l)$^{-5}$.

$$6\,HCHO \rightleftharpoons C_6H_{12}O_6$$

If a 1.0 M solution of glucose were to reach dissociation equilibrium with respect to the above equation, what would be the concentration of formaldehyde in the solution?

Ans. 1.6×10^{-4} molar

16.25. $N_2O_4(g) \rightleftharpoons 2\,NO_2(g)$. At 55°C, K_p for this reaction is 0.66 atm. Compute the percent dissociation of N_2O_4 at 55°C and a total pressure of 380 mm. What are the partial pressures of N_2O_4 and NO_2 at equilibrium?   *Ans.* 50%, $p_{N_2O_4} = 0.17$ atm, $p_{NO_2} = 0.33$ atm

16.26. $2\,NOBr(g) \;\rightleftharpoons\; 2\,NO(g) + Br_2(g)$. If nitrosyl bromide (NOBr) is 34% dissociated at 25°C and a total pressure of 0.25 atmosphere, calculate K_p for the dissociation at this temperature.

Ans. 1.0×10^{-2} atm

16.27. $CO(g) + H_2O(g) \;\rightleftharpoons\; CO_2(g) + H_2(g)$. The equilibrium constant for this reaction at 986°C is 0.63. A mixture of 1 mole of water vapor and 3 moles of CO is allowed to come to equilibrium at a total pressure of 2 atmospheres.

(a) How many moles of H_2 are present at equilibrium?

(b) What is the partial pressure of each gas in the equilibrium mixture?

Ans. (a) 0.68 mole, (b) $p_{CO} = 1.16$ atm, $p_{H_2O} = 0.16$ atm, $p_{CO_2} = p_{H_2} = 0.34$ atm

16.28. $\frac{1}{2}SnO_2(solid) + H_2(gas) \;\rightleftharpoons\; H_2O(steam) + \frac{1}{2}Sn(molten)$. Find the appropriate equilibrium constants (K_p) for this reaction from the following analytical results:

1. At 650°C the equilibrium steam-hydrogen mixture was 38% by volume of H_2.
2. At 800°C the equilibrium steam-hydrogen mixture was 22% by volume of H_2.

Would you recommend higher or lower temperatures for more efficient reduction of tin?

Ans. K_p at 650°C = 1.6, at 800°C = 3.5; higher

16.29. In the preparation of quicklime from limestone, the reaction is

$$CaCO_3(s) \;\rightleftharpoons\; CaO(s) + CO_2(g)$$

Between 850°C and 950°C the equilibrium constant, K_p, is given by the empirical equation $\log K_p$ (in atm) $= 7.282 - 8500/T$, where T is the absolute temperature. If the reaction is carried out in quiet air, what temperature is required for the complete decomposition of the limestone?

Ans. 894°C

16.30. The moisture content of a gas is often expressed in terms of the *dew point*. The dew point is the temperature to which the gas must be cooled before the gas becomes saturated with water vapor. At this temperature, water or ice (depending on the temperature) will be deposited on a solid surface.

The efficiency of $CaCl_2$ as a drying agent was measured by a dew point experiment. Air at 0°C was allowed to pass slowly over large trays containing $CaCl_2$. The air was then passed through a glass vessel through which a copper rod was sealed. The rod was cooled by immersing the emergent part of it in a dry ice bath. The temperature of the rod inside the glass vessel was measured by a thermocouple. As the rod was cooled slowly the temperature at which the first crystals of frost were deposited was observed to be −43°C. The vapor pressure of ice at this temperature is 0.07 mm. Assuming that the $CaCl_2$ owes its desiccating properties to the formation of $CaCl_2 \cdot 2\,H_2O$, calculate K_p at 0°C for the reaction

$$CaCl_2 \cdot 2\,H_2O(s) \;\rightleftharpoons\; CaCl_2(s) + 2\,H_2O(g)$$

Ans. 8×10^{-9} atm²

16.31. Equilibrium constants, K_p, are given for the following reactions at 0°C.

$SrCl_2 \cdot 6\,H_2O\,(s)$	$\rightleftharpoons \quad SrCl_2 \cdot 2\,H_2O\,(s) + 4\,H_2O\,(g)$	6.89×10^{-12} atm⁴
$Na_2HPO_4 \cdot 12\,H_2O\,(s)$	$\rightleftharpoons \quad Na_2HPO_4 \cdot 7\,H_2O\,(s) + 5\,H_2O\,(g)$	5.25×10^{-13} atm⁵
$Na_2SO_4 \cdot 10\,H_2O\,(s)$	$\rightleftharpoons \quad Na_2SO_4\,(s) + 10\,H_2O\,(g)$	4.08×10^{-25} atm¹⁰

The vapor pressure of water at 0°C is 4.58 mm.

(a) Calculate the pressure of water vapor in equilibrium at 0°C with $SrCl_2 \cdot 6\,H_2O$, $Na_2HPO_4 \cdot 12\,H_2O$, and $Na_2SO_4 \cdot 10\,H_2O$. Express the pressures in mm Hg.

(b) Which is the most effective drying agent at 0°C: $SrCl_2 \cdot 2\,H_2O$, $Na_2HPO_4 \cdot 7\,H_2O$, or Na_2SO_4?

(c) At what relative humidities will $Na_2SO_4 \cdot 10\,H_2O$ be efflorescent when exposed to the air at 0°C?

(d) At what relative humidities will Na_2SO_4 be deliquescent (i.e., absorb moisture) when exposed to the air at 0°C?

Ans. (a) 1.23 mm, 2.66 mm, 2.77 mm   (c) less than 60% relative humidity
      (b) $SrCl_2 \cdot 2\,H_2O$ (d) above 60%

Chapter 17

Ionic Equilibrium

REACTIONS THAT ARE NEARLY COMPLETE

In general, when two highly ionized substances are mixed in solution and the possible products are also highly ionized soluble substances, then practically no effective chemical change occurs. An example would be the mixing of solutions of NaCl and KNO_3.

When certain other electrolytes are mixed in solution, an extensive chemical change occurs in the following three cases.

(1) When one of the possible products is slightly soluble and precipitates as a solid. For example, the reaction between solutions containing equivalent amounts of $AgNO_3$ and NaCl goes almost to completion because AgCl is very slightly soluble and precipitates, thus removing from solution nearly all the Ag^+ and Cl^- ions.

(2) When one of the possible products is volatile and forms a gaseous product. The liberation of the gaseous product removes from solution the ions forming it. For example, the reaction of NaCl with an equivalent amount of concentrated H_2SO_4 approaches completion at elevated temperatures because HCl is nearly insoluble and escapes as a gas, thus removing from solution nearly all the H^+ and Cl^- ions.

(3) When one of the possible products, though soluble, is only slightly ionized. For example, in the neutralization reaction between solutions containing equivalent amounts of NaOH and HCl, the H_2O formed is only slightly ionized and therefore removes from solution nearly all the H^+ and OH^- ions.

Calculations involving reactions of type (3) are the subject of this chapter, and calculations for reactions of type (1) will be discussed in the next chapter.

IONIC DISSOCIATION IN AQUEOUS SOLUTIONS

In aqueous solutions, as in gaseous systems, many compounds on dilution tend to dissociate into simpler particles. In water these particles are often oppositely charged ions, stabilized and partly insulated from each other by a firmly attached layer of water molecules. This tendency toward hydration, or solvation in general, seems to be universal, since not only ions but neutral molecules and even the inert gases show evidence of hydration.

The powerful mutual attraction of oppositely charged ions, even when hydrated, causes some pairing off to form the original substance in many cases, so that a *dynamic* equilibrium exists between the dissociating molecule and its resulting ions. Therefore the principles of equilibrium are valid and are particularly applicable to systems of weak acids and bases, such as acetic acid and ammonia.

Important limitation. These interpretations concerning equilibrium do not apply as simply to systems of strong electrolytes which are very soluble, such as aqueous solutions of HCl, H_2SO_4, NaOH, and $CaCl_2$. Strong electrolytes when dissolved in water exist as separate (hydrated) ions, and the high concentrations therefore produce enormous electrical fields which abnormally influence the *effective concentration* or *activity* of the ions. The principles of equilibrium will apply simply as long as the concentrations of ions remain very low. For solutions containing larger ionic concentrations, the same laws may be applied if proper correction is made for the electrical interactions of the ions. Such corrections will not be made in this book. Instead, examples will be chosen for which the numerical solutions should be correct to within 10% even without activity corrections. In general, 10% is about the limit of accuracy expected in this and in the following chapter.

THE HYDRONIUM ION, H_3O^+

Once the universal extent of dynamic hydration or solvation of dissolved materials is emphasized, it is unnecessary to replace the symbol H^+ in all aqueous reactions by H_3O^+ $(H^+ \cdot H_2O)$. The symbol H_3O^+ may be used to advantage when it is desired to stress the similarity between water and ammonia, or to stress the contrast between gaseous and aqueous hydrohalogenides. The general use of H_3O^+, however, is too apt to obscure the important fact that *all ions in water are extensively hydrated.*

THE IONIZATION CONSTANT

The ionization constant is the equilibrium constant for an ionic dissociation, automatically defined by the equilibrium constant equation in terms of the corresponding molar concentrations. Thus the ionization constant is equal to the product of the ionic concentrations divided by the concentration of the undissociated substance. All concentrations are conventionally expressed in *moles per liter,* but the concentration units are normally not explicitly written.

Ionization constants change significantly with temperature. Unless otherwise indicated, the temperature is understood to be 25°C. Also, water will be understood as the solvent in the absence of specific statements to the contrary.

The ionization constant of a weak acid is usually designated K_a. The equilibrium for acetic acid, $HC_2H_3O_2$, may be written as follows:

(1) $HC_2H_3O_2 \rightleftharpoons H^+ + C_2H_3O_2^-$ $K_a = \dfrac{[H^+] \times [C_2H_3O_2^-]}{[HC_2H_3O_2]}$

Or, as often given to stress the Brönsted-Lowry interpretations,

(2) $HC_2H_3O_2 + H_2O \rightleftharpoons H_3O^+ + C_2H_3O_2^-$ $K_a = \dfrac{[H_3O^+] \times [C_2H_3O_2^-]}{[HC_2H_3O_2]}$

(The $[H_2O]$ in dilute solutions may be considered constant, and the usual equilibrium constant includes only the variable quantities.)

These two constants K_a are identical. It must be understood that the acetate ion is also hydrated $(C_2H_3O_2^- \cdot x\,H_2O)$, and that the hydronium ion (H_3O^+) is further hydrated $(H_3O^+ \cdot y\,H_2O)$.

Calculations involving the above equilibrium constant are of interest in three cases.

(1) Solution of acetic acid in water.

When one acetic acid molecule ionizes, one H^+ and one $C_2H_3O_2^-$ are formed. Since the ionization of the acid is the only source of the two ions in solution (the number of H^+ furnished by the water is negligible), then $[H^+] = [C_2H_3O_2^-]$.

(2) Solution of acetic acid plus acetate salt in water.

The acetate salts are completely ionized. If sodium acetate is added to a solution of acetic acid in water, the number of acetate ions resulting from the ionization of the acetic acid is negligible compared with the number of acetate ions added directly as a salt. Then, $[C_2H_3O_2^-]$ = concentration of added salt.

(3) Solution of acetic acid plus stronger acid in water.

If a stronger acid is added to a solution of acetic acid in water, very often the number of H^+ resulting from the ionization of the acetic acid is negligible compared with the number resulting from the stronger acid. Then $[H^+]$ is obtained from the properties of the stronger acid alone.

Cases (2) and (3) are called the *common ion effect*, because the added ion in each case is the same as one of the products of the ionization of the acetic acid. In either case, the effect of the added common ion is to lower the extent of ionization of the acetic acid. This is in accord with Le Chatelier's Principle.

This same type of equilibrium constant applies to the ionization of all weak acids. It does not apply to dilute solutions of such strong acids as HCl, HI, HNO_3, H_2SO_4, and $HClO_4$, which are practically 100% ionized. It applies to the ionization of weak bases like NH_3, in which cases the symbol K_b is used, but not to the strong bases such as NaOH, KOH, and $Ba(OH)_2$, whose ionization is essentially complete. In the case of NH_3, the equilibrium is between the neutral molecule and the products of ionization, NH_4^+ and OH^-.

$$NH_3 + H_2O \; \rightleftharpoons \; NH_4^+ + OH^- \qquad K_b = \frac{[NH_4^+] \times [OH^-]}{[NH_3]}$$

Water is not included in the equilibrium constant equation because $[H_2O]$ is practically constant in dilute aqueous solutions.

An alternative representation is sometimes used:

$$NH_4OH \; \rightleftharpoons \; NH_4 + OH^- \qquad K_b = \frac{[NH_4^+] \times [OH^-]}{[NH_4OH]}$$

The two constants, K_b, are identical. The two formulations differ only in the choice of the convention used to represent the un-ionized ammonia, NH_3 or NH_4OH.

IONIZATION CONSTANT OF WATER

The equilibrium between molecular water and its ions (again disregarding hydration) may be represented by

$$H_2O \; \rightleftharpoons \; H^+ + OH^- \qquad K_w = [H^+] \times [OH^-]$$

Here again, $[H_2O]$ is considered constant and is not written as one of the variable terms in the equilibrium constant equation. Because of the great importance of H_2O and its ions, the numerical value of this equilibrium constant should be remembered. At 25°C the value of the water constant, $K_w = [H^+] \times [OH^-]$, is 1.00×10^{-14}.

NEUTRAL, ACID, AND BASIC SOLUTIONS

In pure water at 25°C,

$$[H^+] = [OH^-] = \sqrt{1.00 \times 10^{-14}} = 1.00 \times 10^{-7} \text{ mole/liter}$$

Hence a neutral solution may be defined as one in which $[H^+] = [OH^-] = 1.00 \times 10^{-7}$ mole/liter. (At 0°C in pure water $[H^+] = [OH^-] = 0.34 \times 10^{-7}$ mole/liter.)

An acid solution is one in which the $[H^+]$ is greater than 10^{-7} mole/liter (as 10^{-2}, 10^{-1}, etc.), or one in which the $[OH^-]$ is less than 10^{-7} mole/liter (as 10^{-10}, 10^{-11}, etc.).

A basic solution is one in which the $[H^+]$ is less than 10^{-7} mole/liter, or one in which the $[OH^-]$ is greater than 10^{-7} mole/liter.

pH AND pOH

The acidity or alkalinity of a solution is expressed often by the common logarithm of the reciprocal of the hydrogen ion concentration in moles per liter. This value is called the pH of the solution. Thus by definition

$$pH = \log \frac{1}{[H^+]} = -\log [H^+] \qquad or \qquad [H^+] = 10^{-pH}$$

For example, if $[H^+] = 10^{-5}$, then $pH = \log \frac{1}{[H^+]} = \log \frac{1}{10^{-5}} = \log 10^5 = 5$. Hence if the $[H^+]$ of a solution is expressed as a simple power of 10, the pH value of the solution is equal to the negative of the exponent. The smaller the pH, the greater the acidity.

Similarly, $\qquad pOH = \log \frac{1}{[OH^-]} \qquad$ and $\qquad pH + pOH = 14.00$ (at 25°C)

For example, if the pH of a solution is 2.6, its $pOH = 14 - 2.6 = 11.4$. The smaller the pOH, the greater the alkalinity.

The following table illustrates the above relations.

$[H^+]$, moles/liter		$[OH^-]$, moles/l	pH	pOH	
1	$= 10^0$	10^{-14}	0	14	Strongly acidic
0.1	$= 10^{-1}$	10^{-13}	1	13	
0.001	$= 10^{-3}$	10^{-11}	3	11	
0.00001	$= 10^{-5}$	10^{-9}	5	9	Weakly acidic
0.0000001	$= 10^{-7}$	10^{-7}	7	7	Neutral
0.000000001	$= 10^{-9}$	10^{-5}	9	5	Weakly basic
0.00000000001	$= 10^{-11}$	10^{-3}	11	3	
0.0000000000001	$= 10^{-13}$	10^{-1}	13	1	
0.00000000000001	$= 10^{-14}$	1	14	0	Strongly basic

pK NOTATION

For convenience pK_a, defined as $\log \frac{1}{K_a}$ or $-\log K_a$, is often used to express the strength of an acid or base. Thus an acid whose ionization constant is 10^{-4} has a pK_a of 4. Similarly, $pK_b = -\log K_b$; for any equilibrium constant, $pK = -\log K$.

HYDROLYSIS

Just as neutral molecules can act as acids or bases, so can ions. In the discussion of the ionization of acetic acid, the reaction

$$HC_2H_3O_2 \rightleftharpoons H^+ + C_2H_3O_2^-$$

was written as a reversible reaction. If acetic acid is dissolved in water, a small proportion of it will ionize according to the above equation. Similarly, if sodium acetate (a strong electrolyte) is dissolved in water, some of the acetate ions will react with the hydrogen ions in the water to form undissociated acetic acid. This is the reverse reaction of the equation written above. When hydrogen ions of the water are used up by such a reaction, more water will ionize, so that the product of the concentrations of H^+ and OH^- will again equal K_w. The situation can be represented by the two equations:

$$(1) \qquad C_2H_3O_2^- \;+\; H^+ \;\rightleftharpoons\; HC_2H_3O_2$$

$$(2) \qquad\qquad\quad H_2O \;\rightleftharpoons\; H^+ \;+\; OH^-$$

The net effect is given by the sum of these two equations:

$$(3) \qquad C_2H_3O_2^- \;+\; H_2O \;\rightleftharpoons\; HC_2H_3O_2 \;+\; OH^-$$

Equation (3) shows that a solution of sodium acetate in water is basic, because an excess of OH^- is produced. The acetate ion, $C_2H_3O_2^-$, is said to be the conjugate base of acetic acid, because the ion can be converted into the acid simply by attaching an H^+. (Similarly, $HC_2H_3O_2$ is said to be the conjugate acid of $C_2H_3O_2^-$.) In this case, the H^+ came from the water.

To calculate the extent of reaction (3), we can start with the known equilibrium conditions for reactions (1) and (2):

$$(1) \qquad K_{a\,(HC_2H_3O_2)} \;=\; \frac{[H^+] \times [C_2H_3O_2^-]}{[HC_2H_3O_2]}$$

$$(2) \qquad K_w \;=\; [H^+] \times [OH^-]$$

$[H^+]$ can be eliminated by dividing the second equation by the first:

$$\frac{K_w}{K_{a\,(HC_2H_3O_2)}} \;=\; \frac{[OH^-] \times [HC_2H_3O_2]}{[C_2H_3O_2^-]}$$

The right-hand side of this equation has exactly the correct form for the equilibrium constant for reaction (3) above. (The concentration of H_2O is conventionally not included in the equilibrium constant because it remains practically constant.) Reaction (3) is called a hydrolysis reaction, and K_w/K_a is called K_h, the hydrolysis constant.

A similar type of hydrolysis reaction is undergone by any negative ion that can act as a base in water solution to form the conjugate acid. Some examples are CN^-, $H_2BO_3^-$, SCN^-, NO_2^-. Since hydrolysis is a reverse of acid dissociation, the tendency toward hydrolysis runs counter to the tendency of the conjugate acid toward ionization. The weaker the acid, the greater the difficulty of removing a proton, and the easier for its anion, or conjugate base, to attach a proton from water (that is, hydrolyze). This relationship appears mathematically as the inverse proportionality between the ionization constant of the acid and the hydrolysis constant of the conjugate base. Acetic acid is a *moderately weak* acid, and the acetate ion hydrolyzes to a *slight* extent. HCN is a *very weak* acid and the cyanide ion, CN^-, hydrolyzes to a *great* extent. Chloride ion, on the other hand, does not undergo the hydrolysis reaction at all because its conjugate acid, HCl, is a *strong* acid and cannot form in water solution.

An analogous reaction will be shown by positive ions which can act as acids in water solution to form the conjugate base. Thus, a solution of ammonium chloride would be acidic because of the reaction

$$NH_4^+ \;\rightleftharpoons\; NH_3 \;+\; H^+$$

The equilibrium constant for this reaction can be shown to have the form

$$K_h \;=\; \frac{[H^+] \times [NH_3]}{[NH_4^+]} \;=\; \frac{K_w}{K_{b\,(NH_3)}}$$

where $K_{b\,(NH_3)}$ is the ionization constant for the weak base NH_3. K_h for positive ion hydrolysis is often called K_a, the dissociation constant of the ion as an acid. This is particularly true for acid cations such as the heavy metal cations that are much better known than their conjugate, positively charged bases. Thus the equilibrium constant for the reaction of Fe^{3+},

$$Fe^{3+} + H_2O \rightleftharpoons Fe(OH)^{2+} + H^+ \qquad K_a = \frac{[Fe(OH)^{2+}] \times [H^+]}{[Fe^{3+}]}$$

is usually called the acid ionization constant for Fe^{3+}, rather than the hydrolysis constant. The above reaction is sometimes written with the hydrated forms of ions to show that ferric ion, like the neutral acids, demonstrates its acidity by loss of a proton.

$$Fe(H_2O)_6^{3+} \rightleftharpoons Fe(H_2O)_5(OH)^{2+} + H^+$$

The two formulations of this equation are equivalent.

BUFFER SOLUTIONS

If the $[H^+]$ and pH of a solution are not appreciably affected by the addition of small amounts of acids and bases, the solution is said to be buffered. A solution will have these properties if it contains relatively large amounts of both a weak acid and a weak base. If a small amount of a strong acid is added to this solution, most of the added H^+ will combine with an equivalent amount of the weak base of the buffer to form the conjugate acid of that weak base; thus the $[H^+]$ and pH of the solution remain almost constant. If a small amount of a strong base is added to the buffer solution, most of the OH^- will combine with an equivalent amount of the weak acid of the buffer to form the conjugate base of that weak acid. In this way, the $[H^+]$ and pH of the buffer solution are not appreciably affected by the addition of small amounts of acid or base.

Any pair of weak acid and base can be used to form a buffer solution, as long as each can form its conjugate base or acid in water solution. A particularly simple case of the buffer solution is one in which the weak acid and weak base are conjugates of each other. Thus acetic acid may be chosen as the weak acid and acetate ion as the weak base. Since relatively large amounts of each are needed, it would not be satisfactory to use just a solution of acetic acid in water, in which the concentration of the acetate ion is relatively small. An acetic acid-acetate buffer can be made in one of the following three ways.

(1) Dissolve relatively large amounts of acetic acid and an acetate salt (sodium or potassium, for example) in water.

(2) Dissolve a relatively large amount of acetic acid in water. *Partially* neutralize the acid by adding some strong base, like sodium hydroxide. The amount of acetate formed will be equivalent to the amount of strong base added. The amount of acetic acid left in solution will be the starting amount minus the amount converted to acetate.

(3) Dissolve a relatively large amount of an acetate salt in water. *Partially* neutralize the acetate by adding some strong acid, like HCl. The amount of acetic acid formed will be equivalent to the amount of strong acid added. The amount of acetate ion left in solution will be the starting amount minus the amount converted to acetic acid.

The ratio of acetic acid to acetate ion in solution can be chosen so as to give a desired $[H^+]$ or pH for the buffer solution, according to a rearranged form of the ionization equilibrium equation for acetic acid:

$$[H^+] = K_{a\,(HC_2H_3O_2)} \times \frac{[HC_2H_3O_2]}{[C_2H_3O_2^-]}$$

Usually the ratio is kept within the limits of 10 and 0.1.

WEAK POLYPROTIC (POLYBASIC) ACIDS

If multiple ionization is possible, as in H_2S and H_2CO_3, each stage of ionization has its own equilibrium constant. Subscripts are usually used to distinguish the various constants.

Primary ionization:　　　$H_2S \rightleftharpoons H^+ + HS^-$　　$K_1 = \dfrac{[H^+] \times [HS^-]}{[H_2S]}$

Secondary ionization:　　$HS^- \rightleftharpoons H^+ + S^{2-}$　　$K_2 = \dfrac{[H^+] \times [S^{2-}]}{[HS^-]}$

The secondary ionization constant of polyprotic acids is always smaller than the primary (K_2 less than K_1), the tertiary, if there is one, is smaller than the secondary, and so on.

It should be clear that $[H^+]$ means the actual concentration of hydrogen ions *in solution*. In an aqueous mixture containing several acids, the different acids all contribute to the hydrogen ion concentration in solution, but there is only one value of $[H^+]$ in any given solution and this value must simultaneously satisfy the equilibrium conditions for all the different acids. Although it might seem very complex to solve a problem in which several equilibria are involved, simplifications can be made when all sources but one make only a negligible contribution (less than 10%, for the purposes of this book) to the total concentration of a particular ion.

In the case of polyprotic acids, K_1 is often so much greater than K_2 that only the K_1 equilibrium need be considered to compute $[H^+]$ in a solution of the acid. Examples where this assumption may and may not be made will be given in specific problems.

Another problem of interest is the calculation of the concentration of the divalent ion in a solution of a weak polyprotic acid when the $[H^+]$ is determined by a stronger acid present in the solution. This is an example of the common ion effect. Just as in the case of the monoprotic acid, so here the $[H^+]$ can be fixed by adding to the solution an amount of a strong acid such that the total $[H^+]$ is essentially that due to the strong acid. In such a case the concentration of the divalent ion can best be calculated by multiplying the K_1 equilibrium equation by the K_2 equation. Again illustrating with H_2S, we find

$$K_1 \times K_2 = \frac{[H^+] \times [HS^-]}{[H_2S]} \times \frac{[H^+] \times [S^{2-}]}{[HS^-]} = \frac{[H^+]^2 \times [S^{2-}]}{[H_2S]}$$

TITRATION AND INDICATORS

When a base is added in small increments to a solution of acid, the pH of the solution rises on each addition of base. When the pH is plotted against the amount of base added, the steepest rise occurs at the equivalence point, when the acid is exactly neutralized. This region of steepest rise is called the *end point*, and the whole process of base addition and determination of the end point is called *titration*. The graph showing the change of pH during the titration is called a *titration curve*. Several curves are shown in Fig. 17-1, for the titration of 50.0 cc of a 0.100 M acid, either HCl or $HC_4H_7O_3$ (β-hydroxybutyric acid), with 0.100 M base, either NaOH or NH_3. Note that although the end point always occurs on the addition of exactly 50.0 cc of base, the curves differ in their steepness and also in the pH value at the end point. The curve has its steepest rise for the titration of a strong acid (HCl) with a strong base (NaOH). The shallowest rise at the end point occurs for the titration of a weak acid ($HC_4H_7O_3$) with a weak base (NH_3).

The points along a titration curve can be calculated by methods previously discussed in this chapter. Four parts of the curve can be distinguished.

1. Starting point.   0% neutralization.

In the case of the strong acid, $[H^+]$ in the initial solution is simply the molarity of the acid.

In the case of the weak acid $[H^+]$ is calculated by the same methods used to compute the extent of ionization of any weak acid in terms of its ionization constant and molarity.

2. Approach to the end point.   5% to 95% neutralization.

For the case of the strong acid the neutralization reaction,

$$H^+ + OH^- \longrightarrow H_2O$$

or

$$H^+ + NH_3 \longrightarrow NH_4^+$$

may be assumed to go to completion to the extent of the amount of base added.   The amount of unreacted H^+ is then the difference between the initial amount of H^+ and the amount neutralized.   Allowance is made for the dilution effect of increasing the total volume of the solution on addition of base.

For the case of the weak acid, the neutralization reaction may be written,

$$HC_4H_7O_3 + OH^- \longrightarrow C_4H_7O_3^- + H_2O$$

or

$$HC_4H_7O_3 + NH_3 \longrightarrow C_4H_7O_3^- + NH_4^+$$

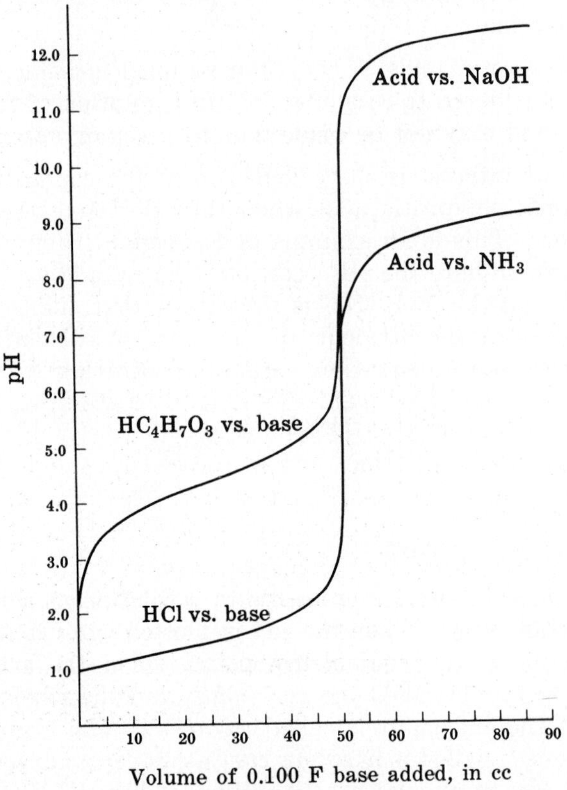

Volume of 0.100 F base added, in cc

Each curve represents the titration of 50.0 cc of 0.100 M acid, either HCl or $HC_4H_7O_3$, with base, either NaOH or NH_3. Note that at low pH the shape of the curve is determined by the strength of the acid, at high pH by the strength of the base, and at pH values near the end point by the strength of both the acid and the base.

Fig. 17-1

In either case the amount of the hydroxybutyrate ion, $C_4H_7O_3^-$, is equal to the amount of base added. The amount of unionized acid, $HC_4H_7O_3$, is the difference between the initial amount and the amount neutralized. Then,

$$[H^+] = K_{a\,(HC_4H_7O_3)} \frac{[HC_4H_7O_3]}{[C_4H_7O_3^-]}$$

3. **The endpoint.   100% neutralization.**

The pH at the end point is the same as for a solution of the salt containing the ions remaining at neutralization, NaCl, NH_4Cl, $NaC_4H_7O_3$ or $NH_4C_4H_7O_3$, as the case may be. NaCl solutions are neutral and have a pH of 7.00. The other three salts hydrolyze, and the pH can be evaluated by solving the hydrolysis equilibria. The two titrations with NH_3 have end points at pH values below 7 because of the dominance of NH_4^+ hydrolysis (NH_4^+ hydrolyzes more than $C_4H_7O_3^-$), while the $HC_4H_7O_3 - NaOH$ titration has at the end point a pH value greater than 7 because of the hydrolysis of the hydroxybutyrate ion.

4. **Extension beyond the end point.   Over 105% neutralization.**

With NaOH titrations, the excess OH^- beyond that needed for neutralization accumulates in the solution. The $[OH^-]$ is computed in terms of this excess and the total volume of solution. $[H^+]$ can then be computed from the K_w relationship.

$$[H^+] = \frac{K_w}{[OH^-]}$$

With NH_3 titrations, excess NH_3 beyond that needed for neutralization accumulates. The amount of NH_4^+ formed in neutralization is equimolar to the amount of acid initially present. Then $[OH^-]$ can be computed from the K_b equilibrium for NH_3.

$$[OH^-] = K_{b\,(NH_3)} \frac{[NH_3]}{[NH_4^+]}$$

Points within 5% of the starting point or of the end point can be computed by the same equilibria, but some of the simplifying assumptions made above are no longer valid.

Polyprotic acids like H_3PO_4 may have two or more distinct end points, corresponding to neutralization of the first, second, and subsequent hydrogens. In such a case each end point would occur at a different pH value.

Titrations can also be carried out in the reverse direction, by adding acid to base. The calculations of points on the curves are done by similar methods.

The end point of a titration, the region of steepest rise in the titration curve, can be determined experimentally if an instrument is available for measuring the pH after each addition of base. A simpler device is to insert into the solution a small amount of an *indicator* substance which has the property of undergoing a pH-dependent color change at the pH value of the end point. Such indicators are indeed known. They themselves are weak acids (or bases conjugate to weak acids) whose conjugate acid and basic forms are differently colored.

Although the electronic rearrangements responsible for the differences in color accompanying the ionization of these substances cannot be discussed here, the equilibria between the two forms of the indicator can be understood easily in terms of acid-base equilibria. If the acid dissociation of the indicator is written,

$$HIn \rightleftharpoons H^+ + In^-$$

then
$$[H^+] = K_{a\,(indicator)} \frac{[HIn]}{[In^-]}$$

If the indicator is so strongly colored that only a small amount need be used, too small to produce an appreciable contribution to the acidity or basicity of the solution, then the ratio of the two forms, HIn and In$^-$, reflects the [H$^+$] pre-existing in the solution before the addition of the indicator.

The most obvious color change occurs when the ratio changes from a value just below *one* to a value just above *one* (or vice versa). In other words, an indicator will have its color change at [H$^+$] values close to K_a for the indicator, or at pH values close to pK_a for the indicator. A variety of indicators is known, each with its own pK_a. A suitable indicator for any particular acid-base titration is one whose pK_a is in the sharply rising portion of the titration curve.

COMPLEX IONS

The combination of anions with protons to form acids is just one type of a more comprehensive class of reactions in which negative ions or neutral molecules combine with a positive ion to form a complex ion. The charge of the complex is equal to the algebraic sum of the charges of the positive ion and the attaching groups, or ligands. Thus Fe^{3+} combines with Cl$^-$ to form $FeCl^{2+}$, Ni^{2+} combines with $6 NH_3$ to form $Ni(NH_3)_6^{2+}$, and Al^{3+} combines with $6 F^-$ to form AlF_6^{3-}.

In some cases these complex ions are so stable that salts can be prepared from them which show no appreciable amounts of the separate constituents. An example is the ferricyanide ion, $Fe(CN)_6^{3-}$, solutions of which are quite different in analytical properties from those of Fe^{3+} or of CN$^-$. More commonly, the complex ion is not so stable and in solution is partly dissociated into its components. In such a case there is an equilibrium constant which regulates the allowable simultaneous values of the concentrations of the various species. An example is $FeBr^{2+}$, which may be formed or dissociated easily by slight modifications of experimental conditions.

$$Fe^{3+} + Br^- \rightleftharpoons FeBr^{2+} \qquad K_1 = \frac{[FeBr^{2+}]}{[Fe^{3+}][Br^-]}$$

The constant K_1 is called a stability constant; the larger its value, the stabler the complex. The subscript 1 refers to the complexation of *one* ligand per cation.

In some cases several ligands may be complexed, one at a time. A separate equilibrium equation may be written for each *successive* ligand addition, such as the following.

$$Cd^{2+} + CN^- \rightleftharpoons CdCN^+ \qquad K_1 = \frac{[CdCN^+]}{[Cd^{2+}][CN^-]}$$

$$CdCN^+ + CN^- \rightleftharpoons Cd(CN)_2 \qquad K_2 = \frac{[Cd(CN)_2]}{[CdCN^+][CN^-]}$$

Analogous equations can be written for the addition of a third and fourth cyanide, with constants K_3 and K_4. In addition to the stepwise formation equilibria, we sometimes write a single overall equation for the formation of a complex containing several ligands from the free cation and ligands.

$$Cd^{2+} + 4 CN^- \rightleftharpoons Cd(CN)_4^{2-} \qquad K_{overall} = \frac{[Cd(CN)_4^{2-}]}{[Cd^{2+}][CN^-]^4}$$

It can easily be shown that $K_{overall} = K_1 \times K_2 \times K_3 \times K_4$.

Equilibria between a complex ion and its components are sometimes written in reverse.

$$Cd(CN)_4^{2-} \rightleftharpoons Cd^{2+} + 4 CN^- \qquad K_d = \frac{[Cd^{2+}][CN^-]^4}{[Cd(CN)_4^{2-}]}$$

The *dissociation constant, K_d,* is the reciprocal of the *overall stability constant, $K_{overall}$.*

Solved Problems

IONIZATION OF ACIDS AND BASES

17.1. At 25°C, a 0.010 M ammonia solution is 4.3% ionized. Calculate (a) the concentration of the OH^- and NH_4^+ ions, (b) the concentration of molecular ammonia, (c) the ionization constant of aqueous ammonia. (d) Calculate $[OH^-]$ after 0.0090 mole of NH_4Cl is added to one liter of the above solution. (e) Calculate $[OH^-]$ of a solution prepared by dissolving 0.010 mole of NH_3 and 0.0050 mole of HCl per liter.

$$NH_3 + H_2O \rightleftharpoons NH_4^+ + OH^-$$

The label on a solution refers to the stoichiometric or weight composition and does not indicate the concentration of any particular component of an ionic equilibrium. Thus the designation 0.010 M NH_3 means that the solution might have been made by dissolving 0.010 mole of ammonia in enough water to make a liter of solution. It does not mean that the concentration of unionized ammonia in solution, $[NH_3]$, is 0.010 M. Rather, the sum of the ionized and the unionized ammonia is equal to 0.010 mole per liter.

(a) $[NH_4^+] = [OH^-] = 0.043 \times 0.010 \text{ mole}/l = 0.00043 \text{ mole}/l$

(b) $[NH_3] = 0.010 \text{ mole}/l - 0.00043 \text{ mole}/l = 0.010 \text{ mole}/l$ (approximately)

(c) $K_b = \dfrac{[NH_4^+] \times [OH^-]}{[NH_3]} = \dfrac{0.00043 \times 0.00043}{0.010} = 1.8 \times 10^{-5}$

(d) Since the base is so slightly ionized, we may assume that (1) all the $[NH_4^+]$ is derived from the NH_4Cl and that (2) the $[NH_3]$ at equilibrium is the same as the stoichiometric molarity of the base.

$$K_b = \frac{[NH_4^+] \times [OH^-]}{[NH_3]} \qquad [OH^-] = \frac{K_b \times [NH_3]}{[NH_4^+]} = \frac{(1.8 \times 10^{-5})(0.010)}{0.0090}$$

$$= 2.0 \times 10^{-5} \text{ mole}/l$$

The addition of NH_4Cl represses the ionization of NH_3, thus reducing greatly the $[OH^-]$ of the solution.

Assumptions made to simplify the solution of a problem should always be verified after the problem has been solved. In this case we assumed that practically all the NH_4^+ came from the added NH_4Cl. In addition to this, there is a small amount of NH_4^+ resulting from the dissociation of the NH_3. From the chemical equation listed in the statement of the problem the amount of NH_4^+ coming from the NH_3 must equal the amount of OH^-, which we now know is 2.0×10^{-5} M. The correct total $[NH_4^+]$ is then the sum of the contributions of NH_4Cl and NH_3, $.0090 + 2.0 \times 10^{-5}$. This sum is indeed equal to .0090, well within our 10% allowance. We have thus justified our assumption.

The student may wonder why we must make assumptions at all. This problem could have been solved by a more complete analysis as follows, without having to make the assumption.

	NH_3	+ H_2O	$\rightleftharpoons$	NH_4^+	+ OH^-
Moles/l at start from NH_3	.010			0	0
Moles/l at start from NH_4^+	0			.0090	0
Change by reaction	$-x$			$+x$	$+x$
Moles/l at equilibrium	$.010 - x$			$.0090 + x$	x

$$[OH^-] = x = \frac{K_b \times [NH_3]}{[NH_4^+]} = \frac{1.8 \times 10^{-5}(.010 - x)}{.0090 + x}$$

This is a quadratic equation in x and can be solved by the usual methods. In the solution, $x = \dfrac{-b \pm \sqrt{b^2 - 4ac}}{2a}$, the acceptable root requires the same sign for the square root as the sign of b. The value of the square root, however, is so close to the value of b, that the square root must be evaluated to 4 significant figures in order that x be known to even 2 figures. There are mathematical methods of solving the problem by approximations without having to evaluate the square root to 4 significant figures, but these mathematical approximations are essentially equivalent to the *chemical* approximations we had originally made.

In this case, failure to make the chemical approximation led to mathematical complications. Because of this, and because of the desirability of applying chemical intuition to problem analysis, we will make simplifying approximations in this book in advance, wherever they seem to make chemical sense. After solving the problem, we will always check the validity of our assumption.

(e) Since HCl is a strong acid, the 0.005 mole of HCl will react completely with 0.005 mole of NH_3 to form 0.005 mole NH_4^+. Of the original 0.010 mole of NH_3, only half will remain as unionized ammonia.

$$[OH^-] = \frac{K_b \times [NH_3]}{[NH_4^+]} = \frac{(1.8 \times 10^{-5})(0.005)}{0.005} = 1.8 \times 10^{-5} \text{ mole}/l$$

Check of assumption: The amount of NH_4^+ contributed by dissociation of NH_3 must be equal to $[OH^-]$, or 1.8×10^{-5} mole/l. This is indeed small compared with the .005 mole/l formed by neutralization of NH_3 with HCl.

17.2. Calculate the molarity of an acetic acid solution which is 2.0% ionized. K_a for $HC_2H_3O_2$ is 1.8×10^{-5} at 25°C.

$$HC_2H_3O_2 \rightleftharpoons H^+ + C_2H_3O_2^-$$

Let x = molarity of acetic acid solution.
Then $[H^+] = [C_2H_3O_2^-] = 0.020x$, and $[HC_2H_3O_2] = x - 0.020x = x$ (approx.).

$$\frac{[H^+] \times [C_2H_3O_2^-]}{[HC_2H_3O_2]} = K_a \quad \text{or} \quad \frac{(0.020x)(0.020x)}{x} = 1.8 \times 10^{-5}$$

Solving, $(0.020)^2 x = 1.8 \times 10^{-5}$ and $x = \dfrac{1.8 \times 10^{-5}}{(0.020)^2} = \dfrac{1.8 \times 10^{-5}}{4 \times 10^{-4}} = 0.045$ M.

17.3. Calculate the percent ionization of a 1.00 M solution of hydrocyanic acid, HCN. K_a of HCN is 4.8×10^{-10}.

$$HCN \rightleftharpoons H^+ + CN^-$$

Since H^+ and CN^- are present in the solution only as a result of the ionization, their concentrations must be equal.

Let $x = [H^+] = [CN^-]$. Then $[HCN] = 1.0 - x$. Let us assume that x will be very small compared with 1.0 and that $[HCN] = 1.0$ mole/l within the allowed 10% error. Then

$$K_a = \frac{[H^+] \times [CN^-]}{[HCN]} \quad \text{or} \quad 4.8 \times 10^{-10} = \frac{x^2}{1.0}, \quad \text{and} \quad x = 2.2 \times 10^{-5} \text{ mole}/l$$

$$\text{Percent ionized} = \frac{\text{ionized HCN}}{\text{total HCN}} \times 100 = \frac{2.0 \times 10^{-5} \text{ mole}/l}{1.0 \text{ mole}/l} \times 10^2 = 0.0020\%$$

Check of assumption: x ($= 2.2 \times 10^{-5}$) is indeed very small compared with 1.0.

17.4. The $[H^+]$ in a 0.072 M solution of benzoic acid is 2.1×10^{-3} mole/l. Compute K_a for the acid on the basis of this information.

$$HC_7H_5O_2 \rightleftharpoons H^+ + C_7H_5O_2^-$$

Since the hydrogen ion and benzoate ion come only from the ionization of the acid, their concentrations must be equal.

$[H^+] = [C_7H_5O_2^-] = 2.1 \times 10^{-3}$ mole/l. $[HC_7H_5O_2] = 0.072 - 2.1 \times 10^{-3} = 0.070$ mole/l.

$$K_a = \frac{[H^+] \times [C_7H_5O_2^-]}{[HC_7H_5O_2]} = \frac{(2.1 \times 10^{-3})^2}{0.070} = 6.3 \times 10^{-5}$$

17.5. The ionization constant of formic acid, HCO_2H, is 1.8×10^{-4}. What is the percentage ionization of a 0.0010 M solution of this acid?

Let $x = [H^+] = [HCO_2^-]$; then $[HCO_2H] = 0.0010 - x$.

Let us assume, as in Problem 17-3, that the percentage ionization is less than 10% and that the formic acid concentration, $(0.0010 - x)$, may be approximated by 0.0010. Then

$$K_a = \frac{[H^+][HCO_2^-]}{[HCO_2H]} = \frac{x^2}{.0010} = 1.8 \times 10^{-4} \quad \text{and} \quad x = 4.2 \times 10^{-4}$$

In checking the assumption, we see that x is not negligible compared with 0.0010. Therefore the previous assumption and the solution based on it must be rejected and the full quadratic form of the equation must be solved.

$$\frac{x^2}{.0010 - x} = 1.8 \times 10^{-4}$$

Solving, $x = 3.4 \times 10^{-4}$. (We reject the negative root, -5.2×10^{-4}).

$$\text{Percent ionized} = \frac{\text{ionized HCO}_2\text{H}}{\text{total HCO}_2\text{H}} \times 100 = \frac{3.4 \times 10^{-4}}{0.0010} \times 100 = 34\%$$

This exact solution (34% ionization) shows that the solution based on the original assumption was in error by almost 25%.

17.6. What concentration of acetic acid is needed to give a $[H^+]$ of 3.5×10^{-4} mole/l? K_a is 1.8×10^{-5}.

Let x = molarity of acetic acid. $[H^+] = [C_2H_3O_2^-] = 3.5 \times 10^{-4}$ and $[HC_2H_3O_2] = x - 3.5 \times 10^{-4}$.

$$\frac{[H^+][C_2H_3O_2^-]}{[HC_2H_3O_2]} = \frac{(3.5 \times 10^{-4})^2}{x - 3.5 \times 10^{-4}} = 1.8 \times 10^{-5} \quad \text{and} \quad x = 7.1 \times 10^{-3} \text{ mole/}l$$

17.7. A 0.100 M solution of an acid (specific gravity 1.010) is 4.5% ionized. Compute the freezing point of the solution. The molecular weight of the acid is 300.

To calculate the molality of the solution, we must find the number of moles of acid dissolved in one kilogram of water.

Weight of one liter of solution = 1000 ml × 1.010 g/ml = 1010 g
Weight of solute in the solution = 0.100 mole × 300 g/mole = 30 g
Hence, weight of water in one liter of solution = 1010 g − 30 g = 980 g

$$\text{Molality of solution} = \frac{0.100 \text{ mole of acid}}{0.980 \text{ kg of water}} = 0.102 \text{ } m$$

If the acid were not ionized at all, the freezing point lowering would be $(1.86 \times 0.102) = 0.190°C$. Because of ionization, the total number of dissolved particles is greater than 0.102 mole per kg of solvent. The freezing point depression is determined by the total number of dissolved particles, regardless of whether they are charged or uncharged.

Let α = fraction ionized. For every mole of acid added to the solution, there will be $(1 - \alpha)$ moles of unionized acid at equilibrium, α moles of H^+, and α moles of anion base conjugate to the acid, or a total of $(1 + \alpha)$ moles of dissolved particles. Hence the molality with respect to all dissolved particles is $(1 + \alpha)$ times the molality computed without regard to ionization.

Freezing point depression = $(1 + \alpha) \times 0.190°C = 1.045 \times 0.190°C = 0.199°C$. The freezing point of the solution is $-0.199°C$.

17.8. A 0.0100 M solution of chloroacetic acid, $HC_2H_2O_2Cl$, is also 0.0020 F in sodium chloroacetate $NaC_2H_2O_2Cl$. K_a for chloroacetic acid is 1.4×10^{-3}. What is $[H^+]$ in the solution?

We could proceed as in part (d) of Problem 17-1 and assume that the concentration of chloroacetate ion can be approximated by the formality of the sodium salt, 0.0020, and that the extent of dissociation of the acid is small.

$[H^+] = x$, $[C_2H_2O_2Cl^-] = 0.0020 + x = 0.0020$ (approx.), $[HC_2H_2O_2Cl] = 0.0100 - x = 0.0100$ (approx.).

$$x = [H^+] = K_a \frac{[HC_2H_2O_2Cl]}{[C_2H_2O_2Cl^-]} = 1.4 \times 10^{-3} \times \frac{.0100}{.0020} = 7.0 \times 10^{-3}$$

Check: We see that the assumption led to a self-inconsistency. The value of x obtained, 7.0×10^{-3}, can be neglected in comparison with neither 0.0020 nor 0.0100.

We must start again, without making the simplifying assumptions.

$$x = [H^+] = 1.4 \times 10^{-3} \times \frac{.0100 - x}{.0020 + x}$$

The solution of the resulting quadratic equation gives the proper value for $[H^+]$, 2.4×10^{-3} mole/l.

Chloroacetic acid is apparently a sufficiently strong acid that the common ion effect does not repress its ionization to a residual value small enough to be neglected. A hint of this result might have been taken from the relatively large value of K_a.

17.9. Calculate $[H^+]$ and $[C_2H_3O_2^-]$ in a solution that is 0.100 M in $HC_2H_3O_2$ and 0.050 M in HCl. K_a for $HC_2H_3O_2$ is 1.8×10^{-5}.

The HCl contributes so much more H^+ than the $HC_2H_3O_2$ that we can take $[H^+]$ as equal to the molarity of the HCl, 0.050 mole/l.

Then if $[C_2H_3O_2^-] = x$, we have $[HC_2H_3O_2] = 0.100 - x = 0.100$ (approx.) and

$$[C_2H_3O_2^-] = \frac{[HC_2H_3O_2] \times K_a}{[H^+]} = \frac{0.100 \times 1.8 \times 10^{-5}}{0.05} = 3.6 \times 10^{-5}$$

Check of assumptions: (1) Contribution of acetic acid to $[H^+]$, x, is indeed small compared with .050. (2) x is indeed small compared with 0.100.

17.10. Calculate $[H^+]$, $[C_2H_3O_2^-]$, and $[CN^-]$ in a solution that is both 0.100 M in $HC_2H_3O_2$ ($K_a = 1.8 \times 10^{-5}$) and 0.200 M in HCN ($K_a = 4.8 \times 10^{-10}$).

This problem is similar to the preceding one in that one of the acids, acetic, completely dominates the other in terms of contribution to the total $[H^+]$ of the solution. We base this assumption on the fact that K_a for $HC_2H_3O_2$ is much greater than for HCN; we will check the assumption after solving the problem. We will proceed by treating the acetic acid as if the HCN were not present.

Let $[H^+] = [C_2H_3O_2^-] = x$; then $[HC_2H_3O_2] = 0.100 - x = 0.100$ (approx.) and

$$\frac{[H^+] \times [C_2H_3O_2^-]}{[HC_2H_3O_2]} = \frac{x^2}{0.100} = 1.8 \times 10^{-5} \quad \text{from which} \quad x = 1.3 \times 10^{-3} \text{ mole}/l$$

Check of assumption: x is indeed small compared with 0.100.

Now we treat the HCN equilibrium established at a value of $[H^+]$ determined by the acetic acid, 1.3×10^{-3} mole/l.

Let $[CN^-] = y$; then $[HCN] = 0.200 - y = 0.200$ (approx.) and

$$y = [CN^-] = \frac{K_a \times [HCN]}{[H^+]} = \frac{4.8 \times 10^{-10} \times 0.200}{1.3 \times 10^{-3}} = 7 \times 10^{-8}$$

Check of assumptions: (1) y is indeed small compared with 0.200. (2) The amount of H^+ contributed by HCN ionization, equal to the amount of CN^- formed (7×10^{-8}), is indeed small compared with the amount of H^+ contributed by $HC_2H_3O_2$.

17.11. Calculate $[H^+]$ in a solution that is 0.100 M HCOOH ($K_a = 1.8 \times 10^{-4}$) and 0.100 M HOCN ($K_a = 2.2 \times 10^{-4}$).

This is a problem in which two weak acids both contribute to $[H^+]$, neither contributing such a preponderant amount that the other's share can be neglected.

	HCOOH	$\rightleftharpoons$	H^+	+	HCO_2^-	HOCN	$\rightleftharpoons$	H^+	+	OCN^-
Initial moles/l	0.100		0		0	0.100		0		0
Change by reaction	$-x$		$+x$		$+x$	$-y$		$+y$		$+y$
Moles/l at equil.	$0.100 - x$		$x+y$		x	$0.100 - y$		$x+y$		y
Approx. moles/l at equil.	0.100		$x+y$		x	0.100		$x+y$		y

The final line in the above tabulation is based on the assumption that x and y are both small compared with 0.100.

$$\frac{x(x+y)}{0.100} = 1.8 \times 10^{-4} \qquad\qquad \frac{y(x+y)}{0.100} = 2.2 \times 10^{-4}$$

Dividing the HOCN equation by the HCOOH equation,

$$\frac{y(x+y)}{0.100} \times \frac{0.100}{x(x+y)} = \frac{2.2 \times 10^{-4}}{1.8 \times 10^{-4}}, \qquad \frac{y}{x} = 1.22, \qquad y = 1.22x$$

Subtracting the HCOOH equation from the HOCN equation,

$$\frac{y(x+y) - x(x+y)}{0.100} = 2.2 \times 10^{-4} - 1.8 \times 10^{-4} \quad \text{or} \quad \frac{y^2 - x^2}{0.100} = 0.4 \times 10^{-4}$$

Substitute $y = 1.22x$ into the last equation and solve to obtain $x = 2.8 \times 10^{-3}$. Then $y = 1.22x = 3.4 \times 10^{-3}$ and $[H^+] = x + y = 6.2 \times 10^{-3}$.

Check of assumptions: The values of x and y are less than 10% of 0.100. (They are not much less than 10%. In this case we are just about at the limit of error we allowed ourselves in setting up the ground rules for this chapter.)

IONS OF WATER, pH

17.12. Calculate the $[H^+]$ and the $[OH^-]$ in 0.100 M $HC_2H_3O_2$ which is 1.3% ionized.

$$[H^+] = 0.013 \times 0.100 = 0.0013 = 1.3 \times 10^{-3} \text{ mole}/l$$

$$[OH^-] = \frac{1.00 \times 10^{-14}}{[H^+]} = \frac{1.00 \times 10^{-14}}{1.3 \times 10^{-3}} = 0.77 \times 10^{-11} = 7.7 \times 10^{-12} \text{ mole}/l$$

Note that $[H^+]$ is computed as if the $HC_2H_3O_2$ were the only contributor, whereas $[OH^-]$ is based on the ionization of water. Obviously, if water ionizes to supply OH^-, it must supply an equal amount of H^+ at the same time. Implied in this solution is the assumption that water's contribution to $[H^+]$ (7.7×10^{-12}) is negligible compared with that of the $HC_2H_3O_2$. This assumption will be valid in all but the most dilute, or weakest, acid solutions. In computing $[OH^-]$, however, water is the only source and it therefore cannot be overlooked.

17.13. Determine the $[OH^-]$ and the $[H^+]$ in 0.010 M ammonia solution which is 4.2% ionized.

$$[OH^-] = 0.042 \times 0.0100 = 4.2 \times 10^{-4} \text{ mole}/l$$

$$[H^+] = \frac{1.00 \times 10^{-14}}{[OH^-]} = \frac{1.00 \times 10^{-14}}{4.2 \times 10^{-4}} = 0.24 \times 10^{-10} = 2.4 \times 10^{-11} \text{ mole}/l$$

Here we have made the assumption that the contribution of water to $[OH^-]$ (equal to $[H^+]$, or 2.4×10^{-11}) is negligible compared with that of NH_3. K_w is used to compute $[H^+]$, since water is the only supplier of H^+. In general, $[H^+]$ for acidic solutions can usually be computed without regard to the water equilibrium; then K_w is used to compute $[OH^-]$. Conversely, $[OH^-]$ for basic solutions can usually be computed without regard to the water equilibrium; then K_w is used to compute $[H^+]$.

17.14. Express the following H^+ concentrations in terms of pH:

(a) 1×10^{-3}, (b) 5.4×10^{-9} mole/l.

(a) $\text{pH} = \log \dfrac{1}{[H^+]} = \log \dfrac{1}{10^{-3}} = \log 10^3 = 3$

(b) $\text{pH} = \log \dfrac{1}{[H^+]} = \log \dfrac{1}{5.4 \times 10^{-9}} = \log \dfrac{10^9}{5.4} = \log 10^9 - \log 5.4 = 9 - 0.73 = 8.27$

Another Method. $\text{pH} = \log \dfrac{1}{5.4 \times 10^{-9}} = -\log (5.4 \times 10^{-9}) = -\log 5.4 - \log 10^{-9}$

$$= -0.73 - (-9.00) = 9.00 - 0.73 = 8.27$$

17.15. Calculate the pH values of the following solutions, assuming complete ionization: (a) 4.9×10^{-4} N acid, (b) 0.0016 N base.

(a) $[H^+] = 4.9 \times 10^{-4}$ mole/l

$$pH = \log \frac{1}{[H^+]} = \log \frac{1}{4.9 \times 10^{-4}} = \log \frac{10^4}{4.9} = \log 10^4 - \log 4.9 = 4 - 0.69 = 3.31$$

(b) $[H^+] = \dfrac{10^{-14}}{[OH^-]} = \dfrac{10^{-14}}{1.6 \times 10^{-3}} = \dfrac{10^{-11}}{1.6}$ mole/l

$$pH = \log \frac{1}{[H^+]} = \log \frac{1.6}{10^{-11}} = \log 1.6 + \log 10^{11} = 0.20 + 11 = 11.20$$

Another Method. $pOH = \log \dfrac{1}{[OH^-]} = \log \dfrac{1}{1.6 \times 10^{-3}} = \log 10^3 - \log 1.6$
$$= 3.00 - 0.20 = 2.80$$

$$pH = 14.00 - pOH = 14.00 - 2.80 = 11.20$$

17.16. Change the following pH values to $[H^+]$ values: (a) 4, (b) 3.6.

(a) Directly from pH: $[H^+] = 10^{-pH} = 10^{-4}$ mole/l

(b) Directly from pH: $[H^+] = 10^{-pH} = 10^{-3.6}$ mole/l. Expanding,

$$pH = \log \frac{1}{[H^+]} = 3.6, \quad \frac{1}{[H^+]} = \text{antilog } 3.6 = 4.0 \times 10^3, \quad \text{and} \quad [H^+] = 2.5 \times 10^{-4} \text{ mole/}l$$

Or: $\log [H^+] = -pH = -3.6 = +0.4 - 4.0;$ then
$$[H^+] = \text{antilog } 0.4 \times \text{antilog} (-4) = 2.5 \times 10^{-4} \text{ mole/}l.$$

Note that a negative number like -3.6 cannot be found in tables of logarithms, nor can -0.6. Only positive mantissas appear in the printed logarithm tables. The positive mantissa was achieved by adding and subtracting the next higher integer to the negative number: $-3.6 = 4.0 - 3.6 - 4.0 = 0.4 - 4.0$. Then 0.4 can be found to be the logarithm of 2.5.

17.17. What is the pH of (a) 5.0×10^{-8} M HCl, (b) 5.0×10^{-10} M HCl?

(a) If we were to consider only the contribution of the HCl to the acidity of the solution, $[H^+]$ would be 5.0×10^{-8} and the pH would be greater than 7. This obviously cannot be true because a solution of a pure acid, no matter how dilute, cannot be less acid than pure water alone. It is necessary in this problem to take into account something that we have omitted in all previous problems dealing with acid solutions, the contribution of water to the total acidity. A complete analysis of the water equilibrium is required.

	H_2O	$\rightleftharpoons$	H^+	$+$	OH^-
Moles/l from HCl			5.0×10^{-8}		
Change by ionization of H_2O			x		x
Moles/l at equilibrium			$(5.0 \times 10^{-8} + x)$		x

$$K_w = [H^+] \times [OH^-] = (5.0 \times 10^{-8} + x)x = 1.00 \times 10^{-14}$$

from which $x = 0.78 \times 10^{-7}$. Then $[H^+] = (5.0 \times 10^{-8} + x) = 1.28 \times 10^{-7}$ and pH $= -\log (1.28 \times 10^{-7}) = 6.89.$

(b) Although the method of (a) could be used here, the problem can be simplified by noting that the HCl is so dilute as to make only a negligible contribution to $[H^+]$ as compared with the ionization of water. We may therefore write directly: $[H^+] = 1.00 \times 10^{-7}$ and pH $= 7.00$.

HYDROLYSIS

17.18. Calculate the extent of hydrolysis in a 0.100 F solution of NH_4Cl. K_b for NH_3 is 1.8×10^{-5}.

$$NH_4^+ \rightleftharpoons NH_3 + H^+$$

$$K_h \;=\; \frac{[NH_3] \times [H^+]}{[NH_4^+]} \;=\; \frac{K_w}{K_b} \;=\; \frac{1.00 \times 10^{-14}}{1.8 \times 10^{-5}} \;=\; 5.6 \times 10^{-10}$$

According to the reaction equation, equal amounts of NH_3 and H^+ are formed. Let $x = [NH_3] = [H^+]$. Then $[NH_4^+] = 0.100 - x = 0.100$ mole/l (approx.).

$$K_h \;=\; \frac{[NH_3] \times [H^+]}{[NH_4^+]} \qquad \text{or} \qquad 5.6 \times 10^{-10} = \frac{x^2}{0.100}$$

Solving, $x = 7.5 \times 10^{-6}$ mole/l. Checking the approximation, x is very small compared with 0.100.

$$\text{Fraction hydrolyzed} \;=\; \frac{\text{amount hydrolyzed}}{\text{total amount}} \;=\; \frac{7.5 \times 10^{-6}\text{ mole}/l}{0.100\text{ mole}/l} \;=\; 7.5 \times 10^{-5} \;=\; 0.0075\%.$$

17.19. Calculate $[OH^-]$ in a 1.00 F solution of NaOCN. K_a for HOCN is 2.2×10^{-4}.

$$OCN^- + H_2O \;\rightleftharpoons\; HOCN + OH^-$$

$$K_h \;=\; \frac{[OH^-] \times [HOCN]}{[OCN^-]} \;=\; \frac{K_w}{K_a} \;=\; \frac{1.00 \times 10^{-14}}{2.2 \times 10^{-4}} \;=\; 4.5 \times 10^{-11}$$

Since the source of OH^- and $HOCN$ is the hydrolysis reaction, they must exist in **equal** concentrations.

Let $x = [OH^-] = [HOCN]$. Then $[OCN^-] = 1.00 - x = 1.00$ mole/l (approx.).

$$K_h \;=\; \frac{[OH^-] \times [HOCN]}{[OCN^-]}, \qquad 4.5 \times 10^{-11} = \frac{x^2}{1.00}, \qquad \text{and} \qquad x = [OH^-] = 6.7 \times 10^{-6} \text{ mole}/l$$

Check of approximation: x is very small compared with 1.00.

17.20. The acid ionization (hydrolysis) constant of Zn^{2+} is 2.2×10^{-10}. (a) Calculate the pH of a 0.0010 F solution of $ZnCl_2$. (b) What is the basic dissociation constant of $Zn(OH)^+$?

(a) $Zn^{2+} + H_2O \rightleftharpoons Zn(OH)^+ + H^+ \qquad K_a = \dfrac{[Zn(OH)^+] \times [H^+]}{[Zn^{2+}]} = 2.2 \times 10^{-10}$

Let $x = [Zn(OH)^+] = [H^+]$. Then $[Zn^{2+}] = 0.0010 - x = 0.0010$ (approx.).

$$\frac{x^2}{0.0010} = 2.2 \times 10^{-10} \qquad \text{and} \qquad x = [H^+] = 5 \times 10^{-7}$$

$$\text{pH} \;=\; -\log(5 \times 10^{-7}) \;=\; -\log 5 - \log 10^{-7} \;=\; -0.7 - (-7) \;=\; +6.3$$

Check of approximation: x is very small compared with 0.0010.

(b) Zn^{2+}, as an acid, is conjugate to the base $Zn(OH)^+$. For the basic dissociation,

$$Zn(OH)^+ \;\rightleftharpoons\; Zn^{2+} + OH^- \qquad \text{and} \qquad K_b = \frac{K_w}{K_{a\,(Zn^{2+})}} = \frac{1.00 \times 10^{-14}}{2.2 \times 10^{-10}} = 4.5 \times 10^{-5}$$

17.21. Calculate the extent of hydrolysis and the pH of 0.0100 F $NH_4C_2H_3O_2$. K_a for $HC_2H_3O_2$ is 1.8×10^{-5}, and K_b for NH_3 is 1.8×10^{-5}.

This is a case where both cation and anion hydrolyze.

For NH_4^+, $K_h = \dfrac{K_w}{K_{b\,(NH_3)}} = \dfrac{1.00 \times 10^{-14}}{1.8 \times 10^{-5}} = 5.6 \times 10^{-10}$

For $C_2H_3O_2^-$, $K_h = \dfrac{K_w}{K_{a\,(HC_2H_3O_2)}} = \dfrac{1.00 \times 10^{-14}}{1.8 \times 10^{-5}} = 5.6 \times 10^{-10}$

By coincidence, the hydrolysis constants for these two ions are identical. The production of H^+ by NH_4^+ hydrolysis must therefore exactly equal the production of OH^- by $C_2H_3O_2^-$ hydrolysis. The solution is thus neutral, $[H^+] = [OH^-] = 1.00 \times 10^{-7}$, and the pH is 7.00.

For NH_4^+ hydrolysis, $\dfrac{[NH_3] \times [H^+]}{[NH_4^+]} = K_h = 5.6 \times 10^{-10}$

Let $x = [NH_3]$. Then $.0100 - x = [NH_4^+]$ and

$$\frac{x}{.0100 - x} = \frac{5.6 \times 10^{-10}}{1.00 \times 10^{-7}} = 5.6 \times 10^{-3} \quad \text{or} \quad x = 5.6 \times 10^{-5}$$

$$\text{Percent } NH_4^+ \text{ hydrolyzed} = \frac{5.6 \times 10^{-5}}{.0100} \times 100 = 0.56\%.$$

The percentage hydrolysis of acetate ion must also be 0.56%, because K_h is the same.

17.22. Calculate the pH in a 0.100 F solution of NH_4OCN. K_b for NH_3 is 1.8×10^{-5} and K_a for HOCN is 2.2×10^{-4}.

As in the previous problem, both cation and anion hydrolyze. Since NH_3 is a weaker base than HOCN is an acid, however, NH_4^+ hydrolyzes more than OCN^-, and the pH of the solution is less than 7. In order to preserve electrical neutrality, there cannot be an appreciable difference between $[NH_4^+]$ and $[OCN^-]$. (The balancing of a slight difference could be accounted for by $[H^+]$ or $[OH^-]$.) Thus $[NH_3]$ must be practically equal to $[HOCN]$ and we will indeed assume that they are equal.

Let $x = [NH_3] = [HOCN]$; then $0.100 - x = [NH_4^+] = [OCN^-]$.

For NH_4^+, $K_h = \dfrac{[NH_3][H^+]}{[NH_4^+]} = \dfrac{K_w}{K_b} = \dfrac{1.00 \times 10^{-14}}{1.8 \times 10^{-5}} = 5.6 \times 10^{-10}$ and

$$[H^+] = 5.6 \times 10^{-10} \times \frac{0.100 - x}{x} \tag{1}$$

For OCN^-, $K_h = \dfrac{[HOCN][OH^-]}{[OCN^-]} = \dfrac{K_w}{K_a} = \dfrac{1.00 \times 10^{-14}}{2.2 \times 10^{-4}} = 4.5 \times 10^{-11}$ and

$$[OH^-] = 4.5 \times 10^{-11} \times \frac{0.100 - x}{x} \tag{2}$$

Dividing equation (1) by equation (2), the x terms cancel:

$$\frac{[H^+]}{[OH^-]} = \frac{5.6 \times 10^{-10}}{4.5 \times 10^{-11}} = 12.4 \tag{3}$$

$[H^+]$ and $[OH^-]$ must satisfy the K_w relationship:

$$[H^+] \times [OH^-] = K_w = 1.00 \times 10^{-14} \tag{4}$$

Multiplying (3) by (4), we obtain

$$[H^+]^2 = 12.4 \times 10^{-14}, \quad [H^+] = 3.5 \times 10^{-7}, \quad \text{and} \quad pH = -\log[H^+] = 6.46$$

Check of assumption: Our assumption that $[NH_4^+] = [OCN^-]$ was based on the principle of electroneutrality. It is correct if $[H^+]$ and $[OH^-]$ are both small compared with the shift in the concentrations of the other ions. We shall therefore solve for x, the decrease in either $[NH_4^+]$ or $[OCN^-]$ accompanying the hydrolysis. From equation (1),

$$x = [NH_3] = \frac{5.6 \times 10^{-10} \times (0.100 - x)}{[H^+]} = \frac{5.6 \times 10^{-10} \times (0.100 - x)}{3.5 \times 10^{-7}} = 1.6 \times 10^{-4}$$

Both $[H^+]$ and $[OH^-]$ are small compared with x.

POLYPROTIC ACIDS

17.23. Calculate the H^+ concentration of 0.10 M H_2S solution. K_1 and K_2 for H_2S are respectively 1.0×10^{-7} and 1.2×10^{-13}.

Most of the H^+ results from the primary ionization: $H_2S \rightleftharpoons H^+ + HS^-$.

Let $x = [H^+] = [HS^-]$. Then $[H_2S] = 0.10 - x = 0.10$ mole/l (approximately).

$$K_1 = \frac{[H^+] \times [HS^-]}{[H_2S]}, \quad 1.0 \times 10^{-7} = \frac{x^2}{0.10}, \quad \text{and} \quad x = 1.0 \times 10^{-4} \text{ mole/}l$$

Check of assumptions: (1) x is indeed small compared with 0.10. (2) Using the above value of $[H^+]$ and $[HS^-]$, find the extent of the second dissociation.

$$[S^{2-}] \;=\; \frac{K_2 \times [HS^-]}{[H^+]} \;=\; \frac{1.2 \times 10^{-13} \times 1.0 \times 10^{-4}}{1.0 \times 10^{-4}} \;=\; 1.2 \times 10^{-13}$$

The extent of the second dissociation is so small that it does not appreciably lower $[HS^-]$ or raise $[H^+]$ as calculated from the first dissociation only.

17.24. Calculate the concentration of $C_8H_4O_4^{2-}$ (a) in a 0.010 M solution of $H_2C_8H_4O_4$, (b) in a solution which is 0.010 M with respect to $H_2C_8H_4O_4$ and 0.020 M with respect to HCl. The ionization constants for $H_2C_8H_4O_4$, phthalic acid, are:

$$H_2C_8H_4O_4 \;\rightleftharpoons\; H^+ + HC_8H_4O_4^- \qquad K_1 = 1.1 \times 10^{-3}$$
$$HC_8H_4O_4^- \;\rightleftharpoons\; H^+ + C_8H_4O_4^{2-} \qquad K_2 = 3.9 \times 10^{-6}$$

(a) If there were no second dissociation, the $[H^+]$ could be computed on the basis of the K_1 equation.

$$\frac{x^2}{0.010 - x} = 1.1 \times 10^{-3} \qquad \text{and} \qquad x = [H^+] = [HC_8H_4O_4^-] = 2.8 \times 10^{-3} \text{ mole}/l$$

If we assume that the second dissociation does not appreciably affect $[H^+]$ or $[HC_8H_4O_4^-]$, then

$$[C_8H_4O_4^{2-}] \;=\; \frac{K_2 \times [HC_8H_4O_4^-]}{[H^+]} \;=\; \frac{3.9 \times 10^{-6} \times 2.8 \times 10^{-3}}{2.8 \times 10^{-3}} \;=\; 3.9 \times 10^{-6}$$

Check of assumption: The extent of the second dissociation, 3.9×10^{-6}, is indeed small compared with the extent of the first dissociation, 2.8×10^{-3}.

(b) The $[H^+]$ in solution may be assumed to be essentially that contributed by the HCl. Also, this large common ion concentration represses the ionization of the phthalic acid, so that we assume that $[H_2C_8H_4O_4] = 0.010$ mole/l. The most convenient equation to use is the $K_1 \times K_2$ equation, since all the concentrations for this equation are known but one.

$$\frac{[H^+]^2 \times [C_8H_4O_4^{2-}]}{[H_2C_8H_4O_4]} \;=\; K_1 \times K_2 \;=\; (1.1 \times 10^{-3})(3.9 \times 10^{-6}) \;=\; 4.3 \times 10^{-9}$$

$$[C_8H_4O_4^{2-}] \;=\; \frac{(4.3 \times 10^{-9})[H_2C_8H_4O_4]}{[H^+]^2} \;=\; \frac{(4.3 \times 10^{-9})(0.010)}{(0.020)^2} \;=\; 1.1 \times 10^{-7} \text{ mole}/l$$

Check of assumptions: Solving for the first dissociation,

$$[HC_8H_4O_4^-] \;=\; \frac{K_1 \times [H_2C_8H_4O_4]}{[H^+]} \;=\; \frac{1.1 \times 10^{-3} \times .010}{.020} \;=\; 6 \times 10^{-4}$$

The amount of H^+ contributed by this dissociation, 6×10^{-4} mole/l, is indeed less than 10% of the amount contributed by HCl (.020 mole/l). The amount of H^+ contributed by the second dissociation is still less.

17.25. Calculate the extent of hydrolysis of 0.005 F K_2CrO_4. The ionization constants of H_2CrO_4 are $K_1 = 0.18$, $K_2 = 3.2 \times 10^{-7}$.

Just as in the ionization of polyprotic acids, so in the hydrolysis of their salts, the reaction proceeds in successive stages. The extent of the second stage is generally very small compared with the first. This is particularly true in this case, where H_2CrO_4 is practically a strong acid with respect to its first ionization. The equation of interest is

$$CrO_4^{2-} + H_2O \;\rightleftharpoons\; HCrO_4^- + OH^-$$

which indicates that the conjugate acid of the hydrolyzing CrO_4^{2-} is $HCrO_4^-$. As the ionization constant for $HCrO_4^-$ is K_2, the hydrolysis constant for the reaction is K_w/K_2.

$$\frac{[OH^-] \times [HCrO_4^-]}{[CrO_4^{2-}]} \;=\; K_h \;=\; \frac{K_w}{K_2} \;=\; \frac{1.0 \times 10^{-14}}{3.2 \times 10^{-7}} \;=\; 3.1 \times 10^{-8}$$

Let $x = [OH^-] = [HCrO_4^-]$. Then $[CrO_4^{2-}] = 0.005 - x = 0.005$ mole/l (approx.) and

$$\frac{x^2}{0.005} = 3.1 \times 10^{-8} \qquad \text{or} \qquad x = 1.2 \times 10^{-5}$$

Fractional hydrolysis $= \dfrac{1.2 \times 10^{-5}}{0.005} = 2.4 \times 10^{-3} = 0.24\%$.

Check of assumption: x is indeed small compared with 0.005.

17.26. What is the pH of a 0.0010 F solution of Na_2S? The ionization constants for H_2S are $K_1 = 1.0 \times 10^{-7}$ and $K_2 = 1.2 \times 10^{-13}$.

As in the previous problem, the first stage of hydrolysis, leading to HS^-, is predominant.

$$S^{2-} + H_2O \rightleftharpoons HS^- + OH^-$$

$$K_h = \frac{[HS^-] \times [OH^-]}{[S^{2-}]} = \frac{K_w}{K_2} = \frac{1.00 \times 10^{-14}}{1.2 \times 10^{-13}} = 0.083$$

Because of the large value for K_h, it cannot be assumed that $[S^{2-}]$ at equilibrium is approximately 0.001 mole/l, the stoichiometric concentration. In fact, the hydrolysis is so extensive that the opposite assumption can safely be made, that is, if $x = [S^{2-}]$, then $[HS^-] = [OH^-] = 0.0010 - x = 0.0010$ mole/l (approximately). This is tantamount to an assumption of almost 100% hydrolysis.

$$\frac{0.001 \times 0.001}{x} = 0.083 \quad \text{and} \quad x = [S^{2-}] = \frac{10^{-6}}{0.083} = 1.2 \times 10^{-5} \text{ mole}/l$$

The residual unhydrolyzed $[S^{2-}]$ is just about 1% of the original. Thus the assumption that $[OH^-]$ is approximately 0.0010 was justified.

$$pOH = -\log 0.0010 = -(-3.00) = 3.00 \qquad pH = 14.00 - pOH = 11.00$$

Additional check: Consider the second stage of hydrolysis.

$$HS^- + H_2O \rightleftharpoons H_2S + OH^- \qquad K_h = \frac{K_w}{K_1} = \frac{1.00 \times 10^{-14}}{1.0 \times 10^{-7}} = 1.0 \times 10^{-7}$$

Solve for $[H_2S]$ by assuming the values of $[OH^-]$ and $[HS^-]$ already obtained.

$$[H_2S] = \frac{K_h \times [HS^-]}{[OH^-]} = \frac{1.0 \times 10^{-7} \times 1.0 \times 10^{-3}}{1.0 \times 10^{-3}} = 1.0 \times 10^{-7}$$

The extent of the second hydrolysis, 1.0×10^{-7} mole/l, is indeed negligible compared with the first, 1.0×10^{-3} mole/l.

17.27. Calculate $[H^+]$, $[H_2PO_4^-]$, $[HPO_4^{2-}]$, and $[PO_4^{3-}]$ in 0.0100 M H_3PO_4. K_1, K_2, and K_3 are 7.1×10^{-3}, 6.2×10^{-8}, and 4.4×10^{-13} respectively.

We begin by assuming that H^+ comes principally from the first stage of dissociation, and that the concentration of any anion formed by one stage of an ionization is not appreciably lowered by the succeeding stage of ionization.

$$H_3PO_4 \rightleftharpoons H^+ + H_2PO_4^- \qquad K_1 = 7.1 \times 10^{-3}$$

Let $[H^+] = [H_2PO_4^-] = x$. Then $[H_3PO_4] = 0.0100 - x$, and $\dfrac{x^2}{.0100 - x} = 7.1 \times 10^{-3}$ or $x = 0.0056$ mole/l.

We next take the above solutions for $[H^+]$ and $[H_2PO_4^-]$ to solve for $[HPO_4^{2-}]$.

$$H_2PO_4^- \rightleftharpoons H^+ + HPO_4^{2-} \qquad K_2 = 6.2 \times 10^{-8}$$

$$[HPO_4^{2-}] = \frac{K_2 \times [H_2PO_4^-]}{[H^+]} = \frac{6.2 \times 10^{-8} \times .0056}{.0056} = 6.2 \times 10^{-8}$$

Check of assumption: The extent of the second dissociation, 6.2×10^{-8} mole/l, indeed is small compared with the first, 5.6×10^{-3} mole/l.

Next we take the above solutions for $[H^+]$ and $[HPO_4^{2-}]$ to solve for $[PO_4^{3-}]$.

$$HPO_4^{2-} \rightleftharpoons H^+ + PO_4^{3-} \qquad K_3 = 4.4 \times 10^{-13}$$

$$[PO_4^{3-}] = \frac{K_3 \times [HPO_4^{2-}]}{[H^+]} = \frac{4.4 \times 10^{-13} \times 6.2 \times 10^{-8}}{5.6 \times 10^{-3}} = 4.9 \times 10^{-18} \text{ mole}/l$$

Check of assumption: The depletion of HPO_4^- by the third stage, 4.9×10^{-18} mole/l, is an insignificant fraction of the amount present as a result of the second step, 6.2×10^{-8} mole/l.

17.28. Nitrilotriacetic acid has acid dissociation constants 9.3×10^{-4}, 8.5×10^{-4}, and 2.0×10^{-11}. Abbreviating the neutral acid as H_3X, what are $[H^+]$, $[H_2X^-]$, and $[HX^{2-}]$ in a 0.0100 M solution of H_3X?

Although the third dissociation is negligible in comparison with the first two, the second is not negligible compared with the first. In principle, the problem could be attacked by solving for the K_1 and K_2 equations simultaneously. Instead of this, we will take a more chemical intuitive approach, equivalent mathematically to a method of successive approximations. We first assume that only the first dissociation takes place, as in the previous problems in this chapter dealing with polyprotic acids. Let $[H^+] = [H_2X^-] = x$; then $[H_3X] = 0.0100 - x$ and

$$\frac{[H^+] \times [H_2X^-]}{[H_3X]} = \frac{x^2}{.0100 - x} = K_1 = 9.3 \times 10^{-4} \quad \text{or} \quad x = 2.6 \times 10^{-3} \text{ mole}/l$$

Next we treat the second stage of dissociation as if we started from the result of the first dissociation.

$$H_2X^- \rightleftharpoons H^+ + HX^{2-}$$

Let $[HX^{2-}] = y$; then $[H^+] = 2.6 \times 10^{-3} + y$, $[H_2X^-] = 2.6 \times 10^{-3} - y$ and

$$\frac{[H^+] \times [HX^{2-}]}{[H_2X^-]} = \frac{(2.6 \times 10^{-3} + y) \times y}{2.6 \times 10^{-3} - y} = K_2 = 8.5 \times 10^{-4}$$

In solving this equation, we may not neglect y in comparison with 2.6×10^{-3}. (If we did, y would appear to be $8.5, \times 10^{-4}$, which is obviously more than 10% of 2.6×10^{-3}.) The solution of the full quadratic equation in y gives:

$$[HX^{2-}] = y = 5.5 \times 10^{-4} \text{ mole}/l, \quad [H^+] = 2.6 \times 10^{-3} + y = 3.2 \times 10^{-3} \text{ mole}/l,$$

$$[H_2X^-] = 2.6 \times 10^{-3} - y = 2.0 \times 10^{-3} \text{ mole}/l$$

It is now necessary to substitute the corrected values into the K_1 equation to be sure that this condition is still satisfied.

$$\frac{[H^+] \times [H_2X^-]}{[H_3X]} = \frac{3.2 \times 10^{-3} \times 2.0 \times 10^{-3}}{.0100 - 2.6 \times 10^{-3}} = 8.6 \times 10^{-4}$$

(sufficiently close to K_1, 9.3×10^{-4}). The student may complete the verification by proving that the third dissociation has no appreciable effect on the concentrations already evaluated.

17.29. What is the pH of $0.0100\,M$ NaHCO$_3$? K_1 and K_2 for H_2CO_3 are 4.5×10^{-7} and 5.7×10^{-11}.

This has similarities to Problem 17-22 in that there is one reaction tending to make the solution acidic (the K_2 acid dissociation of HCO_3^-) and another reaction tending to make the solution basic (the hydrolysis of HCO_3^-).

$$HCO_3^- \rightleftharpoons H^+ + CO_3^{2-} \qquad\qquad K_2 = 5.7 \times 10^{-11} \qquad\qquad (1)$$

$$HCO_3^- + H_2O \rightleftharpoons OH^- + H_2CO_3 \qquad K_h = \frac{K_w}{K_1} = \frac{1.00 \times 10^{-14}}{4.5 \times 10^{-7}} = 2.2 \times 10^{-8} \qquad (2)$$

(Note that the hydrolysis constant for reaction (2) is related to K_1, because both hydrolysis and the K_1 equilibrium involve H_2CO_3 and HCO_3^-.) We see that the equilibrium constant for (2) is greater than that for (1); thus the pH is certain to exceed 7.

We assume that after self-neutralization both $[H^+]$ and $[OH^-]$ will be so small as to have no appreciable effect on the ionic charge balance. Therefore electrical neutrality can be preserved only by maintaining a fixed total anionic charge among the various carbonate species. That is, for every negative charge removed by converting HCO_3^- to H_2CO_3 another negative charge must be created by converting HCO_3^- to CO_3^{2-}. This leads to the following conditions:

$$[H_2CO_3] = [CO_3^{2-}] = x, \qquad [HCO_3^-] = 0.0100 - 2x = 0.0100 \text{ (approx.)}$$

$$\frac{[H^+] \times [CO_3^{2-}]}{[HCO_3^-]} = \frac{[H^+] \times x}{0.0100} = 5.7 \times 10^{-11} \qquad\qquad (3)$$

$$\frac{[OH^-] \times [H_2CO_3]}{[HCO_3^-]} = \frac{[OH^-] \times x}{.0100} = 2.2 \times 10^{-8} \qquad\qquad (4)$$

Multiplying (3) by (4) and considering that $[H^+] \times [OH^-] = 1.00 \times 10^{-14}$, we obtain

$$\frac{(1.00 \times 10^{-14})x^2}{(.0100)^2} = (5.7 \times 10^{-11})(2.2 \times 10^{-8}) \quad \text{or} \quad x = 1.12 \times 10^{-4}$$

Provisional check: $2x$ is indeed small compared with 0.0100.

We return now to equation *(3)*.

$$[H^+] = \frac{5.7 \times 10^{-11} \times [HCO_3^-]}{[CO_3^{2-}]} = \frac{5.7 \times 10^{-11} \times 0.0100}{1.12 \times 10^{-4}} = 5.1 \times 10^{-9} \text{ mole}/l$$

$$pH = -\log[H^+] = -(\log 5.1 + \log 10^{-9}) = -(0.71 - 9) = 8.29$$

Final check: Both $[H^+]$ and $[OH^-]$ are small compared with x, the shift in the concentrations of the other ions.

BUFFER SOLUTIONS AND TITRATION CURVES

17.30. A buffer solution of pH 8.50 is desired. *(a)* Starting with 0.0100 gram-formula weight of KCN and the usual inorganic reagents of the laboratory, how would you prepare one liter of the buffer solution? K_a for HCN is 4.8×10^{-10}. *(b)* How much would the pH change after the addition of 5×10^{-5} mole $HClO_4$ to 100 cc of the buffer? *(c)* How much would the pH change after the addition of 5×10^{-5} mole NaOH to 100 cc of the buffer?

(a) To find the desired $[H^+]$: $\log[H^+] = -pH = -8.50 = 0.50 - 9.00$.   Then

$$[H^+] = \text{antilog } 0.50 \times \text{antilog } (-9.00) = 3.2 \times 10^{-9} \text{ mole}/l$$

The buffer solution could be prepared by mixing CN^- (weak base) with HCN (weak acid) in the proper proportions so as to satisfy the ionization constant equilibrium for HCN.

For the equilibrium $HCN \rightleftharpoons H^+ + CN^-$, $K_a = 4.8 \times 10^{-10} = \dfrac{[H^+] \times [CN^-]}{[HCN]}$.

Then the required ratio $\dfrac{[CN^-]}{[HCN]} = \dfrac{K_a}{[H^+]} = \dfrac{4.8 \times 10^{-10}}{3.2 \times 10^{-9}} = 0.150$.

This ratio of CN^- to HCN can be attained if some of the CN^- is neutralized with a strong acid, like HCl, to form an equivalent amount of HCN. The total cyanide available for both forms is 0.0100 mole.

Let $x = [HCN]$; then $[CN^-] = 0.0100 - x$.

Substituting in $\dfrac{[CN^-]}{[HCN]} = 0.150$, $\dfrac{0.0100 - x}{x} = 0.150$; $x = 0.0087$ and $0.0100 - x = 0.0013$.

The buffer solution can be prepared by dissolving 0.0100 gfw KCN and 0.0087 mole HCl in enough water to make up one liter of solution.

(b) 100 cc of the buffer contains

$$(0.0087 \text{ mole}/l)(0.100 \; l) = 8.7 \times 10^{-4} \text{ mole HCN}$$

and

$$(0.0013 \text{ mole}/l)(0.100 \; l) = 1.3 \times 10^{-4} \text{ mole } CN^-$$

The addition of 5×10^{-5} mole of strong acid will convert more CN^- to HCN. The resulting amount of HCN will be $(8.7 \times 10^{-4} + 0.5 \times 10^{-4})$ mole and the resulting amount of CN^- will be $(1.3 \times 10^{-4} - 0.5 \times 10^{-4})$ mole. Only the ratio of the two concentrations is needed.

$$[H^+] = \frac{K_a \times [HCN]}{[CN^-]} = \frac{4.8 \times 10^{-10} \times 9.2}{0.8} = 5.5 \times 10^{-9} \text{ mole}/l$$

$$pH = -\log[H^+] = 9 - 0.74 = 8.26$$

The drop in pH caused by the addition of the acid is $(8.50 - 8.26)$, or 0.24 pH unit.

(c) The addition of 5×10^{-5} mole of strong base will convert an equivalent amount of HCN to CN^-.

HCN: resulting amount $= 8.7 \times 10^{-4} - 0.5 \times 10^{-4} = 8.2 \times 10^{-4}$ mole

CN^-: resulting amount $= 1.3 \times 10^{-4} + 0.5 \times 10^{-4} = 1.8 \times 10^{-4}$ mole

$$[H^+] = \frac{K_a \times [HCN]}{[CN^-]} = \frac{4.8 \times 10^{-10} \times 8.2}{1.8} = 2.2 \times 10^{-9} \text{ mole}/l$$

$$pH = -\log[H^+] = 9 - 0.34 = 8.66$$

The rise in pH caused by the addition of base is $(8.66 - 8.50)$, or 0.16 pH unit.

17.31. If 0.00010 mole H_3PO_4 is added to a solution buffered at pH 7.00, what are the relative proportions of the four forms: H_3PO_4, $H_2PO_4^-$, HPO_4^{2-}, PO_4^{3-}? K_1, K_2, and K_3 for phosphoric acid are respectively 7.1×10^{-3}, 6.2×10^{-8}, 4.4×10^{-13}.

Since the solution was previously well buffered, we can assume that the pH is not changed by addition of the phosphoric acid. Then if $[H^+]$ is fixed, the ratio of two of the desired concentrations can be calculated from each of the ionization constant equations.

$$\frac{[H^+] \times [H_2PO_4^-]}{[H_3PO_4]} = K_1 \qquad \frac{[H^+] \times [HPO_4^{2-}]}{[H_2PO_4^-]} = K_2 \qquad \frac{[H^+] \times [PO_4^{3-}]}{[HPO_4^{2-}]} = K_3$$

$$\frac{[H_3PO_4]}{[H_2PO_4^-]} = \frac{[H^+]}{K_1} \qquad \frac{[H_2PO_4^-]}{[HPO_4^{2-}]} = \frac{[H^+]}{K_2} \qquad \frac{[HPO_4^{2-}]}{[PO_4^{3-}]} = \frac{[H^+]}{K_3}$$

$$= \frac{1.0 \times 10^{-7}}{7.1 \times 10^{-3}} \qquad = \frac{1.0 \times 10^{-7}}{6.2 \times 10^{-8}} \qquad = \frac{1.0 \times 10^{-7}}{4.4 \times 10^{-13}}$$

$$= 1.4 \times 10^{-5} \qquad\qquad = 1.6 \qquad\qquad = 2.3 \times 10^{5}$$

Since the ratio $\dfrac{[H_3PO_4]}{[H_2PO_4^-]}$ is very small and the ratio $\dfrac{[HPO_4^{2-}]}{[PO_4^{3-}]}$ is very large, practically all of the material will exist as $H_2PO_4^-$ and HPO_4^{2-}. The sum of the amounts of these two ions will be practically equal to 0.00010 mole; and if the total volume of solution is one liter, the sum of these two ion concentrations will be 0.00010 mole/l.

Let $x = [HPO_4^{2-}]$; then $[H_2PO_4^-] = 0.00010 - x$.

Substituting in $\dfrac{[H_2PO_4^-]}{[HPO_4^{2-}]} = 1.6$, we obtain $\dfrac{0.00010 - x}{x} = 1.6$.

Solving, $x = [HPO_4^{2-}] = 3.8 \times 10^{-5}$ mole/l.　　$[H_2PO_4^-] = 0.00010 - x = 6.2 \times 10^{-5}$ mole/l.

$[H_3PO_4] = (1.4 \times 10^{-5})[H_2PO_4^-] = (1.4 \times 10^{-5})(6.2 \times 10^{-5}) = 9 \times 10^{-10}$ mole/l.

$[PO_4^{3-}] = \dfrac{[HPO_4^{2-}]}{2.3 \times 10^5} = \dfrac{3.8 \times 10^{-5}}{2.3 \times 10^5} = 1.7 \times 10^{-10}$ mole/l.

17.32. K_a for $HC_2H_3O_2$ is 1.8×10^{-5}. A 40.0 cc sample of 0.0100 M $HC_2H_3O_2$ is titrated with 0.0200 F NaOH. Calculate the pH after the addition of　(a) 3.0 cc,　(b) 10.0 cc, (c) 20.0 cc,　(d) 30.0 cc of the NaOH solution.

We can keep track of the changing amounts of the various species and the increasing volume by a tabulation such as the following.

		a	b	c	d
Amount of base added	0 cc	3.0 cc	10.0 cc	20.0 cc	30.0 cc
Total volume	.040 l	.043 l	.050 l	.060 l	.070 l
Moles $HC_2H_3O_2$, before neutralization (.040 $l \times$ .0100 mole/l)	4.0×10^{-4}	4.0×10^{-4}	4.0×10^{-4}	4.0×10^{-4}	4.0×10^{-4}
Moles OH^- added (Vol. NaOH in $l \times$ .0200 mole/l)	0.0×10^{-4}	0.6×10^{-4}	2.0×10^{-4}	4.0×10^{-4}	6.0×10^{-4}
Moles $C_2H_3O_2^-$ formed	0.0×10^{-4}	0.6×10^{-4}	2.0×10^{-4}	4.0×10^{-4}	4.0×10^{-4}
Moles $HC_2H_3O_2$ remaining	4.0×10^{-4}	3.4×10^{-4}	2.0×10^{-4}	x	y
Moles OH^- excess					2.0×10^{-4}

Note that the amount of acetic acid neutralized (amount of $C_2H_3O_2^-$) follows the amount of OH^- added up to complete neutralization. Additional OH^-, having no more acid to neutralize, accumulates in the solution. Up to the end point, the amount of $HC_2H_3O_2$ remaining is obtained by simply subtracting the amount of $C_2H_3O_2^-$ from the initial amount of $HC_3H_3O_2$. At the end point and beyond, however, $[HC_2H_3O_2]$ cannot be set equal to zero but must be solved from the ionic equilibrium conditions.

(a) and (b). Here absolute concentrations of conjugate acid and base are not needed; only their **ratio** is required.

$$\begin{array}{ccc} & (a) & (b) \\ [H^+] = \dfrac{K_a \times [HC_2H_3O_2]}{[C_2H_3O_2^-]} & \dfrac{1.8 \times 10^{-5} \times 3.4}{0.6} & \dfrac{1.8 \times 10^{-5} \times 2.0}{2.0} \\ & = 1.0 \times 10^{-4} & = 1.8 \times 10^{-5} \\ pH = -\log{[H^+]} & 4.00 & 4.74 \end{array}$$

(c) $[C_2H_3O_2^-] = \dfrac{4.0 \times 10^{-4}\ \text{mole}}{.060\ l} = 6.7 \times 10^{-3}$ mole/l. The solution at the end point is the same as 6.7×10^{-3} F $NaC_2H_3O_2$. Consider the hydrolysis of $NaC_2H_3O_2$.

$$C_2H_3O_2^- + H_2O \rightleftharpoons HC_2H_3O_2 + OH^-$$

Let $[HC_2H_3O_2] = [OH^-] = x$, $[C_2H_3O_2^-] = 6.7 \times 10^{-3} - x = 6.7 \times 10^{-3}$ (approx.); then

$$\dfrac{[HC_2H_3O_2][OH^-]}{[C_2H_3O_2^-]} = \dfrac{x^2}{6.7 \times 10^{-3}} = K_h = \dfrac{1.00 \times 10^{-14}}{1.8 \times 10^{-5}} = 5.5 \times 10^{-10}, \quad x = 1.9 \times 10^{-6}$$

Check of assumption: x is indeed small compared with 6.7×10^{-3}.

$$pOH = -\log{[OH^-]} = -\log{(1.9 \times 10^{-6})} = 6 - 0.28 = 5.72. \quad pH = 14.00 - 5.72 = 8.28.$$

(d) From the excess OH^- beyond that needed to neutralize all the acetic acid, we know the following:

$$[OH^-] = \dfrac{2.0 \times 10^{-4}\ \text{mole}}{.070\ l} = 2.9 \times 10^{-3}\ \text{mole}/l$$

$$pOH = -\log{[OH^-]} = 3 - 0.46 = 2.54. \quad pH = 14.00 - 2.54 = 11.46.$$

17.33. Calculate a point on the titration curve for the addition of 2.0 cc of 0.0100 F NaOH to 50.0 cc of 0.0100 M chloroacetic acid, $HC_2H_2O_2Cl$. $K_a = 1.4 \times 10^{-3}$.

Our simplifying assumption, made in the previous problem, breaks down here. If the amount of chloroacetate ion formed were equivalent to the amount of NaOH added, we would have the following:

Total volume $= .0520\ l$. Amount OH^- added $= .0020\ l \times .01$ mole/$l = 2.0 \times 10^{-5}$ mole.

$$[C_2H_2O_2Cl^-] = \dfrac{2.0 \times 10^{-5}\ \text{mole}}{.0520\ l} = 3.8 \times 10^{-4}\ \text{mole}/l \tag{1}$$

$$[HC_2H_2O_2Cl] = \dfrac{.0500\ l \times .0100\ \text{mole}/l}{.0520\ l} - 3.8 \times 10^{-4}\ \text{mole}/l \tag{2}$$

$$= .0096 - .0004 = 9.2 \times 10^{-3}\ \text{mole}/l$$

$$[H^+] = \dfrac{K_a \times [HC_2H_2O_2Cl]}{[C_2H_2O_2Cl^-]} = \dfrac{1.4 \times 10^{-3} \times 9.2 \times 10^{-3}}{3.8 \times 10^{-4}} = 3.4 \times 10^{-2} \tag{3}$$

This answer is obviously ridiculous. $[H^+]$ cannot possibly exceed the initial molarity of the acid. Apparently the amount of chloroacetate ion is greater than the equivalent amount of base added. This fact is related to the relatively strong acidity of the acid and to the appreciable ionization of the acid even before the titration begins. Mathematically, this is taken into account by an equation of electroneutrality, according to which there must be equal numbers of cationic and anionic charges in the solution.

$$[H^+] + [Na^+] = [C_2H_2O_2Cl^-] + [OH^-] \tag{4}$$

It is safe to drop the OH^- term in equation (4) in this case because it is so much smaller than $[C_2H_2O_2Cl^-]$. $[Na^+]$ is obtained from the amount of NaOH added and the total volume of solution.

$$[Na^+] = \dfrac{2.0 \times 10^{-5}\ \text{mole}}{.052\ l} = 3.8 \times 10^{-4}\ \text{mole}/l \tag{5}$$

Chloroacetate ion can be computed from equations (4) and (5) with the neglect of $[OH^-]$.

$$[C_2H_2O_2Cl^-] = 3.8 \times 10^{-4} + [H^+] \tag{6}$$

The neutral acid concentration is then the total molarity of acid (including its ion) minus $[C_2H_2O_2Cl^-]$.

$$[HC_2H_2O_2Cl] = \dfrac{.0100 \times .050}{.052} - 3.8 \times 10^{-4} - [H^+] = 9.2 \times 10^{-3} - [H^+] \tag{7}$$

Note that (*6*) and (*7*) differ from (*1*) and (*2*) only in the inclusion of the $[H^+]$ terms.

Now we may return to the ionization equilibrium for the acid.

$$[H^+] \;=\; \frac{K_a \times [HC_2H_2O_2Cl]}{[C_2H_2O_2Cl^-]} \;=\; \frac{(1.4 \times 10^{-3})(9.2 \times 10^{-3} - [H^+])}{3.8 \times 10^{-4} + [H^+]}$$

The solution of the quadratic equation is $[H^+] = 2.8 \times 10^{-3}$, and $pH = -\log[H^+] = 2.55$.

The complication treated in this problem occurs whenever $[H^+]$ or $[OH^-]$ during partial neutralization cannot be neglected in comparison with the concentrations of other ions in solution. This is likely to be the case near the beginning of the titration of a moderately strong weak acid or near the end of a titration of a strong acid with a moderately strong weak base.

17.34. An acid-base indicator has a K_a of 3.0×10^{-5}. The acid form of the indicator is red and the basic form is blue. (*a*) By how much must the pH change in order to change the indicator from 75% red to 75% blue? (*b*) For which of the titrations shown in Fig. 17-1 would this indicator be a suitable choice?

(*a*)
$$[H^+] \;=\; \frac{K_a \times [\text{acid}]}{[\text{base}]}$$

75% red: $[H^+] \;=\; \dfrac{3.0 \times 10^{-5} \times 75}{25} \;=\; 9 \times 10^{-5};\quad pH = 4.05.$

75% blue: $[H^+] \;=\; \dfrac{3.0 \times 10^{-5} \times 25}{75} \;=\; 1.0 \times 10^{-5};\quad pH = 5.00.$

Change in $pH = 5.00 - 4.05 = 0.95.$

(*b*) The indicator changes its color in the pH range 4 to 5. The two HCl titration curves in Fig. 17-1 are rising steeply at pH 4 to 5; thus the indicator would be suitable for them. For the $HC_4H_7O_3$ titrations, however, this pH range occurs nowhere near the end point; an indicator is needed which changes its color at higher pH values.

COMPLEX IONS

17.35. A liter of solution was prepared containing 0.0050 mole per liter of silver in the +I oxidation state and 1.00 mole per liter of NH_3. What is the concentration of free Ag^+ in the solution at equilibrium? K_d for $Ag(NH_3)_2^+$ is 5.9×10^{-8}.

Most of the silver, approximately 0.0050 mole, will be in the form of the complex ion, $Ag(NH_3)_2^+$. The concentration of free NH_3 at equilibrium is practically unchanged from 1.00 mole/liter since only 0.010 mole of NH_3 would be used up to form 0.0050 mole of complex.

$$Ag(NH_3)_2^+ \;\rightleftharpoons\; Ag^+ + 2\,NH_3$$

$$K_d = \frac{[Ag^+] \times [NH_3]^2}{[Ag(NH_3)_2^+]}, \qquad 5.9 \times 10^{-8} = \frac{x(1.00)^2}{0.0050}, \qquad x = [Ag^+] = 3.0 \times 10^{-10}\ \text{mole}/l$$

17.36. K_1 for the complexation of NH_3 with Ag^+ is 1.8×10^3. (*a*) With reference to the preceding problem, what is the concentration of $Ag(NH_3)^+$? (*b*) What is K_2 for this system?

(*a*) K_1 refers to the following:

$$Ag^+ + NH_3 \;\rightleftharpoons\; Ag(NH_3)^+ \qquad K_1 = \frac{[Ag(NH_3)^+]}{[Ag^+][NH_3]}$$

$$[Ag(NH_3)^+] = K_1 \times [Ag^+] \times [NH_3] = (1.8 \times 10^3)(3.0 \times 10^{-10})(1.00)$$
$$= 5.4 \times 10^{-7}\ \text{mole}/l$$

Numerical values for $[Ag^+]$ and $[NH_3]$ are taken from the preceding problem.

This problem is actually a check on an assumption made in the previous problem, where it was assumed that practically all the dissolved silver was in the complex, $Ag(NH_3)_2^+$. If $[Ag(NH_3)^+]$ had turned out to be greater than about 5×10^{-4} mole/l, the assumption would have been shown to be incorrect.

(b) K_1, K_2, and K_d are interrelated.

$$K_2 = \frac{[Ag(NH_3)_2{}^+]}{[Ag(NH_3)^+][NH_3]} = \frac{\dfrac{[Ag(NH_3)_2{}^+]}{[Ag^+][NH_3]^2}}{\dfrac{[Ag(NH_3)^+]}{[Ag^+][NH_3]}} = \frac{\dfrac{1}{K_d}}{K_1}$$

$$= \frac{1}{K_1 K_d} = \frac{1}{(1.8 \times 10^3)(5.9 \times 10^{-8})} = 9.4 \times 10^3$$

17.37. How much NH_3 should be added to a solution of 0.0010 F $Cu(NO_3)_2$ in order to reduce $[Cu^{2+}]$ to 10^{-12} mole per liter? K_d for $Cu(NH_3)_4{}^{2+}$ is 4.7×10^{-15}. Neglect the amount of copper in complexes containing fewer than 4 ammonias per copper.

$$Cu(NH_3)_4{}^{2+} \rightleftharpoons Cu^{2+} + 4\,NH_3 \qquad K_d = \frac{[Cu^{2+}] \times [NH_3]^4}{[Cu(NH_3)_4{}^{2+}]} = 4.7 \times 10^{-15}$$

Since the sum of the concentrations of copper in the complex and in the free ionic state must equal 0.0010, and since the amount of the free ion is very small, the concentration of the complex is 0.0010 mole/l.

Let $x = [NH_3]$. Then $\dfrac{10^{-12}\,(x^4)}{10^{-3}} = 4.7 \times 10^{-15}$, $x^4 = 4.7 \times 10^{-6}$, and $x = 0.047$.

The concentration of NH_3 at equilibrium is 0.047 mole/l. The amount of NH_3 used up in forming 0.001 mole/l of complex is 0.004 mole/l. Hence the total amount of NH_3 to be added is $0.047 + 0.004 = 0.051$ mole/l.

17.38. The stability constants for complexation of Cl^- with Fe^{3+} are 30, 4.5, and 0.10 for K_1, K_2, and K_3 respectively. What is the principal iron-containing species in 0.0100 F $FeCl_3$? Assume that the solution has been mildly acidified to prevent hydrolysis.

This differs from the previous problems in that here there is not a large excess of the complexing agent which would drive the complexation reactions to completion. We may start by imagining that the salt is completely ionized, giving .0100 M Fe^{3+} and .0300 M Cl^-. Then we assume that only the first stage of complexation is significant.

	Fe^{3+}	+	Cl^-	$\rightleftharpoons$	$FeCl^{2+}$
Concn. if there is no complex	.0100		.0300		.000
Change by complexation	$-x$		$-x$		$+x$
Concn. at equilibrium	$.0100 - x$		$.0300 - x$		x

$$\frac{x}{(.0100 - x)(.0300 - x)} = K_1 = 30$$

Solving, $x = 0.0043 = [FeCl^{2+}]$, $.0300 - x = 0.0257 = [Cl^-]$, $.0100 - x = 0.0057$ mole/$l = [Fe^{3+}]$.

Check of assumption: We must now prove that the complexation of a second Cl^- is not extensive enough to modify the above values significantly. We substitute the above values into the K_2 equation.

	$FeCl^{2+}$	+	Cl^-	$\rightleftharpoons$	$FeCl_2{}^+$
Concn. without a 2nd complexation	.0043		.0257		0
Change by 2nd complexation	$-y$		$-y$		$+y$
Concn. at equilibrium	$.0043 - y$		$.0257 - y$		y

$$\frac{y}{(.0043 - y)(.0257 - y)} = K_2 = 4.5, \qquad y = .0004 \text{ mole}/l = [FeCl_2{}^+]$$

This value for $[FeCl_2{}^+]$ is just barely less than 10% of the previous value for $[FeCl^{2+}]$. Within our allowed error, then, we need not go back to the K_1 equation to correct the extent of the first complexation, in which we had assumed that $[Fe^{3+}] + [FeCl^{2+}] = 0.0100$. (A simultaneous treatment of the first two stages of complexation would give the following values: $[Fe^{3+}] = .0054$, $[FeCl^{2+}] = .0041$, $[FeCl_2{}^+] = .00046$, $[Cl^-] = .0250$. Usually it is not worthwhile to make such a small correction.) We have thus proved that a majority of the iron is in the form of Fe^{3+}, almost as much in $FeCl^{2+}$, and an order of magnitude less in $FeCl_2{}^+$. It can be shown similarly that the amount in un-ionized $FeCl_3$ is still less.

Supplementary Problems

ACIDS AND BASES

> *Note*: Many of the problems at the end of this section involve multiple equilibria. They may be omitted in elementary treatments of ionic equilibrium.

17.39. Calculate the ionization constant of formic acid, HCO_2H, which ionizes 4.2% in 0.10 M solution. *Ans.* 1.8×10^{-4}

17.40. A solution of acetic acid is 1.0% ionized. Determine the molarity and the $[H^+]$ of the solution. K_a of $HC_2H_3O_2$ is 1.8×10^{-5}. *Ans.* 0.18 M, 1.8×10^{-3} mole/l

17.41. The ionization constant of ammonia in water is 1.8×10^{-5}. Determine (a) the degree of ionization and (b) the OH^- concentration of a 0.08 M solution of NH_3. *Ans.* 1.5%, 1.2×10^{-3} mole/l

17.42. Chloroacetic acid, a monoprotic acid, has a K_a value of 1.4×10^{-3}. Compute the freezing point of a 0.10 M solution of this acid. Assume that the molarity and the molality are the same in this case. *Ans.* $-0.21°C$

17.43. Calculate $[OH^-]$ in a 0.0100 M solution of aniline, $C_6H_5NH_2$. K_b for the basic dissociation of aniline is 4.2×10^{-10}. What is the $[OH^-]$ in a solution of 0.0100 F aniline hydrochloride, which contains the ion $C_6H_5NH_3^+$? *Ans.* 2.0×10^{-6}, 2.0×10^{-11} mole/l

17.44. Determine $[OH^-]$ of a 0.050 M solution of ammonia to which has been added sufficient NH_4Cl to make the total $[NH_4^+]$ equal to 0.10 mole/l. K_b of ammonia is 1.8×10^{-5}. *Ans.* 9×10^{-6} mole/l

17.45. Calculate $[H^+]$ in a liter of solution in which are dissolved 0.08 mole $HC_2H_3O_2$ and 0.10 gfw $NaC_2H_3O_2$. K_a for $HC_2H_3O_2$ is 1.8×10^{-5}. *Ans.* 1.4×10^{-5} mole/l

17.46. A 0.025 M solution of a monobasic acid had a freezing point of $-0.060°C$. What are K_a and pK_a for the acid? *Ans.* 3.0×10^{-3}, 2.52

17.47. Fluoroacetic acid has a K_a of 2.6×10^{-3}. What concentration of the acid is needed so that $[H^+]$ is 2.0×10^{-3} mole/l? *Ans.* 3.5×10^{-3} M

17.48. What is $[NH_4^+]$ in a solution that is 0.0200 M NH_3 and 0.0100 M KOH? K_b for NH_3 is 1.8×10^{-5}. *Ans.* 3.6×10^{-5} M

17.49. What molarity of NH_3 provides a hydroxide ion concentration of 1.5×10^{-3}? K_b for NH_3 is 1.8×10^{-5}. *Ans.* 0.12 M

17.50. What is $[HCOO^-]$ in a solution that is both 0.015 M HCOOH and 0.020 M HCl? K_a for HCOOH is 1.8×10^{-4}. *Ans.* 1.3×10^{-4} M

17.51. What are $[H^+]$, $[C_3H_5O_3^-]$, and $[OC_6H_5^-]$ in a solution that is 0.030 M $HC_3H_5O_3$ and 0.100 M HOC_6H_5? K_a values for $HC_3H_5O_3$ and HOC_6H_5 are 1.4×10^{-4} and 1.0×10^{-10} respectively. *Ans.* $[H^+] = 2.0 \times 10^{-3}$, $[C_3H_5O_3^-] = 2.0 \times 10^{-3}$, $[OC_6H_5^-] = 5 \times 10^{-9}$ M

17.52. Find the value of $[OH^-]$ in a solution made by dissolving 0.005 mole each of ammonia and pyridine in enough water to make 200 cc of solution. K_b for ammonia and pyridine are 1.8×10^{-5} and 1.5×10^{-9} respectively. What are the concentrations of ammonium and pyridinium ions? *Ans.* $[OH^-] = [NH_4^+] = 7 \times 10^{-4}$ M, [pyridinium ion] $= 5 \times 10^{-8}$ M

17.53. Calculate the pH and pOH values of the following solutions, assuming complete ionization: (a) 0.00345 N acid, (b) 0.000775 N acid, (c) 0.00886 N base. *Ans.* (a) pH = 2.46, pOH = 11.54; (b) 3.11, 10.89; (c) 11.95, 2.05

17.54. Convert the following pH values to $[H^+]$ values: (a) 4, (b) 7, (c) 2.50, (d) 8.26. *Ans.* (a) 10^{-4}, (b) 10^{-7}, (c) 3.2×10^{-3}, (d) 5.5×10^{-9} mole/l

17.55. The $[H^+]$ of an HNO_3 solution is 1×10^{-3} mole/l, and the $[H^+]$ of an NaOH solution is 1×10^{-12} mole/l. Find the formality and pH of each solution.
Ans. HNO_3: 0.001 F, pH = 3; NaOH: 0.01 F, pH = 12

17.56. Compute $[H^+]$ and $[OH^-]$ in a 0.0010 molar solution of a monobasic acid which is 4.2% ionized. What is the pH of the solution? What are K_a and pK_a for the acid?
Ans. $[H^+] = 4.2 \times 10^{-5}$, $[OH^-] = 2.4 \times 10^{-10}$, pH = 4.38, $K_a = 1.8 \times 10^{-6}$, $pK_a = 5.74$

17.57. Compute $[OH^-]$ and $[H^+]$ in a 0.10 N solution of a weak base which is 1.3% ionized. What is the pH of the solution?   *Ans.* $[H^+] = 7.7 \times 10^{-12}$, $[OH^-] = 1.3 \times 10^{-3}$, pH = 11.11

17.58. What is the pH of a solution containing 0.010 mole HCl per liter? Calculate the change in pH if 0.020 gfw $NaC_2H_3O_2$ is added to a liter of this solution. K_a of $HC_2H_3O_2$ is 1.8×10^{-5}.
Ans. Initial pH = 2.0, final pH = 4.74

17.59. Calculate the percent hydrolysis in a 0.0100 F solution of KCN. K_a for HCN is 4.8×10^{-10}.
Ans. 4.5%

17.60. The basic ionization constant for hydrazine, N_2H_4, is 8.7×10^{-7}. What would be the percentage hydrolysis of 0.100 F N_2H_5Cl, a salt containing the acid ion conjugate to hydrazine base?
Ans. 0.034%

17.61. A 0.25 F solution of pyridinium chloride, $C_5H_6N^+Cl^-$, was found to have a pH of 2.89. What is K_b for the basic dissociation of pyridine, C_5H_5N?   *Ans.* 1.5×10^{-9}

17.62. K_a for the acid ionization of Fe^{3+} to $Fe(OH)^{2+}$ and H^+ is 6.8×10^{-3}. What is the maximum pH value which could be used so that at least 95% of the total +III iron in a dilute solution is in the Fe^{3+} form?   *Ans.* 0.9

17.63. A 0.010 M solution of NpO_2NO_3 was found to have a pH of 5.44. What is the hydrolysis constant, K_a, for NpO_2^+, and what is K_b for NpO_2OH?   *Ans.* $K_a = 1.3 \times 10^{-9}$, $K_b = 7.7 \times 10^{-6}$

17.64. Calculate the $[H^+]$ of a 0.050 M H_2S solution. K_1 of H_2S is 1.0×10^{-7}.   *Ans.* 7.1×10^{-5} mole/l

17.65. What is the S^{2-} concentration of a 0.050 M H_2S solution? K_2 of H_2S is 1.2×10^{-13}.
Ans. 1.2×10^{-13} mole/l

17.66. What is $[S^{2-}]$ in a solution that is 0.050 M H_2S and 0.0100 M HCl? Use data from the two previous problems.   *Ans.* 6.0×10^{-18} M

17.67. K_1 and K_2 for oxalic acid, $H_2C_2O_4$, are 5.9×10^{-2} and 6.4×10^{-5}. What is $[OH^-]$ in a 0.005 formal solution of $Na_2C_2O_4$?   *Ans.* 8.8×10^{-7} M

17.68. A buffer solution was prepared by dissolving 0.0200 mole propionic acid and 0.015 gfw sodium propionate in enough water to make a liter of solution. (*a*) What is the pH of the buffer? (*b*) What would be the pH change if 0.010 millimole HCl were added to 10 cc of the buffer? (*c*) What would be the pH change if 1.0×10^{-5} gfw NaOH were added to 10 cc of the buffer? K_a for propionic acid is 1.34×10^{-5}.   *Ans.* (*a*) 4.75, (*b*) −0.05, (*c*) +0.05

17.69. The base imidazole has a K_b of 8.1×10^{-8}. (*a*) In what amounts should 0.0200 M HCl and 0.0200 M imidazole be mixed to make 100 cc of a buffer at pH 7.00? (*b*) If the resulting buffer is diluted to one liter, what is the pH of the diluted buffer?   *Ans.* (*a*) 31 cc acid, 69 cc base; (*b*) 7.00

17.70. In the titration of HCl with NaOH represented in Fig. 17-1, Page 172, calculate the pH after the addition of a total of 20.0, 30.0, and 60.0 cc of NaOH.   *Ans.* 1.37, 1.60, 11.96

17.71. In the titration of β-hydroxybutyric acid, $HC_4H_7O_3$, with NaOH represented in Fig. 17-1, calculate the pH after the addition of a total of 20.0, 30.0, and 70.0 cc of NaOH. pK_a for $HC_4H_7O_3$ is 4.39.
Ans. 4.21, 4.57, 12.22

17.72. In the titration of HCl with NH_3 represented in Fig. 17-1, calculate the total volume of the NH_3 solution needed to bring the pH to 3.00 and to 8.00. K_b for NH_3 is 1.8×10^{-5}.
Ans. 49.0 cc, 52.8 cc

17.73. The dye bromcresol green has a pK_a value of 4.95. For which of the four titrations shown in Fig. 17-1 would bromcresol green be a suitable end point indicator?
Ans. HCl *vs.* NaOH, HCl *vs.* NH_3

17.74. Bromphenol blue is an indicator with a K_a value of 5.84×10^{-5}. What percentage of this indicator is in its basic form at a pH of 4.84?   *Ans.* 80%

17.75. Consider a solution of a monoprotic weak acid of acidity constant K_a. What is the minimum concentration C at which the concentration of the undissociated acid can be written as C within a 10% limit of error? Assume that activity coefficient corrections can be neglected.
Ans. $C = 90K_a$

17.76. What is the percentage ionization of 0.0065 M chloroacetic acid? K_a for this acid is 1.4×10^{-3}.
Ans. 37%

17.77. What concentration of dichloroacetic acid gives a $[H^+]$ of 8.5×10^{-3} mole/l? K_a for the acid is 5.5×10^{-2}.   *Ans.* 9.8×10^{-3} M

17.78. Calculate $[H^+]$ in a 0.200 M dichloroacetic acid solution that is also 0.100 F in sodium dichloro-acetate. K_a for dichloroacetic acid is 5.5×10^{-2}.   *Ans.* 0.053 M

17.79. How much solid sodium dichloroacetate should be added to a liter of 0.10 M dichloroacetic acid to reduce $[H^+]$ to 0.030 M? K_a for dichloroacetic acid is 5.5×10^{-2}. Neglect the increase in volume of the solution on addition of the salt.   *Ans.* 0.10 gfw

17.80. Calculate $[H^+]$ and $[C_2HO_2Cl_2{}^-]$ in a solution that is 0.010 M in HCl and 0.010 M in $HC_2HO_2Cl_2$. K_a for $HC_2HO_2Cl_2$ (dichloroacetic acid) is 5.5×10^{-2}.   *Ans.* 0.0176 M, 0.0076 M

17.81. Calculate $[H^+]$, $[C_2H_3O_2{}^-]$ and $[C_7H_5O_2{}^-]$ in a solution that is 0.0200 M in $HC_2H_3O_2$ and 0.0100 M in $HC_7H_5O_2$. K_a values for $HC_2H_3O_2$ and $HC_7H_5O_2$ are 1.8×10^{-5} and 6.5×10^{-5}, respectively.
Ans. 9.8×10^{-4}, 3.6×10^{-4}, and 6.2×10^{-4} mole/l

17.82. What is the pH of 7.0×10^{-8} M acetic acid? What is the concentration of un-ionized acetic acid? K_a is 1.8×10^{-5}. (*Hint:* Assume essentially complete ionization of the acetic acid in solving for $[H^+]$.)   *Ans.* 6.85, 5.4×10^{-10} M

17.83. Liquid ammonia ionizes to a slight extent. At $-50°$, its ion product, $K_{NH_3} = [NH_4{}^+][NH_2{}^-] = 10^{-30}$. How many amide ions, $NH_2{}^-$, are present per μl of pure liquid ammonia?   *Ans.* 6×10^2

17.84. Calculate the pH of 1.0×10^{-3} M sodium phenolate, $NaOC_6H_5$. K_a for HOC_6H_5 is 1.0×10^{-10}.
Ans. 10.43

17.85. Calculate $[H^+]$ and $[CN^-]$ in 0.0100 F NH_4CN. K_a for HCN is 4.8×10^{-10} and K_b for NH_3 is 1.8×10^{-5}.   *Ans.* 5.2×10^{-10}, 0.0048 mole/l

17.86. Calculate the pH and $[NH_3]$ at the end point in the titration of β-hydroxybutyric acid with NH_3, at the concentrations indicated for Fig. 17-1, Page 172. K_a for the acid is 4.1×10^{-5} and K_b for NH_3 is 1.8×10^{-5}.   *Ans.* 6.82, 1.8×10^{-4} M

17.87. Malonic acid is a dibasic acid having $K_1 = 1.4 \times 10^{-3}$ and $K_2 = 2.2 \times 10^{-6}$. Compute the concentration of the divalent malonate ion in (*a*) 0.0010 M malonic acid, (*b*) a solution that is 0.00010 M in malonic acid and 0.00040 M in HCl.   *Ans.* (*a*) 2.2×10^{-6}, (*b*) 3.4×10^{-7} mole/l

17.88. Compute the pH of a 0.010 M solution of H_3PO_4. K_1 and K_2 for H_3PO_4 are respectively 7.1×10^{-3} and 6.2×10^{-8}.   *Ans.* 2.25

17.89. What is $[H^+]$ in a 0.0060 M H_2SO_4 solution? The first ionization of H_2SO_4 is complete and the second ionization has a K_2 of 1.02×10^{-2}. What is $[SO_4{}^{2-}]$ in the same solution?
Ans. $[H^+] = 9.2 \times 10^{-3}$, $[SO_4{}^{2-}] = 3.2 \times 10^{-3}$

17.90. Citric acid is a polyprotic acid with pK_1, pK_2, and pK_3 equal to 3.13, 4.76, and 6.40, respectively. Calculate the concentrations of H^+, the monovalent anion, the divalent anion, and the trivalent anion in 0.0100 M citric acid.   *Ans.* 2.4×10^{-3}, 2.4×10^{-3}, 1.7×10^{-5}, 2.9×10^{-9} mole/l

17.91. If 0.0010 mole of citric acid is dissolved in a liter of a solution buffered at pH 5.00 (without changing the volume), what will be the equilibrium concentrations of citric acid, its monovalent anion, the divalent anion, and the trivalent anion? Use pK values from the preceding problem. *Ans.* 4.8×10^{-6}, 3.5×10^{-4}, 6.2×10^{-4}, 2.5×10^{-5} mole/l

17.92. If 0.00050 gfw $NaHCO_3$ is added to a large volume of a solution buffered at pH 8.00, how much material will exist in each of the three forms: H_2CO_3, HCO_3^-, and CO_3^{2-}? K_1 and K_2 values of H_2CO_3 are respectively 4.5×10^{-7} and 5.7×10^{-11}.
Ans. 1.1×10^{-5}, 4.9×10^{-4}, 2.8×10^{-6} mole, respectively

17.93. pK_1 and pK_2 for pyrophosphoric acid are 1.52 and 2.36 respectively. Neglecting the third and fourth dissociations of this tetraprotic acid, what would be the concentration of the divalent anion in a 0.050 M solution of the acid?   *Ans.* 3.4×10^{-3} M

17.94. What is $[CO_3^{2-}]$ in a 0.0010 F Na_2CO_3 solution after the hydrolysis reactions have come to equilibrium? K_1 and K_2 for H_2CO_3 are 4.5×10^{-7} and 5.7×10^{-11} respectively.
Ans. 6.6×10^{-4} mole/l

17.95. Calculate the pH of 0.050 F NaH_2PO_4 and of 0.0020 F Na_3PO_4. K_1, K_2, and K_3 of H_3PO_4 are 7.1×10^{-3}, 6.2×10^{-8}, and 4.4×10^{-13} respectively.   *Ans.* 4.7, 11.27

17.96. A buffer solution of pH 6.70 can be prepared by employing solutions of NaH_2PO_4 and Na_2HPO_4. If 0.0050 gfw NaH_2PO_4 is weighed out, how much Na_2HPO_4 must be used to make one liter of the solution? Use K values from the preceding problem.   *Ans.* 0.0016 gfw

17.97. How much NaOH must be added to a liter of 0.010 M H_3BO_3 to make a buffer solution of pH 10.10? H_3BO_3 is a monoprotic acid with $K_a = 5.8 \times 10^{-10}$.   *Ans.* 0.0089 gfw

17.98. A buffer solution was prepared by dissolving 0.050 mole formic acid and 0.060 gfw sodium formate in enough water to make a liter of solution. K_a for formic acid is 1.8×10^{-4}.
(a) Calculate the pH of the solution.
(b) If this solution were diluted to ten times its volume, what would be the pH?
(c) If the solution in (b) were diluted again to another ten times its volume, what would be the pH?
Ans. (a) 3.82, (b) 3.84, (c) 3.99

17.99. The amino acid glycine, NH_2CH_2COOH is basic because of its NH_2- group and acidic because of its $-COOH$ group. By a process formally equivalent to hydrolysis, glycine can acquire an additional proton to form $^+NH_3CH_2COOH$. The resulting cation may be considered to be a diprotic acid, since one proton from the $-COOH$ group and one proton from the $^+NH_3-$ group may be lost. The pK_a values for these processes are 2.35 and 9.77 respectively. In a 0.0100 M solution of neutral glycine, what is the pH and what percent of the glycine is in the cationic form at equilibrium?
Ans. 6.14, 0.016%

COMPLEX IONS

17.100. A 0.0010 gfw sample of solid NaCl was added to a liter of 0.010 F $Hg(NO_3)_2$. Calculate the $[Cl^-]$ equilibrated with the newly formed $HgCl^+$. K_1 for $HgCl^+$ formation is 4.7×10^6. Neglect the K_2 equilibrium.   *Ans.* 2×10^{-8} mole/l

17.101. What is the $[Cd^{2+}]$ in a liter of solution prepared by dissolving 0.001 gfw $Cd(NO_3)_2$ and 1.5 moles NH_3. K_d for the dissociation of $Cd(NH_3)_4^{2+}$ into Cd^{2+} and 4 NH_3 is 5.5×10^{-8}. Neglect the amount of cadmium in complexes containing less than 4 ammonia groups.
Ans. 1.1×10^{-11} mole/l

17.102. Silver ion forms $Ag(CN)_2^-$ in the presence of excess CN^-. How much KCN should be added to one liter of a 0.0005 M Ag^+ solution in order to reduce $[Ag^+]$ to 9×10^{-19} mole/l? K_d for the complete dissociation of $Ag(CN)_2^-$ into Ag^+ and $2\,CN^-$ is 1.8×10^{-19}.   *Ans.* 0.011 gfw

17.103. Calculate the concentration of free Ni^{2+} at equilibrium in a 10^{-4} F solution of $K_2Ni(CN)_4$. Neglect lower cyanide complexes. K_d for $Ni(CN)_4^{2-}$ is 10^{-22}.   *Ans.* 2×10^{-6} M

17.104. A recent investigation of the complexation of SCN^- with Fe^{3+} led to values of 130, 16, and 1.0 for K_1, K_2, and K_3 respectively. What is the overall formation constant of $Fe(SCN)_3$ from its component ions, and what is the dissociation constant of $Fe(SCN)_3$ into its simplest ions?
Ans. $K_{\text{overall}} = 2.1 \times 10^3$, $K_d = 5 \times 10^{-4}$

17.105. Sr^{2+} forms a very unstable complex with NO_3^-. A solution that was nominally 0.00100 F $Sr(ClO_4)_2$ and 0.050 F KNO_3 was found to have only 75% of its strontium in the uncomplexed Sr^{2+} form, the balance being $Sr(NO_3)^+$. What is K_1 for complexation?   *Ans.* 6.7

17.106. A solution made up to be 0.0100 F $Cd(NO_3)_2$ and 0.0200 M N_2H_4 was found to have an equilibrium $[Cd^{2+}]$ of 3.0×10^{-3} mole/l. Assuming that the only complex formed was $Cd(N_2H_4)^{2+}$, what is K_1 for complex formation?   *Ans.* 1.8×10^2

17.107. Equal volumes of 0.0010 F $Fe(ClO_4)_3$ and 0.10 F KSCN were mixed. Using the data given in Problem 17-104, find the equilibrium percentages of the iron existing as Fe^{3+}, $FeSCN^{2+}$, $Fe(SCN)_2^+$, and $Fe(SCN)_3$.   *Ans.* 8%, 50%, 40%, 2%

17.108. A solution was made to be 0.00050 M in all forms of Pb(II) and 0.040 M in Cl^-. K_1 and K_2 for complexation of Pb^{2+} with Cl^- are 12.5 and 14.5 respectively. What fractions of Pb(II) are Pb^{2+}, $PbCl^+$, and $PbCl_2$ at equilibrium?   *Ans.* 56%, 28%, 16%

17.109. What is the concentration of free Cd^{2+} in 0.005 F $CdCl_2$? K_1 for chloride complexation of Cd^{2+} is 100. K_2 need not be considered.   *Ans.* 2.8×10^{-3} mole/l

Chapter 18

Solubility Product and Precipitation

SOLUBILITY PRODUCT PRINCIPLE

Consider the equilibrium between solid AgCl and its dissolved ions in a saturated solution:

$$AgCl \ (solid) \ \rightleftharpoons \ Ag^+ + Cl^- \ (in \ solution)$$

The expression for the equilibrium constant is

$$K_{sp} = [Ag^+] \times [Cl^-]$$

where the convention of excluding terms for solids has been honored. This particular case of the equilibrium constant for the solution of ionic substances is called the *solubility product* (K_{sp}).

In a saturated solution of a slightly soluble salt (such as AgCl), the product of the molar concentrations of the dissolved ions (each concentration in moles per liter raised to an appropriate power) is a constant (K_{sp}) at a given temperature.

$$K_{sp} \ of \ BaCO_3 \ = \ [Ba^{2+}] \times [CO_3{}^{2-}]$$
$$K_{sp} \ of \ CaF_2 \ = \ [Ca^{2+}] \times [F^-]^2$$
$$K_{sp} \ of \ Bi_2S_3 \ = \ [Bi^{3+}]^2 \times [S^{2-}]^3$$

As in the previous chapter, the temperature will be taken as 25°C unless a statement is made to the contrary.

APPLICATIONS OF SOLUBILITY PRODUCT TO PRECIPITATION

1. **Precipitation.** The solubility product principle enables us to explain and predict the completeness of precipitation reactions. Whenever the product of the appropriate powers of the concentrations of any two ions in a solution exceeds the value of the corresponding solubility product, the cation-anion combinations will be precipitated until the product of the concentrations of these two ions remaining in solution (raised to their respective powers) again attains the value of the solubility product.

For example: When some NaF is added to a saturated solution of CaF_2, the $[F^-]$ is greatly increased and the product of the ion concentrations may temporarily exceed the value of the solubility product. To restore equilibrium, some Ca^{2+} unites with an equivalent quantity of F^- to form solid CaF_2 until the product of the concentrations of ions remaining in solution, $[Ca^{2+}] \times [F^-]^2$, again attains the value of the solubility product. Note that in this case the final value of $[F^-]$ is much greater than twice that of $[Ca^{2+}]$, since NaF yields a large contribution to the total $[F^-]$.

The ion concentrations that appear in solubility product equations refer only to the simple ions *in solution*, and do not include the material in the precipitate. Additional equilibria may exist between the simple ions and complexes *in solution* in case soluble complexes form; in such cases the normal rules for complex stability constants are followed.

2. **Solution of Precipitates.** Whenever the product of the concentrations of any two ions (raised to the appropriate powers) in a solution is less than the value of the corresponding solubility product, the solution is not saturated. If some of the corresponding solid salt is added to the solution, more of the salt will dissolve.

For example: If hydrochloric acid (which supplies H^+) is added to a solution of $Mg(OH)_2$, the H^+ removes nearly all the OH^- to form H_2O. This greatly decreases the $[OH^-]$, and more $Mg(OH)_2$ may dissolve so that the ion concentration product may again attain the value of K_{sp} for $Mg(OH)_2$.

3. **Prevention of Precipitation.** To prevent the precipitation of a slightly soluble salt, some substance must be added which will keep the concentration of one of the ions so low that the solubility product of the slightly soluble salt is not attained.

For example: H_2S will not precipitate FeS from a strongly acid (HCl) solution of Fe^{2+}. The large $[H^+]$ furnished by the hydrochloric acid represses the ionization of H_2S (common ion effect) and thus reduces the $[S^{2-}]$ to so low a value that the solubility product of FeS is not reached.

Solved Problems

18.1. The solubility of $PbSO_4$ in water is 0.038 gram per liter. Calculate the solubility product of $PbSO_4$.

$$PbSO_4 \ (solid) \ \rightleftharpoons \ Pb^{2+} + SO_4^{2-} \ (in \ solution)$$

The concentrations of the ions must be expressed in moles per liter. To convert 0.038 g/l to moles of ions per liter, divide by the formula weight of $PbSO_4$ (303).

$$0.038 \ g/l \ = \ \frac{0.038 \ g/l}{303 \ g/gfw} \ = \ 1.25 \times 10^{-4} \ gfw/l$$

One gfw of dissolved $PbSO_4$ yields one mole each of Pb^{2+} and SO_4^{2-}. Then 1.25×10^{-4} gfw $PbSO_4$ yields 1.25×10^{-4} mole each of Pb^{2+} and SO_4^{2-}. Substituting these concentration values in the solubility product equation,

$$K_{sp} \ = \ [Pb^{2+}] \times [SO_4^{2-}] \ = \ (1.25 \times 10^{-4})(1.25 \times 10^{-4}) \ = \ 1.6 \times 10^{-8}$$

This method may be applied to any fairly insoluble salt whose ions do not hydrolyze appreciably or form soluble complexes. Sulfides, carbonates, and phosphates, and salts of many of the transition metals like iron, must be treated by taking into account hydrolysis and, in some cases, complexation. Such examples will be given below.

18.2. The solubility of Ag_2CrO_4 in water is 0.024 gram per liter. Determine the solubility product.

$$Ag_2CrO_4 \ \rightleftharpoons \ 2 \ Ag^+ + CrO_4^{2-}$$

To convert 0.024 g/l to moles/l of the ions, divide by the formula weight of Ag_2CrO_4 (332).

$$0.024 \ g/l \ = \ \frac{0.024 \ g/l}{332 \ g/gfw} \ = \ 7.2 \times 10^{-5} \ gfw/l$$

One gfw of dissolved Ag_2CrO_4 yields 2 moles Ag^+ and 1 mole CrO_4^{2-}. Therefore $[Ag^+] = 2(7.2 \times 10^{-5}) = 1.4 \times 10^{-4}$ mole/l and $[CrO_4^{2-}] = (7.2 \times 10^{-5})$ mole/l.

$$K_{sp} \ = \ [Ag^+]^2 \times [CrO_4^{2-}] \ = \ (1.4 \times 10^{-4})^2(7.2 \times 10^{-5}) \ = \ 1.4 \times 10^{-12}$$

18.3. The concentration of the Ag^+ ion in a saturated solution of $Ag_2C_2O_4$ is 2.2×10^{-4} mole per liter. Compute the solubility product of $Ag_2C_2O_4$.

$$Ag_2C_2O_4 \rightleftharpoons 2\,Ag^+ + C_2O_4{}^{2-}$$

The equation indicates that the concentration of $C_2O_4{}^{2-}$ is half that of Ag^+.

$[Ag^+] = 2.2 \times 10^{-4}$ mole/l; then $[C_2O_4{}^{2-}] = \frac{1}{2}(2.2 \times 10^{-4}) = 1.1 \times 10^{-4}$ mole/l.

$$K_{sp} = [Ag^+]^2 \times [C_2O_4{}^{2-}] = (2.2 \times 10^{-4})^2(1.1 \times 10^{-4}) = 5 \times 10^{-12}$$

18.4. The solubility product of $Pb(IO_3)_2$ is 2.5×10^{-13}. What is the solubility of $Pb(IO_3)_2$ (a) in gfw per liter and (b) in grams per liter?

(a) Let $x =$ solubility of $Pb(IO_3)_2$ in gfw/l. Then $[Pb^{2+}] = x$ and $[IO_3^-] = 2x$.

$$[Pb^{2+}] \times [IO_3^-]^2 = K_{sp}$$

Substituting, $\qquad\qquad x \times (2x)^2 = 2.5 \times 10^{-13}$

Then $4x^3 = 2.5 \times 10^{-13}$, $x^3 = 62 \times 10^{-15}$ and $x = 4.0 \times 10^{-5}$ gfw/l.

(b) To convert 4.0×10^{-5} gfw/l to g/l, multiply by the formula weight of $Pb(IO_3)_2$.

Solubility $= 4.0 \times 10^{-5}$ gfw/$l \times 557$ g/gfw $= 0.022$ g/l.

18.5. The $[Ag^+]$ of a solution is 4×10^{-3} mole/l. Calculate the $[Cl^-]$ that must be exceeded before AgCl can precipitate. The solubility product of AgCl at 25°C is 1.8×10^{-10}.

$[Ag^+] \times [Cl^-] = K_{sp}$. Substituting, $(4 \times 10^{-3})[Cl^-] = 1.8 \times 10^{-10}$ or $[Cl^-] = 5 \times 10^{-8}$ mole/l.

Hence a $[Cl^-]$ of 5×10^{-8} mole/l must be exceeded before AgCl precipitates. This problem differs from the previous ones in that the two ions forming the precipitate are furnished to the solution independently. This represents a typical analytical situation, in which some soluble chloride is added to precipitate silver ion present in a solution.

18.6. Calculate the solubility of AgCN in a buffer solution of pH 3.00. K_{sp} for AgCN is 1.2×10^{-16} and K_a for HCN is 4.8×10^{-10}.

In this solution the silver that dissolves remains as Ag^+, but the cyanide that dissolves is converted mostly to HCN on account of the fixed acidity of the buffer. (The complex $Ag(CN_2)^-$ forms appreciably only at higher cyanide ion concentrations.) First calculate the ratio of [HCN] to $[CN^-]$ at this pH.

$$\frac{[H^+] \times [CN^-]}{[HCN]} = K_a \qquad \text{or} \qquad \frac{[HCN]}{[CN^-]} = \frac{[H^+]}{K_a} = \frac{1.0 \times 10^{-3}}{4.8 \times 10^{-10}} = 2.1 \times 10^6$$

Let $x =$ solubility of AgCN, in gfw/l; then

$x = [Ag^+]$, at equilibrium

$x = [CN^-] + [HCN]$.

Very little error is made by neglecting $[CN^-]$ in comparison to [HCN] (1 part in 2.1 million) and equating [HCN] to x.

$$[CN^-] = \frac{[HCN]}{2.1 \times 10^6} = \frac{x}{2.1 \times 10^6}$$

Substituting in the solubility product equation $[Ag^+] \times [CN^-] = 1.2 \times 10^{-16}$,

$$x \times \frac{[HCN]}{2.1 \times 10^6} = \frac{x \times x}{2.1 \times 10^6} = 1.2 \times 10^{-16}$$

from which $x^2 = (1.2 \times 10^{-16})(2.1 \times 10^6) = 2.5 \times 10^{-10}$ and $x = 1.6 \times 10^{-5}$ gfw/l.

18.7. Calculate the NH_4^+ ion concentration (derived from NH_4Cl) needed to prevent $Mg(OH)_2$ from precipitating in a liter of solution which contains 0.010 mole of ammonia and 0.0010 mole of Mg^{2+}. The ionization constant of ammonia is 1.8×10^{-5}. The solubility product of $Mg(OH)_2$ is 1.12×10^{-11}.

(1) To find the maximum $[OH^-]$ that can be present in the solution in order that $Mg(OH)_2$ will not precipitate:

$$[Mg^{2+}] \times [OH^-]^2 = 1.12 \times 10^{-11}$$

$$[OH^-] = \sqrt{\frac{1.12 \times 10^{-11}}{[Mg^{2+}]}} = \sqrt{\frac{1.12 \times 10^{-11}}{0.0010}} = \sqrt{1.12 \times 10^{-8}} = 1.1 \times 10^{-4} \text{ mole}/l$$

$Mg(OH)_2$ will not precipitate if the $[OH^-]$ does not exceed 1.1×10^{-4} mole/l.

(2) To find the $[NH_4^+]$ (derived from NH_4Cl) needed to repress the ionization of NH_3 so that the $[OH^-]$ will not exceed 1.1×10^{-4} mole/l:

$$\frac{[NH_4^+] \times [OH^-]}{[NH_3]} = 1.8 \times 10^{-5}, \qquad \frac{[NH_4^+] (1.1 \times 10^{-4})}{0.010} = 1.8 \times 10^{-5},$$

$$[NH_4^+] = 1.6 \times 10^{-3} \text{ mole}/l$$

(Since 0.010 M ammonia is only slightly ionized, especially in the presence of excess NH_4^+, the $[NH_3]$ may be considered to be 0.010 mole/liter.)

18.8. Given that 2×10^{-4} mole each of Mn^{2+} and Cu^{2+} was contained in one liter of a 0.003 M $HClO_4$ solution, and this solution was saturated with H_2S. Determine whether or not each of these ions, Mn^{2+} and Cu^{2+}, will precipitate as the sulfide. The solubility of H_2S, 0.10 mole/liter, is assumed to be independent of the presence of other materials in the solution. K_{sp} of MnS $= 5 \times 10^{-15}$; of CuS $= 9 \times 10^{-36}$. The K_1 and K_2 values for H_2S are 1.0×10^{-7} and 1.2×10^{-13} respectively.

$[H_2S] = 0.10$ mole/l, since the solution is saturated with H_2S.

$[H^+] = 0.003$ mole/l, since the H_2S contributes negligible H^+ compared with $HClO_4$.

Calculate $[S^{2-}]$ from the combined ionization constants.

$$\frac{[H^+]^2 \times [S^{2-}]}{[H_2S]} = K_1 \times K_2 = (1.0 \times 10^{-7})(1.2 \times 10^{-13}) = 1.2 \times 10^{-20}$$

$$[S^{2-}] = 1.2 \times 10^{-20} \times \frac{[H_2S]}{[H^+]^2} = 1.2 \times 10^{-20} \times \frac{0.10}{(0.003)^2} = 1.3 \times 10^{-16} \text{ mole}/l$$

Whenever the product of the concentrations of two ions (e.g., $[Cu^{2+}] \times [S^{2-}]$) in a solution exceeds the value of K_{sp} (e.g., when $[Cu^{2+}] \times [S^{2-}] > 9 \times 10^{-36}$), the corresponding ionic compound (CuS) will be precipitated until the product of the concentrations of these two ions again attains the value K_{sp}.

First let us tabulate numbers that would describe the solution if no precipitation occurred.

$$[Mn^{2+}] = [Cu^{2+}] = 2 \times 10^{-4} \text{ mole}/l$$

$$[Mn^{2+}] \times [S^{2-}] = (2 \times 10^{-4})(1.3 \times 10^{-16}) = 2.6 \times 10^{-20}$$

$$[Cu^{2+}] \times [S^{2-}] = (2 \times 10^{-4})(1.3 \times 10^{-16}) = 2.6 \times 10^{-20}$$

2.6×10^{-20} is less than the solubility product of MnS, 5×10^{-15}. Hence MnS will not precipitate. The large $[H^+]$ furnished by the HCl represses the ionization of the H_2S (common H^+ ion) and thus reduces the $[S^{2-}]$ to so low a value that the solubility product of MnS is not reached.

2.6×10^{-20} is greater than the solubility product of CuS, 9×10^{-36}. Hence CuS will precipitate.

18.9. In the above problem, how much Cu^{2+} escapes precipitation?

$$[Cu^{2+}] = \frac{K_{sp}}{[S^{2-}]} = \frac{9 \times 10^{-36}}{1.3 \times 10^{-16}} = 7 \times 10^{-20} \text{ mole}/l, \text{ which remains in solution. On a per-}$$

centage basis the amount Cu^{2+} remaining unprecipitated is $\dfrac{7 \times 10^{-20}}{2 \times 10^{-4}} \times 100 = 3 \times 10^{-14}\%$.

18.10. If the solution in the above problem is made neutral by lowering the $[H^+]$ to 10^{-7} mole per liter, will MnS precipitate?

$$[S^{2-}] = 1.2 \times 10^{-20} \times \frac{0.10}{(10^{-7})^2} = 1.2 \times 10^{-7} \text{ mole}/l$$

Then, if no precipitation occurred, $[Mn^{2+}] \times [S^{2-}] = (2 \times 10^{-4})(1.2 \times 10^{-7}) = 2.4 \times 10^{-11}$.

2.4×10^{-11} is greater than the K_{sp} of MnS, 5×10^{-15}. Hence MnS will precipitate.

Lowering the $[H^+]$ to 10^{-7} mole/l increases the ionization of H_2S to such an extent as to furnish a $[S^{2-}]$ sufficient to exceed the solubility product of MnS.

18.11. How much NH_3 must be added to a 0.004 M Ag^+ solution to prevent the precipitation of AgCl when $[Cl^-]$ reaches 0.001 M? K_{sp} for AgCl is 1.8×10^{-10} and K_d for $Ag(NH_3)_2{}^+$ is 5.9×10^{-8}.

Just as acids may be used to lower the concentration of anions in solution, so complexing agents may be used in some cases to lower the concentration of cations. In this problem, the addition of NH_3 converts most of the silver to the complex ion, $Ag(NH_3)_2{}^+$. The upper limit for the uncomplexed $[Ag^+]$ without formation of a precipitate can be calculated from the solubility product.

$$[Ag^+] \times [Cl^-] = 1.8 \times 10^{-10} \qquad [Ag^+] = \frac{1.8 \times 10^{-10}}{[Cl^-]} = \frac{1.8 \times 10^{-10}}{0.001} = 1.8 \times 10^{-7}$$

Enough NH_3 must be added to keep the $[Ag^+]$ below this limiting value of 1.8×10^{-7}. The concentration of $Ag(NH_3)_2{}^+$ at this limit would then be $0.004 - 1.8 \times 10^{-7}$, or practically 0.004.

$$\frac{[Ag^+] \times [NH_3]^2}{[Ag(NH_3)_2{}^+]} = K_d \qquad [NH_3]^2 = \frac{K_d \times [Ag(NH_3)_2{}^+]}{[Ag^+]} = \frac{(5.9 \times 10^{-8})(0.004)}{1.8 \times 10^{-7}}$$

$$= 1.31 \times 10^{-3}, \quad \text{and} \quad [NH_3] = 0.036 \text{ mole}/l$$

The amount of NH_3 that must be added is equal to the sum of the amount of free NH_3 remaining in the solution plus the amount of NH_3 used up in forming 0.004 mole per liter of the complex ion, $Ag(NH_3)_2{}^+$. This sum is $0.036 + 2(0.004) = 0.044$ mole NH_3 to be added per liter.

18.12. What is the solubility of AgSCN in 0.003 M NH_3?

K_{sp} for AgSCN is 1.0×10^{-12}, and K_d for $Ag(NH_3)_2{}^+$ is 5.9×10^{-8}.

We may assume that practically all the dissolved silver will exist as the complex ion, $Ag(NH_3)_2{}^+$. Then, if x is the solubility of AgSCN in gfw/l, $x = [SCN^-] = [Ag(NH_3)_2{}^+]$. The concentration of uncomplexed Ag^+ must be computed from the equilibrium constant for the complex ion dissociation.

$$\frac{[Ag^+] \times [NH_3]^2}{[Ag(NH_3)_2{}^+]} = K_d \qquad [Ag^+] = \frac{K_d \times [Ag(NH_3)_2{}^+]}{[NH_3]^2} = \frac{(5.9 \times 10^{-8})x}{(0.003)^2} = 6.6 \times 10^{-3}\,x$$

This result confirms the validity of the assumption we made that practically all the dissolved silver is in the complex. The ratio of uncomplexed to complexed silver in solution is 6.6×10^{-3}. The above result may now be used in the solubility product equation.

$$[Ag^+] \times [SCN^-] = K_{sp}, \qquad (6.6 \times 10^{-3}\,x)x = 1.0 \times 10^{-12}, \qquad x = 1.2 \times 10^{-5} \text{ gfw}/l$$

The validity of another assumption is confirmed by the answer. If 1.2×10^{-5} mole of complex is formed per liter, the amount of NH_3 used up for complex formation is $2(1.2 \times 10^{-5}) = 2.4 \times 10^{-5}$ mole/l. The concentration of the remaining free NH_3 in solution is practically unchanged from its initial value, 0.003.

18.13. Calculate the simultaneous solubility of CaF_2 and SrF_2. K_{sp} for these two salts are 4.0×10^{-11} and 2.8×10^{-9} respectively.

The two solubilities are not independent of each other because there is a common ion, F^-. We will first assume that most of the F^- in the saturated solution is contributed by the SrF_2 since its K_{sp} is so much larger than that of CaF_2. We can then proceed to solve for the solubility of SrF_2 as if the CaF_2 were not present.

Let x = solubility of SrF_2 = $[Sr^{2+}]$, $2x$ = $[F^-]$. Then

$$4x^3 = K_{sp} = 2.8 \times 10^{-9} \qquad \text{and} \qquad x = 9 \times 10^{-4} \text{ mole/}l, \text{ solubility of } SrF_2$$

The CaF_2 solubility will have to adapt to the concentration of F^- set by the SrF_2 solubility.

$$[Ca^{2+}] = \frac{K_{sp}}{[F^-]^2} = \frac{4.0 \times 10^{-11}}{(2 \times 9 \times 10^{-4})^2} = 1.2 \times 10^{-5} \qquad \text{Solubility of } CaF_2 = 1.2 \times 10^{-5} \text{ gfw/}l$$

Check of assumption: The amount of F^- contributed by the solubility of CaF_2 is twice $[Ca^{2+}]$, or 2.4×10^{-5} mole/l. This is indeed small compared with the amount contributed by SrF_2, $2 \times 9 \times 10^{-4}$, or 1.8×10^{-3} mole/l.

18.14. Calculate the simultaneous solubility of AgSCN and AgBr. The solubility products for these two salts are 1.0×10^{-12} and 5.0×10^{-13}.

It would be dangerous to proceed as in the previous problem by assuming that the more soluble salt, AgSCN in this case, provides all the common ion, Ag^+. The two K_{sp} values do not differ by very much, so that the contribution of AgBr cannot be neglected. Instead we must solve for both equilibria simultaneously.

$$[Ag^+][SCN^-] = 1.0 \times 10^{-12} \tag{1}$$

$$[Ag^+][Br^-] = 5.0 \times 10^{-13} \tag{2}$$

Dividing (1) by (2),
$$\frac{[SCN^-]}{[Br^-]} = 2.0 \tag{3}$$

Because of the need for electrical charge balance,

$$[SCN^-] + [Br^-] = [Ag^+] \tag{4}$$

Dividing (4) by $[Br^-]$,
$$\frac{[SCN^-]}{[Br^-]} + \frac{[Br^-]}{[Br^-]} = \frac{[Ag^+]}{[Br^-]} \tag{5}$$

Substituting from (3),
$$2.0 + 1.0 = 3.0 = \frac{[Ag^+]}{[Br^-]} \tag{6}$$

Substituting (6) into (2),

$$3.0\,[Br^-]^2 = 5.0 \times 10^{-13} \qquad \text{or} \qquad [Br^-] = 4.1 \times 10^{-7}$$

The remaining concentrations can be solved by simple substitutions in (1) and (3):

$$[Ag^+] = 1.2 \times 10^{-6}, \qquad [SCN^-] = 8 \times 10^{-7}$$

The solubilities are 4×10^{-7} gfw AgBr/l and 8×10^{-7} gfw AgSCN/l.

18.15. Calculate the solubility of MnS in pure water. K_{sp} is 5×10^{-15}. K_1 and K_2 for H_2S are 1.0×10^{-7} and 1.2×10^{-13}.

This problem differs from the analogous problems dealing with chromates, oxalates, halides, sulfates, and iodates. The difference lies in the extensive hydrolysis of the sulfide ion.

$$S^{2-} + H_2O \rightleftharpoons HS^- + OH^- \qquad K_h = \frac{K_w}{K_2} = \frac{10^{-14}}{1.2 \times 10^{-13}} = 0.083$$

If x is the solubility of MnS, we cannot equate x to $[S^{2-}]$. Instead, $x = [S^{2-}] + [HS^-] + [H_2S]$. To simplify, we first assume that the first stage of hydrolysis is almost complete and that the second stage proceeds to only a slight extent. In other words, $x = [Mn^{2+}] = [HS^-] = [OH^-]$.

$$[S^{2-}] = \frac{[HS^-][OH^-]}{K_h} = \frac{x^2}{0.083}$$

At equilibrium,

$$[Mn^{2+}][S^{2-}] = \frac{x(x)^2}{.083} = K_{sp} = 5 \times 10^{-15} \qquad \text{or} \qquad x = 7 \times 10^{-6} \text{ gfw/}l$$

Check of assumptions:

(1) $[S^{2-}] = \dfrac{x^2}{0.083} = \dfrac{(7 \times 10^{-6})^2}{.083} = 6 \times 10^{-10}$ mole/l

$[S^{2-}]$ is indeed negligible compared with $[HS^-]$.

(2) $[H_2S] = \dfrac{[H^+][HS^-]}{K_1} = \dfrac{K_w \times [HS^-]}{[OH^-] \times K_1} = \dfrac{10^{-14}x}{10^{-7}x} = 10^{-7}$

$[H_2S]$ is also small compared with $[HS^-]$.

The above approximations would not be valid for sulfides like CuS which are much more insoluble than MnS. First, the water dissociation would begin to play an important role in determining $[OH^-]$. Secondly, the second stage of hydrolysis, producing $[H_2S]$, would not be negligible compared with the first. Even for MnS an additional complication arises because of the complexation of Mn^{2+} with OH^-. The full treatment of sulfide solubilities is a complicated problem because of the multiple equilibria which must be considered.

18.16. The following solutions were mixed: 500 cc of 0.0100 F $AgNO_3$ and 500 cc of a solution that was both 0.0100 F in NaCl and 0.0100 F in NaBr. K_{sp} for AgCl and for AgBr are 1.8×10^{-10} and 5.0×10^{-13}. Calculate $[Ag^+]$, $[Cl^-]$, and $[Br^-]$ in the equilibrium solution.

If there were no precipitation, the diluting effect of mixing would make $[Ag^+] = [Cl^-] = [Br^-] = 0.0050$ M. AgBr is the more insoluble salt and would take precedence in the precipitation process. To find whether AgCl also precipitates, we may assume that it does not. In this case, only Ag^+ and Br^- would be removed by precipitation, and the concentrations of these two remaining ions would remain equal to each other.

$$[Ag^+][Br^-] = [Ag^+]^2 = K_{sp} = 5.0 \times 10^{-13}$$

$$[Ag^+] = [Br^-] = 7.1 \times 10^{-7}$$

We now examine the ion product for AgCl.

$$[Ag^+][Cl^-] = (7.1 \times 10^{-7})(5.0 \times 10^{-3}) = 3.5 \times 10^{-9}$$

Since this ion product exceeds K_{sp} for AgCl, AgCl must also precipitate. In other words, our first assumption was wrong.

Since both halides precipitate, both solubility product requirements must be met simultaneously.

$$[Ag^+][Cl^-] = 1.8 \times 10^{-10} \qquad\qquad (1)$$

$$[Ag^+][Br^-] = 5.0 \times 10^{-13} \qquad\qquad (2)$$

The third equation needed to define the three unknowns is an equation expressing the balancing of positive and negative charges in solution.

$$[Na^+] + [Ag^+] = [Cl^-] + [Br^-] + [NO_3^-]$$

$$.0100 + [Ag^+] = [Cl^-] + [Br^-] + .0050$$

or $\qquad\qquad [Cl^-] + [Br^-] - [Ag^+] = .0050 \qquad\qquad (3)$

Dividing (1) by (2), $\qquad\qquad \dfrac{[Cl^-]}{[Br^-]} = 360$

we see that Br^- plays a negligible role in the total anion concentration of the solution. Also, $[Ag^+]$ must be negligible in equation (3) because of the insolubility of the two silver salts. We thus assume in (3) that $[Cl^-] = .0050$. From (1),

$$[Ag^+] = \dfrac{1.8 \times 10^{-10}}{[Cl^-]} = \dfrac{1.8 \times 10^{-10}}{.005} = 3.6 \times 10^{-8}$$

From (2), $\qquad [Br^-] = \dfrac{5.0 \times 10^{-13}}{[Ag^+]} = \dfrac{5.0 \times 10^{-13}}{3.6 \times 10^{-8}} = 1.4 \times 10^{-5}$

Check of assumptions: Both $[Ag^+]$ and $[Br^-]$ are negligible compared with .0050.

18.17. How much Ag^+ would remain in solution after mixing equal volumes of 0.080 F $AgNO_3$ and 0.080 F HOCN? K_{sp} for AgOCN is 2.3×10^{-7}. K_a for HOCN is 2.2×10^{-4}.

We must consider both the solubility and the acid ionization equilibria.

$$HOCN \rightleftharpoons H^+ + OCN^- \qquad\qquad Ag^+ + OCN^- \rightleftharpoons AgOCN$$

We can think of precipitation as a process which forces HOCN to ionize. H^+ will thus accumulate in the solution as precipitation occurs. To a first approximation, one H^+ must replace every Ag^+ removed by precipitation to provide electrical balance with respect to the NO_3^-. (This approximation is true so long as OCN^- does not add appreciably to the total equilibrium anion concentration and require still more H^+.) To this approximation, the following equations hold. Let

$$x = [Ag^+]$$
$$[NO_3^-] = 0.040 \quad \text{(2-fold dilution of .080 F solution)}$$
$$\text{amt. of } Ag^+ \text{ precipitated} = (.040 - x) \text{ mole}/l$$
$$\text{amt. of } OCN^- \text{ precipitated} = (.040 - x) \text{ mole}/l$$
$$[HOCN] = x - [OCN^-] = x \text{ (approx.)}$$
$$[H^+] = (.040 - x) \text{ mole}/l$$

Then $\qquad \dfrac{[OCN^-]}{[HOCN]} = \dfrac{K_a}{[H^+]} = \dfrac{2.2 \times 10^{-4}}{.040 - x} \qquad$ or $\qquad [OCN^-] = \dfrac{2.2 \times 10^{-4}\, x}{.040 - x}$

and $\qquad [Ag^+][OCN^-] = \dfrac{2.2 \times 10^{-4}\, x^2}{.040 - x} = K_{sp} = 2.3 \times 10^{-7}$ or $\quad x = [Ag^+] = 6.0 \times 10^{-3}$

Check of assumptions: $[OCN^-] = \dfrac{2.2 \times 10^{-4}\, x}{.040 - x} = 3.9 \times 10^{-5}$, small compared with $[NO_3^-]$ and with x.

18.18. Calculate the solubility of AgOCN in 0.0010 M HNO_3.

K_a for HOCN is 2.2×10^{-4} and K_{sp} is 2.3×10^{-7}.

This is a case where no simple approximation can be justified. The dissolved cyanate is not predominantly in a single species at equilibrium, but both $[OCN^-]$ and $[HOCN]$ are of equal orders of magnitude. The student may verify that an assumption that either of these two species may be neglected in comparison with the other leads to an inconsistency.

Let x = solubility of AgOCN in gfw/l. The two equilibria that must be satisfied are:

$$[Ag^+][OCN^-] = K_{sp} = 2.3 \times 10^{-7} \tag{1}$$

and $\qquad \dfrac{[H^+][OCN^-]}{[HOCN]} = K_a = 2.2 \times 10^{-4} \tag{2}$

Mass and charge balance impose the following requirements:

$$x = [Ag^+] = [HOCN] + [OCN^-] \tag{3}$$
(The dissolved anion must remain as OCN^- or be converted to HOCN.)

$$[H^+] = 0.0010 - [HOCN] \tag{4}$$
(H^+ is partly used up to form HOCN.)

Algebraically, the four equations are sufficient for determination of the four unknowns. By substitution, a cubic equation could be obtained in one unknown. Rather than proceed rigorously to the cubic, we will approach the problem by a method of successive approximations. We will employ the following steps:

(a) Estimate x by equation (3): $x = [HOCN] + [OCN^-]$.

(b) Estimate $[OCN^-]$ by equation (1): $[OCN^-] = \dfrac{2.3 \times 10^{-7}}{x}$.

(c) Estimate $[HOCN]$ by (2) with the substitution indicated in (4):

$$[HOCN] = \dfrac{[H^+][OCN^-]}{2.2 \times 10^{-4}} = \dfrac{(0.0010 - [HOCN])[OCN^-]}{2.2 \times 10^{-4}} = \dfrac{0.0010\,[OCN^-]}{2.2 \times 10^{-4} + [OCN^-]}$$

(d) Return to step a and repeat the cycle, each time using the most recent approximate values of the variables. Stop the process when the x values determined in step a for two successive cycles are sufficiently close to each other.

For the first cycle of approximations we must have some provisional values of [HOCN] and [OCN$^-$]. A possible first assumption is that [HOCN] = 0, that [OCN$^-$] = x and that from (1), $x = \sqrt{2.3 \times 10^{-7}} = 4.8 \times 10^{-4}$. A table of the calculations based on this assumption follows.

Approximation	(a) x	(b) [OCN$^-$]	(c) [HOCN]
1.	4.8×10^{-4}	4.8×10^{-4}	6.8×10^{-4}
2.	11.6×10^{-4}	2.0×10^{-4}	4.8×10^{-4}
3.	6.8×10^{-4}	3.4×10^{-4}	6.1×10^{-4}
4.	9.5×10^{-4}	2.4×10^{-4}	5.2×10^{-4}
5.	7.6×10^{-4}	3.0×10^{-4}	5.8×10^{-4}
6.	8.8×10^{-4}	2.6×10^{-4}	5.4×10^{-4}
7.	8.0×10^{-4}	2.9×10^{-4}	5.7×10^{-4}
8.	8.6×10^{-4}	2.7×10^{-4}	5.5×10^{-4}
9.	8.2×10^{-4}	2.8×10^{-4}	5.6×10^{-4}
10.	8.4×10^{-4}	2.7×10^{-4}	5.5×10^{-4}
11.	8.2×10^{-4}		

Note the oscillating nature of the approximate x values and the decreasing amplitude of the oscillations. This is a good sign that we have "zeroed in" on the correct x value. We can take as the solution the average of the last two values between which the solution according to the above method oscillates: $x = 8.3 \times 10^{-4} = [Ag^+]$; [OCN$^-$] = 2.8×10^{-4}; [HOCN] = 5.5×10^{-4}. A final check is obtained by substituting these values into equations (1)-(4).

The choice of the first approximate value of x was somewhat arbitrary. Because of the convergence of the procedure, it would not have mattered if some other value of x had been used, even a guessed value of the right order of magnitude.

Supplementary Problems

18.19. Calculate the solubility products of the following compounds. The solubilities are given in gram-formula weights per liter. (a) BaSO$_4$, 3.2×10^{-5} gfw/l; (b) TlBr, 0.0019 gfw/l; (c) Mg(OH)$_2$, 1.4×10^{-4} gfw/l; (d) Ag$_2$C$_2$O$_4$, 1.1×10^{-4} gfw/l; (e) La(IO$_3$)$_3$, 6.9×10^{-4} gfw/l.
Ans. (a) 1.0×10^{-9}, (b) 3.6×10^{-6}, (c) 1.1×10^{-11}, (d) 5×10^{-12}, (e) 6×10^{-12}

18.20. Calculate the solubility products of the following salts. Solubilities are given in grams per liter. (a) CaC$_2$O$_4$, 0.0055 g/l; (b) BaCrO$_4$, 2.8×10^{-3} g/l; (c) CaF$_2$, 1.7×10^{-2} g/l.
Ans. (a) 1.8×10^{-9}, (b) 1.2×10^{-10}, (c) 4×10^{-11}

18.21. The solubility product of SrF$_2$ at 25°C is 2.8×10^{-9}.
(a) Determine the solubility of SrF$_2$ at 25°C in gfw per liter and in mg per ml.
(b) What is the [Sr^{2+}] and the [F$^-$] (in moles/liter) in a saturated solution of SrF$_2$?
Ans. (a) 9×10^{-4} gfw/l, 0.11 mg/ml
(b) [Sr^{2+}] = 9×10^{-4} mole/l, [F$^-$] = 1.8×10^{-3} mole/l

18.22. What [SO$_4^{2-}$] must be exceeded to produce a RaSO$_4$ precipitate in 500 ml of a solution containing 0.00010 mole of Ra^{2+}? K_{sp} of RaSO$_4$ is 4×10^{-11}. *Ans.* 2×10^{-7} mole/l

18.23. A solution contains a Mg^{2+} ion concentration of 0.001 mole/l. Will Mg(OH)$_2$ precipitate if the OH$^-$ ion concentration of the solution is (a) 10^{-5} mole/l, (b) 10^{-3} mole/l? K_{sp} of Mg(OH)$_2$ is 1.1×10^{-11}. *Ans.* (a) No, (b) Yes

18.24. After solid SrCO$_3$ was equilibrated with a pH 8.50 buffer, the solution was found to contain [Sr^{2+}] = 8×10^{-5} mole/l. What is the solubility product for SrCO$_3$? K_2 for carbonic acid is 5.7×10^{-11}. *Ans.* 1.1×10^{-10}

18.25. Calculate the solubility at 25°C of CaCO$_3$ in a closed vessel containing a solution of pH 8.50. K_{sp} for CaCO$_3$ is 4.8×10^{-9}. K_2 for carbonic acid is 5.7×10^{-11}. *Ans.* 5.2×10^{-4} gfw/l

18.26. How much AgBr could dissolve in one liter of 0.40 M NH$_3$? K_{sp} for AgBr is 5.0×10^{-13}, and K_d for Ag(NH$_3$)$_2^+$ is 5.9×10^{-8}. *Ans.* 1.2×10^{-3} gfw/l

18.27. Solution A was made by mixing equal volumes of 0.0010 M Cd^{2+} and 0.0072 M OH^- solutions as neutral salt and strong base respectively. Solution B was made by mixing equal volumes of 0.0010 M Cd^{2+} and a standard KI solution. What was the concentration of the standard KI solution if the final $[Cd^{2+}]$ in solutions A and B was the same? K_{sp} for $Cd(OH)_2$ is 2.0×10^{-14}, and $K_{overall}$ for formation of CdI_4^{2-} from its simple ions is 4.3×10^6. Neglect the amount of cadmium in all iodide complexes other than CdI_4^{2-}.   *Ans.* 0.9 F KI

18.28. Ag_2SO_4 and $SrSO_4$ are both shaken up with pure water. K_{sp} values for these two salts are 1.7×10^{-5} and 3.0×10^{-7} respectively. Evaluate $[Ag^+]$ and $[Sr^{2+}]$ in the resulting saturated solution. *Ans.* 3.2×10^{-2} M Ag^+, 1.8×10^{-5} M Sr^{2+}

18.29. Calculate $[Ag^+]$ in a solution made by dissolving both Ag_2CrO_4 and $Ag_2C_2O_4$ until saturation is reached with respect to both salts. K_{sp} for these two salts are 1.4×10^{-12} and 5×10^{-12} respectively. *Ans.* 2.3×10^{-4} M

18.30. Calculate $[F^-]$ in a solution saturated with respect to both MgF_2 and SrF_2. K_{sp} values for the two salts are 6.6×10^{-9} and 2.8×10^{-9} respectively.   *Ans.* 2.7×10^{-3} M

18.31. Equal volumes of 0.0200 F $AgNO_3$ and 0.0200 F HCN were mixed. Calculate $[Ag^+]$ at equilibrium. K_{sp} for AgCN is 1.2×10^{-16} and K_a for HCN is 4.8×10^{-10}.   *Ans.* 5×10^{-5} M

18.32. Equal volumes of 0.0100 F $Sr(NO_3)_2$ and 0.0100 F $NaHSO_4$ were mixed. Calculate $[Sr^{2+}]$ and $[H^+]$ at equilibrium. K_{sp} for $SrSO_4$ is 3.0×10^{-7} and K_a for HSO_4^- (the same as K_2 for H_2SO_4) is 1.0×10^{-2}. Take into account the amount of H^+ needed to balance the charge of the SO_4^{2-} remaining in the solution.   *Ans.* 1.0×10^{-3} M Sr^{2+}, 4.3×10^{-3} M H^+

18.33. How much $SrSO_4$ can dissolve in 0.010 M HNO_3? Use data from the previous problem. *Ans.* 8×10^{-4} gfw/l

18.34. Excess solid $Ag_2C_2O_4$ is shaken with (a) 0.0010 M HNO_3, (b) 0.00030 M HNO_3. What is the equilibrium value of $[Ag^+]$ in the resulting solution? K_{sp} for $Ag_2C_2O_4$ is 5×10^{-12}. K_2 for $H_2C_2O_4$ is 6.4×10^{-5}. K_1 is so large that the concentration of free oxalic acid is of no importance in this problem.   *Ans.* (a) 3.8×10^{-4}, (b) 2.5×10^{-4} mole/l

18.35. How much solid $Na_2S_2O_3$ should be added to a liter of water so that .00050 gfw $Cd(OH)_2$ could just barely dissolve? K_{sp} for $Cd(OH)_2$ is 2.0×10^{-14}. K_1 and K_2 for $S_2O_3^{2-}$ complexation are 8.3×10^3 and 3.3×10^2 respectively. (As part of the problem, determine whether CdS_2O_3 or $Cd(S_2O_3)_2^{2-}$ is the predominant species in solution.)   *Ans.* .09 gfw

In the problems below, use the following physical constants for H_2S:
Solubility = 0.10 mole/liter, $K_1 = 1.0 \times 10^{-7}$, $K_2 = 1.2 \times 10^{-13}$.

18.36. What is the maximum possible $[Ag^+]$ in a saturated H_2S solution from which precipitation has not occurred? Solubility product of Ag_2S is 7×10^{-50}.   *Ans.* 8×10^{-19} mole/l

18.37. Determine the $[S^{2-}]$ in a saturated H_2S solution to which enough HCl has been added to produce a $[H^+]$ of 2×10^{-4} mole/l.   *Ans.* 3×10^{-14} mole/l

18.38. Will CoS precipitate in a saturated H_2S solution if the solution contains 0.01 mole/l of Co^{2+} and (a) 0.2 mole/l of H^+? (b) 0.001 mole/l of H^+? K_{sp} of CoS = 6×10^{-21}.   *Ans.* (a) No, (b) Yes

18.39. Given that 0.0010 mole each of Cd^{2+} and Fe^{2+} is contained in one liter of 0.020 M HCl, and this solution is saturated with H_2S. K_{sp} of CdS = 8×10^{-27}; of FeS = 5×10^{-18}.
(a) Determine whether or not each of these ions will precipitate as the sulfide.
(b) How much Cd^{2+} remains in solution at equilibrium?
Ans. (a) Only CdS precipitates, (b) 3×10^{-9} mole/l

18.40. In an attempted determination of the solubility product of Tl_2S, the solubility of this compound in pure CO_2-free water was determined as 6.3×10^{-6} gfw per liter. What is the computed K_{sp}? Assume that the dissolved sulfide hydrolyzes practically completely to HS^-, and that the further hydrolysis to H_2S can be neglected.   *Ans.* 7×10^{-20}

18.41. To a liter of a 0.080 M solution of NH_3 was added 3×10^{-5} mole of Cu^{2+}.
(a) What is the concentration of free Cu^{2+} in the resulting solution? Neglect the amount of copper in complexes other than the tetrammine complex.
(b) How much S^{2-} must be present to begin the precipitation of CuS?
K_{sp} for CuS is 9×10^{-36}, and K_d for $Cu(NH_3)_4^{2+}$ is 4.7×10^{-15}.
Ans. (a) 3×10^{-15} M Cu^{2+}, (b) 3×10^{-21} M S^{2-}

18.42. Calculate the solubility of FeS in pure water. $K_{sp} = 5 \times 10^{-18}$. (*Hint.* The second stage of hydrolysis, producing H_2S, cannot be neglected.)   *Ans.* 7×10^{-7} gfw/l

Chapter 19

Electrochemistry

INTRODUCTION

Not only is matter influenced by electricity, but also many of the important particles of matter *are* electrical in nature. All atomic nuclei are positively charged, and all ions are charged either positively or negatively. In this chapter we deal with two aspects of the connection between chemistry and electricity, (1) *electrolysis*, or the decomposition of matter accompanying the passage of electricity through it, and (2) *galvanic cell action*, or the role of a chemical reaction as an electrical generator. First we review briefly some basic definitions of electrical terms.

ELECTRICAL UNITS

The *coulomb* is the *practical* unit of charge, or quantity of electricity. From the point of view of fundamental particles, the *elementary* unit is the charge of one proton, also equal in magnitude to the charge of one electron. No particle is known whose charge is not an integral multiple of this elementary charge, the value of which is 1.602×10^{-19} coulomb.

The *electrical current* is the rate of transfer of charge. The practical unit, the *ampere*, is a transfer rate of one coulomb per second.

The *potential difference* between two points in a conductor causes the transfer of charge from one point to the other. The *volt* is the practical unit of potential difference.

The *resistance* which a conductor offers to the transfer of charge depends on the material, dimensions, and temperature of the conductor. The *ohm* is the practical unit of resistance.

$$\text{Resistance in } ohms = \text{specific resistance} \times \frac{\text{length in cm}}{\text{cross section area in cm}^2}$$

where *specific resistance* (expressed in ohm-cm) is a constant which depends on the material of which the conductor is made and on the temperature.

The *watt* is the practical unit of electrical *power* (the rate at which electrical work is done). One watt is the power developed by 1 ampere in flowing through a potential difference of 1 volt. One *kilowatt* = 1000 watts.

$$\text{Power in } watts = \text{current in } amperes \times \text{potential difference in } volts$$

The *watt-second*, or *joule*, is the *energy* furnished in 1 second by a current whose rate of expenditure (power) is 1 watt.

$$\text{Energy in } joules = \text{power in } watts \times \text{time in } seconds$$

Electrical energy can be converted into heat. In fact the calorie, the common unit of heat, is defined in terms of the heating effect of a specified amount of electrical energy. One calorie = 4.1840 joules. This relationship is known as the electrical equivalent of heat.

OHM'S LAW

The value of the steady current in a metallic conductor is equal to the potential difference between the ends of the conductor divided by the resistance of the conductor.

$$\text{Current in } amperes \;=\; \frac{\text{potential difference in } volts}{\text{resistance in } ohms}$$

Ohm's law may be applied to the entire circuit or to any part of the circuit. Thus the resistance (R) of a conductor is equal to the potential difference (E) across the conductor divided by the current (I) in the conductor, or $R = E/I$. Ohm's law applies to ionic conductors only if there is no irreversible chemical change. It would thus not apply to the electrolysis of salt solutions in a direct current, but would apply in an alternating current determination of the conductivity of a salt solution with reversible electrodes during which no overall chemical change occurs.

FARADAY'S LAWS OF ELECTROLYSIS

(1) The mass of any substance liberated or deposited at an electrode is proportional to the electrical charge (i.e., the number of coulombs) that has passed through the electrolyte.

(2) The masses of different substances liberated or deposited by the same quantity of electricity (i.e., by the same number of coulombs) are proportional to the equivalent weights of the various substances.

These laws, found empirically by Faraday over half a century prior to the discovery of the electron, can now be shown to be simple consequences of the electrical nature of matter. In any electrolysis, a reduction must occur at the cathode to remove electrons flowing into the electrode and an oxidation must occur at the anode to supply the electrons that leave the electrolysis cell at this electrode. By the principle of continuity of current, electrons must be discharged at the cathode at exactly the same rate at which they are supplied to the anode. By definition of the equivalent weight for oxidation-reduction reactions (that fraction of the formula weight associated with the transfer of one electron), the number of *gram-equivalents* of electrode reaction must be proportional to the amount of charge transported into or out of the electrolytic cell and must, indeed, be equal to the number of *moles of electrons* transported in the circuit. The charge of one mole of electrons is called a *faraday* (abbreviated F) and is equal to 1.602×10^{-19} coulomb per electron $\times$ 6.023×10^{23} electrons per mole, or 9.65×10^4 coulombs per mole of electrons.

The equivalent weight needed for electrolytic calculations can be found by writing the balanced half-reactions for the electrode processes. Thus in the electrolytic reduction of Cu^{2+}, the cathode reaction is:

$$Cu^{2+} + 2e \;\longrightarrow\; Cu$$

The equivalent weight of copper is 1/2 the atomic weight. If a solution of Cu^+ were electrolyzed, the equivalent weight would be the atomic weight of copper because only one electron is captured per copper atom formed,

$$Cu^+ + e \;\longrightarrow\; Cu$$

Special information about the electrode reactions is often needed in order to calculate the equivalent weight for electrolysis, just as in ordinary oxidation-reduction reactions. If a solution containing Fe^{3+} is electrolyzed at low voltages, the electrode reaction for the iron is

$$Fe^{3+} + e \;\longrightarrow\; Fe^{2+}$$

and the equivalent weight of iron is equal to the atomic weight. At higher voltages, the reaction might be

$$Fe^{3+} + 3e \longrightarrow Fe$$

and the equivalent weight of iron would be one-third of the atomic weight.

GALVANIC CELLS

Many oxidation-reduction reactions may be carried out in such a way as to generate electricity. In principle, this may always be done for spontaneous aqueous oxidation-reduction reactions if the oxidizing and reducing agents are not the same (as in the self oxidation-reduction of hydrogen peroxide). Such an arrangement for the production of an electrical current is called a galvanic or electrochemical cell. The experimental requirements are:

(1) The oxidizing and reducing agents are not in physical contact with each other but are contained in separate compartments called half-cells. Each half-cell contains a solution and a metallic conductor (i.e., electrode).

(2) The reducing or oxidizing agent in a half-cell may be either the electrode itself, a solid substance deposited on the electrode, a gas which bubbles around the electrode, or a solute in the solution which bathes the electrode.

(3) The solutions of the two half-cells are connected in some way that allows ions to move between them. Among the possible arrangements to accomplish this are: (a) careful layering of the lighter solution over the heavier one; (b) separation of the two solutions by a porous substance, such as fritted glass, unglazed porcelain, or a fiber permeated with some electrolyte solution; (c) insertion of a connecting electrolyte solution, or salt bridge, between the two solutions.

(4) The potential developed across the two electrodes causes an electrical current to flow if the electrodes are connected to each other by an outside conducting circuit.

STANDARD HALF-CELL POTENTIALS

The reaction occurring in each half-cell may be represented by an ion-electron partial equation of the type described in Chapter 10. The whole cell operation involves a flow of electrons in the external circuit. The electrons generated in the oxidation half-reaction enter the anode, travel through the external circuit to the cathode, and are consumed at the cathode by the reduction half-reaction. From the electrical principle of equal current at all points in a non-branching circuit, the number of electrons generated in the oxidation must exactly balance the number of electrons consumed in the reduction. This demands the same rule for combining two half-reactions into a balanced whole equation that was used in Chapter 10.

In the half-cell of the reducing agent, the oxidation product accumulates during operation of the cell. The reducing agent together with its oxidation product, known as a couple, are thus found in the same compartment during cell operation. Similarly, the other half-cell contains a couple consisting of the oxidizing agent and its reduction product. An arbitrary couple, consisting of both product and reactant of an oxidation-reduction half-reaction, may sometimes be the reducing part of a galvanic cell and sometimes the oxidizing part, depending on what the other couple is. The (Fe^{2+}, Fe^{3+}) couple, for example, takes an oxidizing role when paired against the strongly reducing (Zn, Zn^{2+}) couple, and a reducing role when paired against the strongly oxidizing (Ce^{3+}, Ce^{4+}) couple.

Each couple has an intrinsic ability to generate electrons. This ability can be assigned a numerical value called the *oxidation potential*. When two couples are combined into a whole cell, the couple with the higher oxidation potential provides the reducing agent and generates electrons at its electrode. The other couple provides the oxidizing agent and captures electrons at its electrode. The driving force for the flow of current is the *algebraic difference* between these two oxidation potentials, and is equal numerically to the voltage output of the cell under the limiting operating condition of very low current.

The value of an oxidation potential depends on the intrinsic chemical nature of the particular chemical, on the temperature, and on the concentrations of the various members of the couple. For purposes of reference, half-cell potentials are tabulated for unit concentrations of all the chemicals, defined as one atmosphere pressure for each gas and one mole per liter (or one *molal* in some cases, but not in this book) for every solute appearing in the balanced half-cell reaction. Such reference potentials are called *standard oxidation potentials* and are designated by the symbol E°. The same symbol is used for the standard potential of the whole cell, the value that could be realized in measurement at unit concentrations of all reactants and products. A partial listing of standard oxidation potentials at 25°C is given in Table 19-1. Since only *differences* between two oxidation potentials can be measured by the cell voltage, there is an arbitrary zero point on this scale, the potential of the standard hydrogen electrode, (H_2, H^+).

A couple with a large positive standard oxidation potential, like (Li, Li^+), is strongly reducing. That is, the reduced form has a great tendency to undergo oxidation and thereby transfer electrons to the electrode. Such a standard half-cell is experimentally *negative* with respect to a standard hydrogen electrode since the external circuit receives electrons from the half-cell. Conversely, a couple with a large negative standard oxidation potential, like (F^-, F_2), is strongly oxidizing, captures electrons from the electrode, and is experimentally *positive* with respect to a standard hydrogen electrode.

(A consistent system of half-cell potentials could also be used in which the number tabulated is the reduction potential or the relative tendency of a couple to *capture* electrons. Each couple would appear in such a table of *standard reduction potentials* as the negative of the value appearing in Table 19-1. The corresponding half-cell reaction would be written as a reduction, the reverse of the equation in Table 19-1. In fact, the reduction potentials are recommended for reference compilations by international scientific organizations. Because of long usage, American textbook authors have been slow to adapt to the recommended international convention. The problems in this chapter will be solved under the oxidation potential convention. Values for whole cell potentials and for the experimental sign of any electrode with respect to another must, of course, be independent of the particular convention used. Teachers who wish to use the Stockholm international convention may instruct their students to read Table 19-1 with the corresponding changes in the reaction direction and in the sign indicated above. Additional needed changes in procedure will be indicated below by parenthetical comments headed *Stockholm convention*.)

COMBINATIONS OF COUPLES

The above discussion concerned the combination of two half-cells to form a whole cell. In taking the algebraic difference between the two half-cell oxidation potentials to find the whole cell potential, no attention was paid to the number of electrons in either half-cell reaction. This was justified because the ultimate balanced whole cell reaction is the sum of two half-cell reactions, each multiplied by an integer such that the number of electrons in the *multiplied* reduction equation must equal the number in the *multiplied* oxidation equation.

Table 19-1

STANDARD OXIDATION POTENTIALS AT 25°C	
Reaction	$E°$ (volts)
$Li \rightarrow Li^+ + e$	3.045
$Na \rightarrow Na^+ + e$	2.714
$V \rightarrow V^{2+} + 2e$	1.186
$Zn + 4 NH_3 \rightarrow Zn(NH_3)_4^{2+} + 2e$	1.04
$Zn \rightarrow Zn^{2+} + 2e$	0.763
$Fe \rightarrow Fe^{2+} + 2e$	0.440
$Cd \rightarrow Cd^{2+} + 2e$	0.403
$Tl \rightarrow Tl^+ + e$	0.336
$V^{2+} \rightarrow V^{3+} + e$	0.256
$Ni \rightarrow Ni^{2+} + 2e$	0.250
$Sn \rightarrow Sn^{2+} + 2e$	0.136
$Pb \rightarrow Pb^{2+} + 2e$	0.126
$H_2 \rightarrow 2 H^+ + 2e$	0.000
$Sn^{2+} \rightarrow Sn^{4+} + 2e$	−0.15
$Cu \rightarrow Cu^{2+} + 2e$	−0.337
$Cu \rightarrow Cu^+ + e$	−0.521
$2 I^- \rightarrow I_2 + 2e$	−0.536
$Au + 4 SCN^- \rightarrow Au(SCN)_4^- + 3e$	−0.655
$H_2O_2 \rightarrow O_2 + 2 H^+ + 2e$	−0.682
$Fe^{2+} \rightarrow Fe^{3+} + e$	−0.771
$Ag \rightarrow Ag^+ + e$	−0.799
$2 Br^- \rightarrow Br_2 + 2e$	−1.065
$2 H_2O \rightarrow O_2 + 4 H^+ + 4e$	−1.229
$Mn^{2+} + 2 H_2O \rightarrow MnO_2 + 4 H^+ + 2e$	−1.23
$Tl^+ \rightarrow Tl^{3+} + 2e$	−1.25
$Au \rightarrow Au^{3+} + 3e$	−1.498
$Mn^{2+} + 4 H_2O \rightarrow MnO_4^- + 8 H^+ + 5e$	−1.51
$Ce^{3+} \rightarrow Ce^{4+} + e$	−1.61
$Au \rightarrow Au^+ + e$	−1.69
$2 H_2O \rightarrow H_2O_2 + 2 H^+ + 2e$	−1.776
$Co^{2+} \rightarrow Co^{3+} + e$	−1.81
$2 SO_4^{2-} \rightarrow S_2O_8^{2-} + 2e$	−2.01
$2 F^- \rightarrow F_2 + 2e$	−2.87

There is another way of combining two half-cell reactions in which the electrons do not cancel. This cannot correspond to a whole cell, where the electrons always cancel. The situation is a hypothetical one often used to calculate an unknown half-cell potential on the basis of two known half-cell potentials. In such a case the electron number cannot be left out of consideration. The rule for this case is as follows:

If two oxidation half-reactions are added or subtracted to give a third oxidation half-reaction, the product of $E°$, the standard oxidation potential, times n, the number of electrons in the half-reaction, may be added or subtracted correspondingly to give the $nE°$ value for the resulting half-reaction. (*Stockholm convention*. The same rule applies to reduction half-reactions with standard reduction potentials.)

An example follows:

			$E°$	n	$nE°$
Fe	$\longrightarrow$	$Fe^{2+} + 2e$	0.44	2	0.88
Fe^{2+}	$\longrightarrow$	$Fe^{3+} + e$	-0.77	1	-0.77
Adding: Fe	$\longrightarrow$	$Fe^{3+} + 3e$		3	0.11

Since $nE°$ for the resulting half-reaction is 0.11 volt, and n is 3, $E°$ must be $0.11/3 = 0.04$ volt.

This rule makes it possible to reduce the length of compiled tables, since many half-reactions can be computed even if they are not tabulated.

CONCENTRATION EFFECTS

The exact effect of concentrations upon a potential can be computed from the Nernst equation.

$$E = E° - \frac{RT}{nF} \times 2.303 \log Q$$

$E°$ is the *standard* potential, R is the molar gas constant, T is the absolute temperature, F is the numerical value of the faraday, and Q is the same function of concentrations of products and reactants of the reaction that occurs in the equilibrium constant, the product of the concentrations of all products divided by the product of the concentrations of all reactants, each term raised to a power equal to the coefficient of the corresponding substance in the balanced equation. This equation applies equally well to an oxidation half-reaction, to a reduction half-reaction, or to a whole cell reaction. At 25°C, for potentials measured in volts, the equation becomes

$$E = E° - \frac{0.05916 \log Q}{n}$$

For a half-reaction, n is the number of electrons in the half-equation; for a whole cell reaction, n is the number of electrons in *one* of the multiplied half-equations before cancelling the electrons. This equation is closely related to the laws of chemical equilibrium. Le Chatelier's Principle applies to the potential of a cell in the same sense as it applies to the yield of an equilibrium process. Since Q has product concentrations in the numerator and reactant concentrations in the denominator, an increased concentration of product reduces the potential and an increased concentration of reactant raises the potential.

DIRECTION OF OXIDATION-REDUCTION REACTIONS

Lists like Table 19-1 can be used to predict which oxidation-reduction reactions may occur even when galvanic cells are not involved. Two cases may be considered:

1. Spontaneous Oxidation-reduction Reactions

When chemicals are mixed together, an oxidation-reduction may spontaneously occur in which the oxidation potential of the couple containing the substance to be oxidized is algebraically greater than that of the couple containing the substance to be reduced. (*Stockholm convention.* The reduction potential of the couple containing the substance to be oxidized should be algebraically less than that of the couple containing the substance to be reduced.) In particular, if both reduced and oxidized members of both couples are mixed, all substances being at unit concentrations, any reducing agent can reduce an oxidizing agent occurring lower in the table of standard oxidation potentials but cannot reduce an oxidizing agent that occurs higher. (*Stockholm convention.* Any reducing agent can, at unit concentrations, reduce an oxidizing agent occurring higher in the table of standard reduction potentials.) The same rule of relative position on the table may be applied for general values of the concentrations, but E for each half-cell, as computed by the Nernst equation, replaces $E°$. Usually the qualitative prediction based on $E°$ values is not changed even for moderate deviations from unit concentrations if the two $E°$ values are separated by at least several tenths of a volt.

It should be noted that predictions under this rule indicate what reactions *might* occur but say nothing about the rate at which they *do* occur.

2. Electrode Reactions in Electrolysis

Of all possible oxidations that might occur electrolytically at an anode, the one for which the corresponding oxidation potential is algebraically greatest is the most likely. (*Stockholm convention.* The one with the least value of the reduction potential is most likely.) Correspondingly, the most probable cathodic reduction is that for which the corresponding oxidation potential is algebraically the least. (*Stockholm convention.* The one with the greatest value of the reduction potential is most likely.)

This rule applies strictly only at reversible electrodes in the absence of non-equilibrium polarization effects. Reversible conditions can be approached at very low currents with suitably prepared electrodes.

In applying the rule, the student should keep in mind the following possibilities:

(*a*) A solute molecule or ion may undergo oxidation or reduction.

(*b*) The electrode itself may undergo oxidation at the anode.

(*c*) The solvent may undergo oxidation or reduction. We consider specifically the case of water at 25°C.

   (*1*) The oxidation to molecular oxygen is found in Table 19-1 under the following entry:
$$2\,H_2O \quad \longrightarrow \quad O_2 + 4\,H^+ + 4e \qquad E° = -1.229$$

The unit concentrations to which this $E°$ value applies are 1 atmosphere of oxygen gas and 1 mole per liter of H^+. Although a cell exposed to the air would have O_2 gas at 1/5 atmosphere partial pressure, the effect of the reduced pressure has only a minor effect on E (+.010 volt), which can usually be neglected in making qualitative predictions. A varying $[H^+]$, however, ranging over many powers of 10, can have a profound effect on E. We can calculate E for the above half-cell for neutral solutions, in which $[H^+] = 10^{-7}$, by using the Nernst equation. Assume that the oxygen concentration remains unity (one atmosphere).

$$\begin{aligned}
E &= E° - \frac{0.0592}{4} \log [H^+]^4 [O_2] \\
&= -1.229 - .0148 \log (10^{-7})^4 (1) \\
&= -1.229 - (-28)(.0148) = -.815 \text{ volt}
\end{aligned}$$

It is thus much easier for water to be oxidized in neutral than in acid solutions.

(2) For the reduction of water to H_2, the most appropriate entry in Table 19-1 is

$$H_2 \longrightarrow 2\,H^+ + 2e \qquad E^\circ = 0.000$$

In neutral solutions, where $[H^+] = 10^{-7}$ and the H_2 pressure retains its unit value,

$$
\begin{aligned}
E &= E^\circ - \frac{0.0592}{2} \log \frac{[H^+]^2}{[H_2]} \\
&= 0.000 - .0296 \log (10^{-7})^2 \\
&= 0.000 - (-14)(.0296) = 0.414 \text{ volt}
\end{aligned}
$$

Hydrogen is thus easier to oxidize but water more difficult to reduce in neutral solutions than in acids.

Solved Problems

ELECTRICAL UNITS

19.1. A lamp draws a current of 2.0 amperes. Find the charge in coulombs used by the lamp in 30 seconds.

Charge in coulombs $=$ amperes $\times$ seconds $=$ 2.0 amperes $\times$ 30 seconds $=$ 60 coulombs

19.2. Compute the time required to pass 36,000 coulombs through an electroplating bath using a current of 5 amperes.

Time in seconds $= \dfrac{\text{coulombs}}{\text{amperes}} = \dfrac{36{,}000 \text{ coulombs}}{5 \text{ amperes}} = 7200$ seconds $=$ 2 hours

19.3. What current is in a 55 ohm resistance when connected across a 110 volt line?

Current in amperes $= \dfrac{\text{volts}}{\text{ohms}} = \dfrac{110 \text{ volts}}{55 \text{ ohms}} =$ 2.0 amperes

19.4. An ammeter (A) connected in series with an unknown resistance reads 0.3 ampere. A high-resistance voltmeter (V) connected across the ends of the resistance reads 1.5 volts. Determine the value of the resistance.

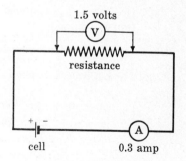

The current is measured by inserting an ammeter *into* the circuit. The voltage is measured by connecting the terminals of a voltmeter *across* two points of the circuit.

Resistance in ohms $= \dfrac{\text{volts}}{\text{amperes}} = \dfrac{1.5 \text{ volts}}{0.3 \text{ ampere}}$
$\qquad\qquad\qquad\quad = 5$ ohms

19.5. What voltage will cause 2.0 amperes to pass through a 60 ohm resistance? How many calories per second will be developed in the resistance when this voltage is applied across its terminals?

Voltage $=$ amperes $\times$ ohms $=$ 2.0 amp $\times$ 60 ohms $=$ 120 volts

Energy/sec $=$ 2.0 amp $\times$ 120 volts $=$ 240 watts $= \dfrac{240 \text{ joules/sec}}{4.18 \text{ joules/cal}} =$ 57 cal/sec

19.6. A dynamo delivers 15 amperes at 120 volts. (*a*) Compute the power in watts and in kilowatts supplied by the dynamo. (*b*) How much electrical energy in kilowatt-hours is supplied by the dynamo in 2 hours? (*c*) What is the cost of this energy at 2¢ per kilowatt-hour?

(*a*) Power = 15 amperes × 120 volts = 1800 watts = 1.8 kilowatts

(*b*) Energy in kilowatt-hours = 1.8 kilowatts × 2 hours = 3.6 kilowatt-hours

(*c*) Cost = 3.6 kw-hr × 2¢/kw-hr = 7.2¢

19.7. A resistance heater was wound around a 50 gram metallic cylinder. A current of .65 amperes was passed through the heater for 24 seconds while the measured voltage drop across the heater was 5.4 volts. The temperature of the cylinder was 22.5°C before the heating period and 29.8°C at the end. If heat losses to the environment can be neglected, what is the specific heat of the cylinder metal?

$$\text{Energy input} = (0.65 \text{ amp})(5.4 \text{ volts})(24 \text{ sec}) = 84 \text{ watt-sec} = 84 \text{ joules}$$

$$= \frac{84 \text{ joules}}{4.18 \text{ joules/cal}} = 20.1 \text{ cal}$$

$$\text{Energy input} = (\text{mass})(\text{specific heat})(\text{temperature rise})$$

$$20.1 \text{ cal} = (50 \text{ g})(\text{specific heat})(29.8° - 22.5°)$$

from which   specific heat = 0.55 cal deg^{-1} g^{-1}.

19.8. How many electrons per second pass through a cross-section of a copper wire carrying 10^{-16} ampere?

The charge passing any cross-section of the conductor = $(10^{-16}$ amp$)(1 \text{ sec}) = 10^{-16}$ coulomb.

$$\text{The number of electrons} = \frac{10^{-16} \text{ coulomb}}{1.6 \times 10^{-19} \text{ coulomb/electron}} = 6 \times 10^2 \text{ electrons}.$$

FARADAY'S LAWS OF ELECTROLYSIS

19.9. Exactly 0.2 faraday is passed through three electrolytic cells in series. One contains silver ion, one zinc ion, and one ferric ion. Assume that the only cathode reaction in each cell is the reduction of the ion to the metal. How many grams of each metal will be deposited?

One faraday liberates 1 g-eq of an element. Hence 0.2 faraday liberates 0.2 g-eq of an element.

Equivalent weight of Ag^+ = 107.9, Zn^{++} = $\frac{65.37}{2}$ = 32.68, Fe^{+++} = $\frac{55.85}{3}$ = 18.62.

$$Ag^+ \text{ deposited} = 0.2 \text{ faraday} \times 107.9 \text{ g/faraday} = 21.58 \text{ g}$$

$$Zn^{++} \text{ deposited} = 0.2 \text{ faraday} \times 32.68 \text{ g/faraday} = 6.54 \text{ g}$$

$$Fe^{+++} \text{ deposited} = 0.2 \text{ faraday} \times 18.62 \text{ g/faraday} = 3.72 \text{ g}$$

19.10. A current of 5.00 amperes flowing for exactly 30 minutes deposits 3.048 grams of zinc at the cathode. Calculate the equivalent weight of zinc from this information.

$$\text{Coulombs used} = 5.00 \text{ amp} \times (30 \times 60) \text{ sec} = 9.00 \times 10^3 \text{ coulombs}$$

$$\text{Number of faradays used} = \frac{9.00 \times 10^3 \text{ coulombs}}{9.65 \times 10^4 \text{ coulombs}/F} = .0933 \ F$$

$$\text{Equivalent weight} = \text{mass deposited by } 1 \ F = \frac{3.048 \text{ g}}{.0933 \ F} = 32.7 \text{ g}/F = 32.7 \text{ g/g-eq}$$

19.11. A certain current liberates 0.504 g of hydrogen in 2 hours. How many grams of oxygen and of copper (from Cu^{++} solution) can be liberated by the same current flowing for the same time?

> Masses of different substances liberated by the same number of coulombs are proportional to their equivalent weights.
>
> Equivalent weight of hydrogen = 1.008, of oxygen = 8.00, of copper = 31.8.
>
> Gram-equivalents of hydrogen in 0.504 g = $\dfrac{0.504\ g}{1.008\ g/g\text{-eq}}$ = $\frac{1}{2}$ g-eq. Then $\frac{1}{2}$ g-eq each of oxygen and copper can be liberated.
>
> $$\text{Mass of oxygen liberated} = \tfrac{1}{2}\ g\text{-eq} \times 8.00\ g/g\text{-eq} = 4.00\ g$$
> $$\text{Mass of copper liberated} = \tfrac{1}{2}\ g\text{-eq} \times 31.8\ g/g\text{-eq} = 15.9\ g$$

19.12. The same quantity of electricity that liberated 2.158 g of silver was passed through a solution of a gold salt and 1.314 g of gold was deposited. The equivalent weight of silver is 107.9. Calculate the equivalent weight of gold. What is the oxidation state of gold in this gold salt?

> Gram-equivalents of Ag in 2.158 g = $\dfrac{2.158\ g}{107.9\ g/g\text{-eq}}$ = 0.02000 g-eq Ag.
>
> Then 1.314 g Au must also represent 0.02000 g-eq.
>
> Equivalent weight of Au = $\dfrac{1.314\ g}{0.02000\ g\text{-eq}}$ = 65.70 g/g-eq.
>
> Oxidation state = number of electrons needed to form one gold atom by reduction
>
> $$= \frac{\text{atomic weight of Au}}{\text{equivalent weight of Au}} = \frac{197.0}{65.7} = 3.$$

19.13. How long would it take to deposit 100 g Al from an electrolytic cell containing Al_2O_3 at a current of 125 amperes? Assume that Al formation is the only cathode reaction.

> Equivalent weight of Al = $\frac{1}{3}$(atomic weight) = $\frac{1}{3}$(27.0) = 9.0 g/g-eq.
>
> Number of g-eq Al = $\dfrac{100\ g}{9.0\ g/g\text{-eq}}$ = 11.1 g-eq. Number of faradays = number of g-eq = 11.1 F.
>
> Time (sec) = $\dfrac{\text{charge (coulombs)}}{\text{current (amperes)}} = \dfrac{(11.1\ F)(9.65 \times 10^4\ coul/F)}{125\ amp} = 8.6 \times 10^3\ \text{sec} = 2.4\ \text{hr.}$

19.14. A current of 15.0 amperes is employed to plate nickel in a $NiSO_4$ bath. Both Ni and H_2 are formed at the cathode. The current efficiency with respect to formation of Ni is 60%. (a) How many grams of nickel are plated on the cathode per hour? (b) What is the thickness of the plating if the cathode consists of a sheet of metal 4.0 centimeters square which is coated on both faces? The density of nickel is 8.9 g/cc. (c) What volume of H_2 (S.T.P.) is formed per hour?

> (a) Total coulombs used = (15.0 amp)(60 × 60 sec) = 5.40×10^4 coulombs.
>
> Number of faradays used = $\dfrac{5.40 \times 10^4\ \text{coulombs}}{9.65 \times 10^4\ \text{coulomb}/F}$ = .560 F.
>
> Number of g-eq Ni deposited = (.560 F)(.60 g-eq Ni/F) = .336 g-eq Ni.
>
> Equivalent weight of Ni = $\frac{1}{2}$(atomic weight) = $\frac{1}{2}$(58.71) = 29.4 g/g-eq.
>
> Mass of Ni deposited = .336 g-eq × 29.4 g/g-eq = 9.9 g.
>
> (b) Area of one face = (4.0 cm)(4.0 cm) = 16 cm². Area of two faces = 32 cm².

$$\text{Volume of 9.9 g Ni} = \frac{\text{mass}}{\text{density}} = \frac{9.9 \text{ g}}{8.9 \text{ g/cm}^3} = 1.11 \text{ cm}^3.$$

$$\text{Thickness of plating} = \frac{\text{volume}}{\text{area}} = \frac{1.11 \text{ cm}^3}{32 \text{ cm}^2} = 0.035 \text{ cm}.$$

(c) Number of g-eq H_2 liberated $= (.560\ F)(.40 \text{ g-eq } H_2/F) = .224 \text{ g-eq } H_2$.

Volume of 1 g-eq ($\frac{1}{2}$ mole) $H_2 = \frac{1}{2}(22.4\ l) = 11.2\ l\ H_2$.

Volume of H_2 liberated $= (.224 \text{ g-eq})(11.2\ l/\text{g-eq}) = 2.51\ l\ H_2$.

19.15. An electrolytic cell, using electrodes of cross sectional area 400 cm² placed 10.0 cm apart, contains a saturated solution of copper sulfate having a specific resistance of 29.2 ohm-cm. Calculate the resistance of the solution between the electrodes.

$$\text{Resistance} = \text{specific resistance} \times \frac{\text{distance between electrodes}}{\text{area of cross section}}$$

$$= 29.2 \text{ ohm-cm} \times \frac{10.0 \text{ cm}}{400 \text{ cm}^2} = 0.73 \text{ ohm}$$

19.16. How many coulombs must be applied to a cell for the electrolytic production of 245 g of $NaClO_4$ from $NaClO_3$? Because of side reactions, the anode efficiency for the desired reaction is 60%.

First it is necessary to know the equivalent weight of $NaClO_4$ for this reaction. The balanced anode reaction equation is

$$ClO_3^- + H_2O \longrightarrow ClO_4^- + 2\,H^+ + 2e$$

$$\text{Equivalent weight of } NaClO_4 = \frac{\text{formula weight}}{\text{number of electrons transferred}} = \frac{122.4}{2} = 61.2.$$

$$\text{Number of g-eq } NaClO_4 = \frac{245 \text{ g}}{61.2 \text{ g/g-eq}} = 4.00 \text{ g-eq } NaClO_4.$$

$$\text{Number of faradays required} = \frac{4.00 \text{ g-eq}}{.60 \text{ g-eq anode product}/F} = 6.7\ F.$$

$$\text{Number of coulombs required} = (6.7\ F)(9.6 \times 10^4 \text{ coulombs}/F) = 6.4 \times 10^5 \text{ coulombs}.$$

GALVANIC CELLS AND ELECTRODE PROCESSES

19.17. What is the standard potential of a cell that uses the (Zn, Zn^{++}) and (Ag, Ag^+) couples? Which couple is negative? Write the equation for the cell reaction occurring at unit concentrations.

The standard oxidation potentials for (Zn, Zn^{++}) and (Ag, Ag^+) are 0.763 and -0.799 volts respectively. The standard potential of the cell is the difference between these two numbers, $0.763 - (-0.799) = 1.562$ volts. The zinc half-cell oxidation potential is higher, and thus zinc is the reducing agent and the negative electrode. The equation is

$$Zn + 2\,Ag^+ \longrightarrow Zn^{++} + 2\,Ag$$

19.18. Can Fe^{3+} oxidize Br^- to Br_2 at unit concentrations?

The (Fe^{2+}, Fe^{3+}) couple has a higher standard oxidation potential, -0.771, than the (Br^-, Br_2) couple, -1.065. Therefore Fe^{2+} can reduce Br_2 but Br^- cannot reduce Fe^{3+}. On the other hand I^-, occurring at a much higher standard half-cell potential, -0.536, is easily oxidized by Fe^{3+} to I_2.

19.19. What is the standard oxidation potential for (MnO_2, MnO_4^-) in acid solution?

The oxidation half-reaction for this couple is

$$MnO_2 + 2\,H_2O \longrightarrow MnO_4^- + 4\,H^+ + 3e$$

which can be written as the difference of two half-reactions whose oxidation potentials are listed in the table. nE° values may be correspondingly subtracted.

				n	E°	nE°
	$Mn^{++} + 4\,H_2O$	$\longrightarrow$	$MnO_4^- + 8\,H^+ + 5e$	5	-1.51	-7.55
Minus	$(Mn^{++} + 2\,H_2O$	$\longrightarrow$	$MnO_2 + 4\,H^+ + 2e$	2	-1.23	$-2.46)$
Subtracting,	$2\,H_2O$	$\longrightarrow$	$MnO_4^- - MnO_2 + 4\,H^+ + 3e$	3		-5.09

Rearranging, $MnO_2 + 2\,H_2O \longrightarrow MnO_4^- + 4\,H^+ + 3e$, the desired reaction, in which $n = 3$.

$$E^\circ \text{ for the desired reaction} = \frac{-5.09}{3} = -1.70$$

19.20. Predict the stabilities at 25° of aqueous solutions of the uncomplexed intermediate oxidation states of (a) thallium and (b) copper.

(a) The question is whether the intermediate state, Tl^+, spontaneously decomposes into the lower and higher states, Tl and Tl^{3+}. The supposed disproportionation reaction,

$$3\,Tl^+ \longrightarrow 2\,Tl + Tl^{3+}$$

could be written in the ion-electron method as

$$2 \times [Tl^+ + e \longrightarrow Tl] \tag{1}$$

$$Tl^+ \longrightarrow Tl^{3+} + 2e \tag{2}$$

In Eq. (1), the (Tl, Tl^+) couple functions as oxidizing agent; in Eq. (2), the (Tl^+, Tl^{3+}) couple functions as reducing agent. The reaction would occur at unit concentrations if E° for the reducing couple is greater than E° for the oxidizing couple. Since -1.25 is less than 0.34, the reaction cannot occur as written. We conclude that Tl^+ does not spontaneously decompose into Tl and Tl^{3+}. On the contrary, the reverse reaction is a spontaneous one.

$$2\,Tl + Tl^{3+} \longrightarrow 3\,Tl^+$$

This means that Tl(III) salts are unstable in solution in the presence of metallic Tl.

(b) The supposed disproportionation of Cu^+ would take the following form:

$$2\,Cu^+ \longrightarrow Cu + Cu^{2+}$$

The ion-electron partial equations are

$$Cu^+ + e \longrightarrow Cu$$

$$Cu^+ \longrightarrow Cu^{2+} + e$$

This process could occur if E° for the supposed reducing couple, (Cu^+, Cu^{2+}), is greater than E° for the oxidizing couple (Cu, Cu^+). Indeed, $+0.184$ (computed by the method of the previous problem) is algebraically greater than -0.521. Therefore Cu^+ is unstable to disproportionation in solution. Compounds of Cu(I) can exist only as extremely insoluble substances or as such stable complexes that only a very small concentration of free Cu^+ can exist in solution.

19.21. What is the potential of the cell containing the (Zn, Zn^{2+}) and (Cu, Cu^{2+}) couples if the Zn^{2+} and Cu^{2+} concentrations are 0.1 and 10^{-9} mole per liter respectively, at $25^\circ C$?

The cell reaction is $Zn + Cu^{2+} \longrightarrow Zn^{2+} + Cu$, with an n value of 2.

$$E = E^\circ - \frac{0.0592 \log Q}{n}$$

E°, the standard cell potential, is equal to the difference between the half-cell potentials, $0.763 - (-0.337) = 1.100$ volts. Q, the concentration function, does not include terms for the solid metals because their concentrations are not variable.

$$E = 1.100 - \frac{0.0592}{2} \log \frac{[Zn^{2+}]}{[Cu^{2+}]} = 1.100 - 0.0296 \log \frac{10^{-1}}{10^{-9}}$$

$$= 1.100 - 0.0296 \log 10^8 = 1.100 - 8(0.0296) = 1.100 - 0.237 = 0.863 \text{ volt}$$

19.22. By how much is the oxidizing power of the (Mn^{2+}, MnO_4^-) couple reduced if the H^+ concentration is reduced from 1 mole per liter to 10^{-4} moles per liter?

The half-cell reaction for oxidation is

$$Mn^{2+} + 4\,H_2O \longrightarrow MnO_4^- + 8\,H^+ + 5e$$

with an n value of 5. Assume that only the H^+ concentration deviates from 1 mole/liter.

$$E = E^\circ - \frac{0.0592}{5} \log \frac{[MnO_4^-]\,[H^+]^8}{[Mn^{2+}]} = -1.51 - 0.0118 \log \frac{(1)(10^{-4})^8}{1}$$

$$= -1.51 - 0.0118 \log 10^{-32} = -1.51 - (-32)(0.0118) = -1.51 + 0.38 = -1.13$$

The position of the couple has moved up the table 0.38 volt (to a position of less oxidizing power) from its standard value.

19.23. In the continued electrolysis of each of the following solutions at pH 7.0 at 25°, predict the main product at each electrode if there are no electrode polarization effects: (a) 1 F $NiSO_4$ with gold electrodes; (b) 1 F $NiBr_2$ with inert electrodes; (c) 1 F Na_2SO_4 with Cu electrodes.

(a) **Anode Reaction**

Three possible oxidation processes, along with their E° values are the following:

$$
\begin{array}{lll}
 & & E^\circ \\
(1) \quad 2\,H_2O & \longrightarrow \quad O_2 + 4\,H^+ + 4e & -1.23 \\
(2) \quad Au & \longrightarrow \quad Au^{3+} + 3e & -1.50 \\
(3) \quad 2\,SO_4^{2-} & \longrightarrow \quad S_2O_8^{2-} + 2e & -2.01 \\
\end{array}
$$

The standard potentials are reasonable values to take in considering (2) and (3). Although the initial concentrations of Au^{3+} and $S_2O_8^{2-}$ are zero, they would increase during prolonged electrolysis if these species were the principal products. In the case of (1), however, the buffering of the solution prevents the buildup of $[H^+]$, and it would be more appropriate to take the E value calculated for pH 7.0 in the last introductory section of this chapter, -0.815 volt. It is apparent that of the three possible anode reactions, (1) has the largest E value and would thus occur most readily.

Cathode Reaction

Two possible processes may be considered, the reverse of the following oxidation half-reactions:

$$
\begin{array}{lll}
 & & E^\circ \\
(4) \quad Ni & \longrightarrow \quad Ni^{2+} + 2e & 0.25 \\
(5) \quad H_2 & \longrightarrow \quad 2\,H^+ + 2e & 0.00 \\
\end{array}
$$

From our rule that the most probable cathode process is that for which the corresponding *oxidation* potential is least, the hydrogen couple is favored on the basis of E° values. Allowing for the effect of the pH 7.0 buffer, however, E for (5) is raised to 0.41 volt, as calculated in the last introductory section of this chapter. The reduction of nickel then becomes the favored process.

In conclusion, the electrode processes to be expected are:

$$
\begin{array}{ll}
\text{Anode:} & 2\,H_2O \longrightarrow O_2 + 4\,H^+ + 4e \\
\text{Cathode:} & Ni^{2+} + 2e \longrightarrow Ni \\
\text{Overall:} & 2\,H_2O + 2\,Ni^{2+} \longrightarrow O_2 + 4\,H^+ + 2\,Ni \\
\end{array}
$$

(b) **Anode Reaction**

The expression "inert electrode" is often used to indicate that we may neglect reaction of the electrode itself, either by virtue of the intrinsically low value for its oxidation potential or by reason of polarization effects related to the preparation of the electrode surface. The remaining possible anode reactions are:

$$
\begin{array}{lll}
 & & E^\circ \\
(6) \quad 2\,Br^- & \longrightarrow \quad Br_2 + 2e & -1.065 \\
(1) \quad 2\,H_2O & \longrightarrow \quad O_2 + 4\,H + 4e & -1.229 \\
\end{array}
$$

When the E value for (1) is computed to -0.815 for pH 7.0, as in (a), oxygen evolution takes precedence. (In practice, "overvoltage" or polarization is more difficult to avoid in the case of reactions involving gases (O_2) as compared with liquids and dissolved solutes, so that electrolysis of $NiBr_2$ at most electrodes would probably lead to Br_2 formation.)

Cathode Reaction

As in (a), Ni reduction would occur.

(c) **Anode Reaction**

In addition to (1) and (3), the reaction of the Cu anode must be considered:

$$\text{(7)} \quad \text{Cu} \quad \longrightarrow \quad \text{Cu}^{2+} + 2e \qquad\qquad \begin{array}{c} E^\circ \\ -0.34 \end{array}$$

Process (7) has the highest E value and would thus take precedence over oxygen evolution.

Cathode Reaction

The new couple to be considered is the sodium couple, the oxidation of which is:

$$\text{(8)} \quad \text{Na} \quad \longrightarrow \quad \text{Na}^+ + e \qquad\qquad \begin{array}{c} E^\circ \\ 2.71 \end{array}$$

This E value is much greater than that of (5), the evolution of H_2 at pH 7.0, 0.41 volt. Therefore hydrogen evolution will occur at the cathode.

19.24. Knowing that K_{sp} for AgCl is 1.8×10^{-10}, calculate E for a silver – silver chloride electrode immersed in 1 M KCl.

The electrode process is a special case of the (Ag, Ag^+) couple, except that silver in the $+\text{I}$ state collects as solid AgCl on the electrode itself. Even solid AgCl, however, has some Ag^+ in equilibrium with it in solution. This $[\text{Ag}^+]$ can be computed from the K_{sp} equation:

$$[\text{Ag}^+] \;=\; \frac{K_{sp}}{[\text{Cl}^-]} \;=\; \frac{1.8 \times 10^{-10}}{1} \;=\; 1.8 \times 10^{-10}$$

This value for $[\text{Ag}^+]$ can be inserted into the Nernst equation for the (Ag, Ag^+) half-reaction.

$$\text{Ag} \quad \longrightarrow \quad \text{Ag}^+ + e \qquad E^\circ = -0.799$$

$$E \;=\; E^\circ - \frac{0.05916}{1} \log [\text{Ag}^+] \;=\; -0.799 - 0.05916 \log 1.8 \times 10^{-10}$$
$$\;=\; -0.799 + 0.576 \;=\; -0.223 \text{ volt}$$

19.25. From data in Table 19-1, calculate the overall stability constant, K_{overall}, of $\text{Zn(NH}_3)_4^{2+}$.

There are two entries in the table for couples connecting the zero and $+\text{II}$ oxidation states of zinc.

$$\begin{array}{llll} & & & E^\circ \\ \text{(1)} & \text{Zn} \longrightarrow \text{Zn}^{2+} + 2e & & 0.76 \\ \text{(2)} \quad \text{Zn} + 4\,\text{NH}_3 \longrightarrow \text{Zn(NH}_3)_4^{2+} + 2e & & 1.04 \end{array}$$

Process (1) refers to the couple in which Zn^{2+} is at unit concentration (1 M). Process (2) refers to the couple in which NH_3 and $\text{Zn(NH}_3)_4^{2+}$ are at unit concentration; the Zn^{2+} concentration to which the E° for (2) refers is that value which satisfies the complex ion equilibrium when the other species are at unit concentration.

$$[\text{Zn}^{2+}] \;=\; \frac{[\text{Zn(NH}_3)_4^{2+}]}{K_{\text{overall}} \times [\text{NH}_3]^4} \;=\; \frac{1}{K_{\text{overall}} \times 1^4} \;=\; \frac{1}{K_{\text{overall}}}$$

In other words, the standard conditions for couple (2) may be thought of as a non-standard case of couple (1) in which $[\text{Zn}^{2+}] = 1/K_{\text{overall}}$. Applying this substitution to the Nernst equation for (1),

$$E = E° - \frac{0.0592}{2} \log [Zn^{2+}] \quad \text{or} \quad 1.04 = 0.76 - 0.0296 \log \frac{1}{K_{overall}}$$

from which $\quad \log K_{overall} = \dfrac{1.04 - 0.76}{0.0296} = 9.5 \quad$ and $\quad K_{overall} = 3 \times 10^9$

19.26. A galvanic cell is set up with a zinc electrode dipping into a solution of zinc sulfate and a copper electrode dipping into a solution of cupric sulfate. The measured potential of the cell is 1.100 volts. Compute the free energy change for the cell reaction

$$Zn + Cu^{2+} \longrightarrow Zn^{2+} + Cu$$

The free energy change of a spontaneous oxidation-reduction reaction is negative and is equal in magnitude to the electrical energy generated in a galvanic cell utilizing the reaction, provided the current drawn from the galvanic cell is very small and there are no irreversible effects at the electrodes.

The faraday, which expresses electrochemical equivalence, is a valid conversion factor for galvanic cells as well as for electrolysis. For every g-at of Zn (2 g-eq) which reacts according to the equation, the number of coulombs passed through the electrical circuit is $2(9.65 \times 10^4)$, since the electrode reactions show that 2 electrons are involved at each electrode for the reaction of one Zn atom with one Cu^{2+}.

$$Zn \longrightarrow Zn^{2+} + 2e$$
$$Cu^{2+} + 2e \longrightarrow Cu$$

Electrical energy in joules $\quad = \quad$ watts $\times$ seconds $\quad = \quad$ volts $\times$ amperes $\times$ seconds
$$= \text{ volts} \times \text{coulombs}$$

Electrical energy per mole $\quad = \quad (1.100 \text{ volts})(2 \times 9.65 \times 10^4 \text{ coulombs})$
$$= 2.123 \times 10^5 \text{ joules}$$
$$= 2.123 \times 10^5 \text{ joules} \times \frac{1 \text{ kcal}}{4184 \text{ joules}} = 50.7 \text{ kcal}$$

Free energy change $\quad = \quad -50.7$ kcal

Supplementary Problems

ELECTRICAL UNITS

19.27. Calculate the current in an electric heater of 80 ohms resistance when it is connected to a 120 volt source. *Ans.* 1.50 amperes

19.28. A current of 1.6 amperes flows through a lamp when it is connected to a 112 volt source. What is the resistance of the lamp? *Ans.* 70 ohms

19.29. What voltage is required to maintain a current of 6 amperes in a resistance of 40 ohms?
Ans. 240 volts

19.30. An ammeter is connected in series with an unknown resistance, and a voltmeter is connected across the terminals of the resistance. The ammeter reads 1.2 amperes and the voltmeter reads 18 volts. Compute the resistance. *Ans.* 15 ohms

19.31. How many coulombs per hour pass through an electroplating bath which uses a current of 5 amperes?   *Ans.* 18,000 coul/hr

19.32. Compute the cost at 5¢ per kilowatt-hour of operating for 8 hours an electric motor which takes 15 amperes at 110 volts.   *Ans.* 66¢

19.33. An electrolytic cell, using electrodes each of 20×15 cm and placed 15 cm apart, contains a solution of silver nitrate having specific resistance 15 ohm-cm. Calculate the resistance of the solution between the electrodes.   *Ans.* 0.75 ohm

19.34. A tank containing 200 liters of water was used as a constant temperature bath. How long would it take to heat the bath from 20°C to 25°C with a 250-watt immersion heater? Neglect the heat capacity of the tank frame and any heat losses to the air.   *Ans.* 4.6 hr

19.35. The specific heat of a liquid was measured by placing 100 g of the liquid in a calorimeter. The liquid was heated by an electrical immersion coil. The heat capacity of the calorimeter together with the coil was previously determined to be 7.50 cal/°C. With the 100 g sample in place in the calorimeter, a current of 0.500 ampere was passed through the immersion coil for exactly 180 seconds. The voltage across the terminals of the coil was measured to be 1.50 volts. The temperature of the sample rose by 0.800°C. Find the specific heat of the liquid.   *Ans.* 0.329 cal/g°C

19.36. The heat of solution of NH_4NO_3 in water was determined by measuring the amount of electrical work needed to compensate for the cooling which would otherwise occur when the salt dissolves. After the NH_4NO_3 was added to the water, electrical energy was provided by passage of a current through a resistance coil until the temperature of the solution reached the value it had prior to the addition of the salt. In a typical experiment, 4.4 g NH_4NO_3 was added to 200 g water. A current of 0.75 ampere was provided through the heater coil, and the voltage across the terminals was 6.0 volts. The current was applied for exactly 5.2 minutes. Calculate ΔH for the solution of 1 gfw NH_4NO_3 in enough water to give the same concentration as was attained in the above experiment.   *Ans.* 6.1 kcal

FARADAY'S LAWS

In the electrolysis problems, assume that the electrode efficiency is 100% for the principal electrode reaction, unless a statement is made in the problem to the contrary.

19.37. What current is required to pass one faraday per hour through an electroplating bath? How many grams of aluminum and of cadmium will be liberated by one faraday?
Ans. 26.8 amp, 8.99 g Al, 56.2 g Cd

19.38. What mass of aluminum is deposited electrolytically in 30 minutes by a current of 40 amperes?
Ans. 6.7 g

19.39. How many amperes are required to deposit on the cathode 5.00 g of gold per hour from a solution containing a salt of gold in the +III oxidation state?   *Ans.* 2.0 amp

19.40. How many hours will it take to produce 100 pounds of electrolytic chlorine from NaCl in a cell that carries 1000 amperes? The anode efficiency for the chlorine reaction is 85%.
Ans. 40.3 hours

19.41. A given quantity of electricity passes through two separate electrolytic cells containing solutions of $AgNO_3$ and $SnCl_2$, respectively. If 2.00 g of silver is deposited in one cell, how many grams of tin are deposited in the other cell?   *Ans.* 1.10 g Sn

19.42. An electrolytic cell contains a solution of $CuSO_4$ and an anode of impure copper. How many kilograms of copper will be refined (i.e., deposited on the cathode) by 150 amperes maintained for 12 hours?   *Ans.* 2.1 kg Cu

19.43. How many hours are required for a current of 3.0 amperes to decompose electrolytically 18 g water?
Ans. 18 hr

19.44. Hydrogen peroxide can be prepared by the successive reactions:

$$2\ NH_4HSO_4 \longrightarrow H_2 + (NH_4)_2S_2O_8$$

$$(NH_4)_2S_2O_8 + 2\ H_2O \longrightarrow 2\ NH_4HSO_4 + H_2O_2$$

The first reaction is an electrolytic reaction, and the second a steam distillation. What current would have to be used in the first reaction to produce enough intermediate to yield 100 g of pure H_2O_2 per hour? Assume 50% anode current efficiency.   *Ans.* 315 amperes

19.45. The electrodes in a lead storage battery are made of Pb and PbO_2. The overall reaction during discharge is

$$Pb + PbO_2 + 2\ H_2SO_4 \longrightarrow 2\ PbSO_4 + 2\ H_2O$$

(a) What is the minimum weight of lead (counting all chemical forms of the element) in a battery case if the battery is designed to deliver 100 ampere-hours? Assume a 25% "coefficient of use." This is the percent of the Pb and PbO_2 in the battery case that actually are available for the electrode reactions. (b) If the average voltage of a storage battery is 2.00 volts under zero load, what is the approximate free energy change for the reaction as written above?
Ans. 6.8 lb lead, -92 kcal

19.46. Neglecting electrode polarization effects, predict the principal product at each electrode in the continued electrolysis at 25° of each of the following: (a) 1 F $Sn(SO_4)_2$ with inert electrodes in 0.1 M H_2SO_4, (b) 1 F LiCl with silver electrodes, (c) 1 F $FeSO_4$ with inert electrodes at pH 7.0, (d) molten NaF.   *Ans.* (a) Sn^{2+} and O_2, (b) H_2 and AgCl, (c) H_2 and Fe^{3+}, (d) Na and F_2

19.47. A galvanic cell was operated under almost ideally reversible conditions at a current of 10^{-16} ampere. (a) At this current how long would it take to deliver a faraday of electricity? (b) How many electrons would be delivered by the cell to a pulsed measuring circuit in 10 milliseconds of operation?
Ans. (a) 3×10^{13} years, (b) 6

GALVANIC CELLS AND OXIDATION-REDUCTION

All these problems refer to conditions at 25°C.

19.48. (a) What is the standard potential of the cell made up of the (Cd, Cd^{2+}) and (Cu, Cu^{2+}) couples? (b) Which couple is positive?   *Ans.* (a) 0.740 volts, (b) (Cu, Cu^{2+})

19.49. What is the standard potential of a cell containing the (Sn, Sn^{2+}) and (Br^-, Br_2) couples?
Ans. 1.201 volts

19.50. Why are Co^{3+} salts unstable in water?
Ans. Co^{3+} can oxidize H_2O, the principal products being Co^{2+} and O_2.

19.51. If H_2O_2 is mixed with Fe^{2+}, which reaction is more likely, the oxidation of Fe^{2+} to Fe^{3+}, or the reduction of Fe^{2+} to Fe? Write the reaction for each possibility and compute the standard potential of the equivalent electrochemical cell.
Ans. More likely, $H_2O_2 + 2\ H^+ + 2\ Fe^{2+} \longrightarrow 2\ H_2O + 2\ Fe^{3+}$; $E° = 1.01$ volts.
Less likely, $H_2O_2 + Fe^{2+} \longrightarrow Fe + O_2 + 2\ H^+$;   the reverse of this reaction occurs with a standard potential of 1.12 volts.

19.52. What substance can be used to oxidize fluorides to fluorine?
Ans. Fluorides may be oxidized electrolytically, but not chemically by any substance listed in Table 19-1.

19.53. Are Fe^{2+} solutions stable in the air? Why can such solutions be preserved by the presence of iron nails?   *Ans.* O_2 oxidizes Fe^{2+} to Fe^{3+}, but Fe reduces Fe^{3+} to Fe^{2+}.

19.54. What is the standard potential of the (Tl, Tl^{3+}) half-cell?   *Ans.* -0.72 volts

19.55. Which of the following intermediate oxidation states is stable with respect to disproportionation in oxygen-free non-complexing media: gold(I), tin(II)?   *Ans.* tin(II)

19.56. Would H_2O_2 behave as oxidant or reductant with respect to each of the following couples at standard concentrations: (a) (I^-, I_2), (b) $(SO_4^{2-}, S_2O_8^{2-})$, (c) (Fe^{2+}, Fe^{3+})?

    Ans. (a) oxidant, (b) reductant, (c) both; in fact iron salts in either +II or +III states catalyze the self oxidation-reduction of H_2O_2.

19.57. What is the potential of a cell containing two hydrogen electrodes, the negative one in contact with 10^{-8} molar H^+ and the positive in contact with 0.025 molar H^+?   *Ans.* 0.379 volts

19.58. Compute the potential of the (Ag, Ag^+) couple with respect to (Cu, Cu^{2+}) if the concentrations of Ag^+ and Cu^{2+} are 4.2×10^{-6} and 1.3×10^{-3} molar respectively.   *Ans.* 0.229 volt

19.59. What is the minimum concentration of Ag^+ that would remain unreduced by a standard (Fe^{2+}, Fe^{3+}) couple at equilibrium?   *Ans.* 0.33 M

19.60. Copper can reduce zinc ions if the resultant copper ions can be kept at a sufficiently low concentration by the formation of an insoluble salt. What is the maximum $[Cu^{2+}]$ in solution if this reaction is to occur, when $[Zn^{2+}]$ is 1 molar?   *Ans.* 6×10^{-38} molar

19.61. A (Tl, Tl^+) couple was prepared by saturating 0.1 M KBr with TlBr and allowing the Tl^+ from the relatively insoluble bromide to equilibrate. This couple was observed to have a potential of -0.443 volts with respect to a (Pb, Pb^{2+}) couple in which $[Pb^{2+}]$ was 0.1 molar. What is the solubility product of TlBr?   *Ans.* 3.6×10^{-6}

19.62. K_d for complete dissociation of $Ag(NH_3)_2^+$ into Ag^+ and NH_3 is 5.9×10^{-8}. Calculate E° for the following half-reaction by reference to Table 19-1: $Ag + 2NH_3 \longrightarrow Ag(NH_3)_2^+ + e$.   *Ans.* -0.371 volt

19.63. Calculate $K_{overall}$ for formation of $Au(SCN)_4^-$ from Au^{3+} and SCN^-?   *Ans.* 6×10^{42}

19.64. Reference tables give the following entry:
$$3OH^- \longrightarrow HO_2^- + H_2O + 2e \qquad E^\circ = -0.878$$
Combining this information with relevant entries in Table 19-1, find K_1 for the acid dissociation of H_2O_2.   *Ans.* 2×10^{-12}

Chapter 20

Photochemistry and Nuclear Chemistry

LIGHT AND MATTER

Light, like many other forms of energy, can interact with matter in a variety of ways. Only a few are listed here.

1. Each substance absorbs light. The particular wavelengths at which light is absorbed together with a quantitative measure of the extent of the absorption constitute a characteristic property of each substance, the *absorption spectrum*. Cupric salts are blue, for example, because they are strong absorbers of red light, which is the complementary color to blue. The absorption spectrum of a substance includes not only the visible range but the long wavelength infrared and short wavelength ultraviolet as well. In the most general case, the absorbing substance does not undergo any permanent change. The subparticles of the absorbing substance (its molecules, atoms, and electrons) display more energetic internal motions, and after a very brief time following absorption, sometimes as short of 10^{-13} sec at certain wavelength regions and concentrations, the absorbed energy is dissipated throughout the bulk matter and the net result is the same as if an equivalent amount of heat had been absorbed.

2. Some substances display the *photoelectric effect*, in which the impingement of light causes the ejection of electrons from the substance. The use of these electrons in electrical circuits which operate control devices is well known to all students.

3. In certain cases the absorption of light can lead to permanent chemical change. Among the familiar examples of *photochemistry* are the exposure step of the photographic process, the photosynthesis of sugars by green plants, and the production of a sunburn.

4. Light can be emitted by substances energized in a variety of ways. If the source of the energy is an exothermic chemical reaction, the resulting emission is called *chemiluminescence*. A familiar example is the glow of the firefly, which accompanies a biological oxidation reaction. If the energy source is light which the substance first absorbs, the emitted light is called *fluorescence* or *phosphorescence*. In these cases the emitted light is characteristic of the substance and need not be of the same color as the absorbed light. Examples of such photo-induced light emissions are the blue fluorescence of laundry whiteners and the light of lasers.

SOME PROPERTIES OF LIGHT

Two principal aspects of light must be recognized, its wave character and its particle character.

1. **Wave Character of Light.**

A light beam is associated with an electromagnetic disturbance, propagated periodically in the direction of the beam. The distance between two successive maxima in the intensity of the disturbance is called the *wavelength* and is usually given the symbol lambda, λ. The *frequency* is the number of such maxima passing a given point per second and is usually given the symbol nu, ν. The product of the wavelength and the frequency is equal to the velocity of propagation, usually designated by c.

$$c = \lambda \nu$$

The velocity of light in vacuum is the same for all wavelengths, 2.998×10^{10} cm/sec, so that a simple inverse proportionality exists between wavelength and frequency:

$$\lambda = \frac{2.998 \times 10^{10} \text{ cm/sec}}{\nu}$$

The velocity of light in the atmosphere is reduced by less than one part in a thousand below the value given above, so that the above value is acceptable for most laboratory applications.

2. Particle Character of Light.

The energy of light is absorbed, emitted, or converted to other forms of energy in individual units, or *quanta* (singular, *quantum*). The unit itself, often referred to as the particle of light, is called a *photon*. The energy of a photon is proportional to the frequency:

$$\epsilon \text{ (in ergs per quantum)} = h\nu = 6.626 \times 10^{-27} \nu$$

The universal proportionality constant, h, is called Planck's constant (1 joule = 10^7 ergs). For photochemical purposes, we are often interested in the energy per mole of quanta, called an *einstein*. One useful equation, in which the proportionality constant includes h, Avogadro's number, and conversion factors for energy and length, is:

$$E \text{ (in kcal per einstein)} = \frac{2.859 \times 10^5}{\lambda \text{ (in Å)}}$$

NUCLEAR CHEMISTRY

In ordinary chemical reactions, the atoms of the reactant molecules regroup themselves to form the product molecules. In such reactions, the outer electrons of the atoms undergo rearrangements in being transferred wholly or in part from one atom to another. The atomic nuclei, on the other hand, change their relative positions with respect to each other, but are themselves unchanged.

There are reactions in which the nuclei themselves are broken down, and in which the product materials do not contain the same elements as the reactants. The chemistry of such nuclear reactions is called nuclear chemistry. Some nuclear processes, such as radioactivity, consist of the spontaneous disintegration of individual nuclei. These reactions occur at rates that are not affected by any laboratory conditions, such as temperature or pressure. Most of the known nuclear reactions, however, result from the interaction of two nuclei or from the impact of a sub-atomic particle upon a nucleus. Processes from this latter class are very sensitive to the experimental control of the energy and relative positions of the reacting particles.

FUNDAMENTAL PARTICLES

For the purposes of this chapter the fundamental particles occurring in nuclear reactions will be limited to the proton, neutron, the negative electron, and the positive electron, or positron. The charge and mass of each of these particles are important properties and are tabulated below. The charges are in the elementary units defined in Chapter 19, i.e., 1.602×10^{-19} coulomb. -1 is the charge of an electron, and $+1$ is the equal but opposite charge of a proton. The mass units are expressed in u on the atomic weight scale.

Table 20-1

FUNDAMENTAL PARTICLES			
Particle	Symbol	Mass	Charge
Proton	p	1.00728	+1
Neutron	n	1.00867	0
Negative electron, Electron	e, e^- β, β^-	0.0005486	−1
Positive electron, Positron	e^+, β^+	0.0005486	+1

BINDING ENERGIES

The mass of an atom, in general, is not equal to the sum of the masses of its component protons, neutrons, and electrons. If we could imagine a reaction in which free protons, neutrons, and electrons combine to form an atom, we would find that for all nuclides except H^1 the mass of the atom is slightly less than the mass of the component parts and also that a tremendous amount of energy is released when the reaction occurs. The loss in mass is exactly equal to the mass equivalent of the released energy, according to the Einstein equation.

$$E = m \times c^2$$

energy = change in mass × (velocity of light)^2

This energy equivalent of the loss of mass is called the binding energy of the nucleus. When m is expressed in grams and c in cm/sec, E is in ergs. A more convenient unit of energy for nuclear reactions is the electron volt (eV), which is the electrical energy needed to accelerate an electron through a potential difference of one volt. When the mass is expressed in atomic mass units (u) and energy in million electron volts (MeV), this equation becomes

energy (in MeV) = 931.5 × change in mass (in u)

The energy on a molar basis corresponding to a single particle energy of 1 eV is the electrical energy of a mole of electrons (see Chapter 19) accelerated through 1 volt, or 1 volt-faraday, or 9.65×10^4 volt-coulombs, or 9.65×10^4 joules, or $\dfrac{9.65 \times 10^4 \text{ joules}}{4.184 \text{ joules/cal}}$, or 23.06×10^3 cal, or 23.06 kcal.

Some typical nuclidic masses for some of the lighter elements are given in Table 20-2.

Table 20-2

TABLE OF NUCLIDIC MASSES (Atomic Weight Scale)			
$_0n^1$	1.00867	$_4Be^9$	9.01219
$_1H^1$	1.00783	$_6C^{12}$	12.00000
$_1H^2$	2.01410	$_6C^{13}$	13.00335
$_1H^3$	3.01605	$_6C^{14}$	14.00324
$_2He^4$	4.00260	$_6C^{16}$	16.01470
$_2He^6$	6.01890	$_7N^{14}$	14.00307
$_3Li^6$	6.01513	$_7N^{16}$	16.00609
$_3Li^7$	7.01601	$_9F^{18}$	18.00095
$_4Be^7$	7.01693	$_{10}Ne^{18}$	18.00572
$_4Be^8$	8.00531		

NUCLEAR EQUATIONS

The rules for balancing nuclear equations are different from the rules for balancing ordinary chemical equations.

(1) Each particle is assigned a superscript equal to its mass number and a subscript equal to its atomic number or nuclear charge.

(2) A free proton is the nucleus of the hydrogen atom, and is therefore assigned the notation $_1H^1$.

(3) A free neutron is assigned zero atomic number because it has no charge. The mass number of a neutron is 1. The notation for a neutron is $_0n^1$.

(4) An electron (beta$^-$, or β^-) is assigned the mass number zero and the atomic number -1; hence the notation $_{-1}e^0$. Only high speed bombarding electrons or electrons ejected from a nucleus are so designated, and not the ordinary electrons that belong to the electron orbital structure of the atom.

(5) A positron is assigned the mass number zero and the atomic number $+1$; hence the notation $_{+1}e^0$.

(6) An alpha-particle (α-particle) is a helium nucleus, and is therefore represented by the notation $_2He^4$, or $_2\alpha^4$.

(7) Gamma radiation (γ) is a form of light, and has zero mass number and zero charge.

(8) In a balanced equation the sum of the subscripts (atomic numbers), written or implied, must be the same on the two sides of the equation. The sum of the superscripts (mass numbers), written or implied, must also be the same on the two sides of the equation. Thus the equation for the primary radioactivity of Ra^{226} is

$$_{88}Ra^{226} \longrightarrow {}_{86}Rn^{222} + {}_2He^4$$

Many nuclear processes may be indicated by a short-hand notation, in which a light bombarding particle and a light product particle are represented by symbols in parentheses between the symbols for the initial target nucleus and the final product nucleus. The symbols n, p, d, α, β^-, β^+, γ are used to represent neutron, proton, deuteron ($_1H^2$), alpha, electron, positron, and gamma rays respectively. The atomic numbers are commonly omitted because the symbol for any element implies its atomic number. Examples of the corresponding long- and short-hand notation for several reactions follow.

$$_7N^{14} + {}_1H^1 \longrightarrow {}_6C^{11} + {}_2He^4 \qquad N^{14}(p, \alpha)C^{11}$$

$$_{13}Al^{27} + {}_0n^1 \longrightarrow {}_{12}Mg^{27} + {}_1H^1 \qquad Al^{27}(n, p)Mg^{27}$$

$$_{25}Mn^{55} + {}_1H^2 \longrightarrow {}_{26}Fe^{55} + 2 {}_0n^1 \qquad Mn^{55}(d, 2n)Fe^{55}$$

Just as an ordinary *chemical* equation is a shortened version of the complete *thermochemical* equation which expresses both energy and mass balance, so each nuclear equation has associated with it, either written or implied, a term expressing energy balance. The symbol Q is usually used to designate the net energy *released* when all reactant and product particles of matter are at zero velocity. Q is positive for exothermic reactions and negative for endothermic. Q is the energy equivalent of the mass decrease accompanying the reaction.

RADIOCHEMISTRY

Special properties of radioactive nuclides make them useful tracers for following complex processes. *Radiochemistry* is that branch of chemistry which involves the applications of radioactivity to chemical problems.

A radioactive nuclide is converted to another nuclide by one of the following processes, in which there is an overall decrease in mass:

1. Alpha decay.

An α-particle is emitted and the daughter nucleus has an atomic number, Z, two units less, and a mass number, A, four units less than the parent. Thus

$$_{88}\text{Ra}^{226} \longrightarrow {}_{86}\text{Rn}^{222} + \alpha$$

2. β^- decay.

A β^--particle is emitted and the daughter has a Z value one unit greater than the parent, with no change in A. Thus

$$_{14}\text{Si}^{31} \longrightarrow {}_{15}\text{P}^{31} + \beta^-$$

3. β^+ decay.

A β^+-particle is emitted and the daughter has a Z value one unit less than the parent, with no change in A. Thus

$$_{21}\text{Sc}^{40} \longrightarrow {}_{20}\text{Ca}^{40} + \beta^+$$

The emitted positron is itself unstable and is normally consumed, after being slowed down by collisions, by the following annihilation reaction:

$$\beta^+ + \beta^- \longrightarrow 2\gamma$$

4. K-electron capture.

By capturing an orbital electron within its own atom, a nucleus can reduce its Z value by one unit without a change in A. Thus

$$_4\text{Be}^7 \xrightarrow{\;K \text{ capture}\;} {}_3\text{Li}^7$$

The stability of a radioactive nucleus toward spontaneous decay is measured by its *half-life*. The half-life is defined as the time in which half of any large sample of identical nuclei will undergo decomposition. This is a fixed number for each type of nucleus and may be given the symbol T. Thus a collection of radioactive atoms will be reduced to one-fourth ($\frac{1}{2} \times \frac{1}{2}$) its original number in a time equal to $2T$, to one-eighth in $3T$, to one-sixteenth in $4T$, and so on. A more general relationship expressing the time, t, for reduction of a sample to a fraction, f, of its initial value is

$$\log f = -\frac{0.301\, t}{T}$$

The above equation may also be written in the following alternative forms

$$f = 10^{-.301t/T} = e^{-.693t/T}$$

where e is the natural base of logarithms, 2.718..., .301 is $\log_{10} 2$, and .693 is $\log_e 2$.

Radioactivity is measured by observing the high-energy particles produced directly or indirectly as a result of the disintegration process. A unit of radioactivity applicable to any radioactive nucleus is the *curie*. One curie (c) is that amount of a radioactive substance in which the number of nuclear disintegrations per second is 3.70×10^{10}. The terms *millicurie* (mc) and *microcurie* (μc) are also used. The strength of a sample, expressed in curies, depends both on the number of atoms of the radioactive nuclide and on the half-life.

Solved Problems

20.1. Determine the frequency of light of the following wavelengths: (a) 1.0 Å, (b) 5000 Å, (c) 4.4 μ, (d) 89 m.

The basic equation for all these problems is: $\nu = \dfrac{2.998 \times 10^{10} \text{ cm/sec}}{\lambda}$.

(a) $\nu = \dfrac{3.0 \times 10^{10} \text{ cm/sec}}{1.0 \text{ Å} \times 10^{-8} \text{ cm/Å}} = 3.0 \times 10^{18} \text{ sec}^{-1}$

(b) $\nu = \dfrac{2.998 \times 10^{10} \text{ cm/sec}}{5000 \text{ Å} \times 10^{-8} \text{ cm/Å}} = 5.996 \times 10^{14} \text{ sec}^{-1}$

(c) $\nu = \dfrac{3.00 \times 10^{10} \text{ cm/sec}}{4.4 \text{ } \mu \times 10^{-4} \text{ cm/}\mu} = 6.8 \times 10^{13} \text{ sec}^{-1}$

(d) $\nu = \dfrac{3.00 \times 10^{10} \text{ cm/sec}}{89 \text{ m} \times 10^{2} \text{ cm/m}} = 3.4 \times 10^{6} \text{ sec}^{-1}$

The frequency unit sec^{-1} is also called a *cycle*. In (d), for example, the answer could also be listed as 3.4×10^6 cycles or 3.4 megacycles.

20.2. In the photoelectric effect, an absorbed quantum of light results in the ejection of an electron from the absorber. The kinetic energy of the ejected electron is equal to the energy of the light quantum absorbed minus the energy of the longest wavelength quantum that causes the effect. Calculate the kinetic energy of a photoelectric electron produced in cesium by 4000 Å light. The critical (maximum) wavelength for the photoelectric effect in cesium is 6600 Å.

$$\nu_{\text{crit}} = \frac{2.998 \times 10^{10} \text{ cm/sec}}{6600 \text{ Å} \times 10^{-8} \text{ cm/Å}} = 4.54 \times 10^{14} \text{ sec}^{-1}$$

$$\nu = \frac{2.998 \times 10^{10} \text{ cm/sec}}{4 \times 10^{3} \text{ Å} \times 10^{-8} \text{ cm/Å}} = 7.50 \times 10^{14} \text{ sec}^{-1}$$

Kinetic energy of electron $= h\nu - h\nu_{\text{crit}} = h(\nu - \nu_{\text{crit}})$
$$= (6.63 \times 10^{-27} \text{ erg sec})(7.50 - 4.54)10^{14} \text{ sec}^{-1} = 1.96 \times 10^{-12} \text{ erg}$$

20.3. It has been found that gaseous iodine molecules dissociate into separated atoms after absorption of light at wavelengths less than 4995 Å. If each quantum is absorbed by one molecule of I_2, what is the minimum input, in kcal per mole, needed to dissociate I_2 by this photochemical process?

$$E \text{ (in kcal/mole)} = \frac{2.859 \times 10^{5}}{\lambda \text{ (in Å)}} = \frac{2.859 \times 10^{5}}{4.995 \times 10^{3}} = 57.2 \text{ kcal/mole}$$

20.4. A beam of electrons accelerated through 4.64 eV in a tube containing mercury vapor was partly absorbed by the vapor. As a result of absorption, electronic changes occurred within a mercury atom and light was emitted. If the full energy of a single electron was converted into light, what was the wavelength of the emitted light?

We can find the conversion factor linking eV to wavelength of light by combining two factors given earlier in this chapter.

$$\lambda \text{ (in Å)} = \frac{2.859 \times 10^{5}}{E \text{ (in kcal/mole)}} \qquad 1 \text{ eV} = 23.06 \text{ kcal/mole}$$

Then $\lambda \text{ (in Å)} = \dfrac{2.859 \times 10^{5}}{E \text{ (in eV)} \times 23.06} = \dfrac{1.240 \times 10^{4}}{E \text{ (in eV)}} = \dfrac{1.240 \times 10^{4}}{4.64} = 2.67 \times 10^{3} \text{ Å}$

20.5. How many protons, neutrons, and electrons are there in each of the following atoms: (a) He^3, (b) C^{12}, (c) Pb^{206}?

(a) From the atomic weight table, we see that the atomic number of He is 2; therefore the nucleus must contain 2 protons. Since the mass number of this isotope is 3, the sum of the protons and neutrons must equal 3; therefore there is 1 neutron. The number of electrons in the atom is the same as the atomic number, 2.

(b) The atomic number of carbon is 6; hence the nucleus must contain 6 protons. The number of neutrons is equal to $(12-6)$, or 6. The number of electrons is the same as the atomic number, 6.

(c) The atomic number of lead is 82; hence there are 82 protons in the nucleus. The number of neutrons is $(206-82)$, or 124. There are 82 electrons.

20.6. Complete the following nuclear equations.

(a) $_7N^{14} + _2He^4 \longrightarrow _8O^{17} + \cdots$

(b) $_4Be^9 + _2He^4 \longrightarrow _6C^{12} + \cdots$

(c) $_4Be^9 (p, \alpha) \cdots$

(d) $_{15}P^{30} \longrightarrow _{14}Si^{30} + \cdots$

(e) $_1H^3 \longrightarrow _2He^3 + \cdots$

(f) $_{20}Ca^{43} (\alpha, \cdots) _{21}Sc^{46}$

(a) The sum of the subscripts on the left is $(7+2) = 9$. The subscript of the first product on the right is 8. Hence the second product on the right must have a subscript (net charge) of 1.

The sum of the superscripts on the left is $(14+4) = 18$. The superscript of the first product on the right is 17. Hence the second product on the right must have a superscript (mass number) of 1.

The particle with nuclear charge 1 and mass number 1 is the proton, $_1H^1$.

(b) The nuclear charge of the second product particle (its subscript) is $(4+2)-6 = 0$. The mass number of the particle (its superscript) is $(9+4)-12 = 1$. Hence the particle must be the neutron, $_0n^1$.

(c) The reactants, $_4Be^9$ and $_1H^1$, have a combined nuclear charge of 5 and mass number of 10. In addition to the α-particle, a product will be formed of charge $5-2 = 3$, and mass $10-4 = 6$. This is $_3Li^6$, since lithium is the element of atomic number 3.

(d) The nuclear charge of the second particle is $15-14 = +1$. The mass number is $30-30 = 0$. Hence the particle must be the positron, $_{+1}e^0$.

(e) The nuclear charge of the second particle is $1-2 = -1$. Its mass number is $3-3 = 0$. Hence the particle must be a β^-, or negative electron, $_{-1}e^0$.

(f) The reactants, $_{20}Ca^{43}$ and $_2He^4$, have a combined nuclear charge of 22 and mass number 47. The ejected product will have a charge $22-21 = 1$, and mass $47-46 = 1$. This is a proton and should be represented in the parentheses by p.

20.7. What is the total binding energy of C^{12} and what is the average binding energy per nucleon?

Although binding energy is a term referring to the nucleus, it is more convenient to use the mass of the whole atom in calculations. Then $M_n = M - Zm_e$, where M_n, M, and m_e are the nuclear, atomic and electron masses respectively. The binding energy, B.E., can be represented as follows:

$$\text{B.E. for } C^{12} = 6(M_n \text{ for } H^1) + 6(M_n \text{ for } n) - (M_n \text{ for } _6C^{12})$$
$$= 6[(M \text{ for } H^1) - m_e] + 6(M \text{ for } n) - [(M \text{ for } C^{12}) - 6m_e]$$
$$= 6(M \text{ for } H^1) + 6(M \text{ for } n) - (M \text{ for } C^{12})$$

In other words, a mass difference equation can just as well be used in terms of whole atom masses. The Zm_e terms always cancel because there are just as many electrons in the whole atom in question as are needed to make whole hydrogen atoms from the constituent protons. Whole atom masses can, in fact, be used for mass difference calculations in all nuclear reaction types discussed in this chapter except in β^+ processes, where there is a resulting annihilation of two electron masses (one β^+ and one β^-).

The data needed for this problem can be obtained from Table 20-2.

$$\text{Mass of 6 } H^1 \text{ atoms} = 6 \times 1.0078 = 6.0468$$
$$\text{Mass of 6 neutrons} = 6 \times 1.0087 = 6.0522$$
$$\text{Total mass of component particles} = 12.0990$$
$$\text{Mass of } C^{12} = 12.0000$$
$$\text{Loss in mass on formation of } C^{12} = .0990$$

$$\text{Binding energy} = 931 \times 0.0990 \text{ MeV} = 92.2 \text{ MeV}$$

Since there are 12 nucleons (protons and neutrons), the average binding energy per nucleon is (92.2 MeV)/12, or 7.68 MeV.

20.8. Evaluate Q for the $Li^7 (p, n)Be^7$ reaction.

The change of mass for the reaction must be computed.

Reactants		Products	
$_3Li^7$	7.01601	$_0n^1$	1.00867
$_1H^1$	1.00783	$_4Be^7$	7.01693
	8.02384		8.02560

Increase of mass $= 8.02560 - 8.02384 = .00176$.

A corresponding net amount of energy must be consumed, equal to 931×0.00176 MeV, or 1.64 MeV, and $Q = -1.64$ MeV. This energy is supplied as kinetic energy of the bombarding proton, and is *part* of the acceleration requirement for the proton supplied by a cyclotron or some other accelerating device.

20.9. The Q value for the $He^3(n, p)$ reaction is 0.76 MeV. What is the nuclidic mass of He^3?

The reaction is
$$_2He^3 + _0n^1 \longrightarrow _1H^1 + _1H^3.$$

The mass loss must be $0.76/931 = 0.00082\, u$.

The mass balance can be calculated on the basis of whole atoms.

Reactants		Products	
He^3	x	H^1	1.00783
n	1.00867	H^3	3.01605
	$x + 1.00867$		4.02388

Then $(x + 1.00867) - 4.02388 = 0.00082$ or $x = 3.01603$.

20.10. Calculate the maximum kinetic energy of the β^- emitted in the radioactive decay of He^6.

The process referred to is
$$_2He^6 \longrightarrow _3Li^6 + \beta^-$$

In computing the mass change during this process, only the whole atomic masses of He^6 and Li^6 need be considered.

$$\text{Mass of } He^6 = 6.01890$$
$$\text{Mass of } Li^6 = 6.01513$$
$$\text{Loss in mass} = .00377$$

$$\text{Energy equivalent} = 931 \times .00377 \text{ MeV} = 3.51 \text{ MeV}$$

The maximum kinetic energy of the β^--particle is 3.51 MeV.

20.11. Consider the two nuclides of mass number 7: Li^7 and Be^7. Which of the two is stabler? How does the unstable nuclide decay into the stable one?

Be⁷ has a larger mass than Li⁷. Thus $_4Be^7$ can decay spontaneously into $_3Li^7$, but not vice versa. There are two types of decay process in which Z is decreased by one unit without a change in mass number A: β^+ emission and K capture. These two processes have different mass balance requirements.

Assume that the process is β^+ emission.

$$_4Be^7 \longrightarrow \beta^+ + {}_3Li^7$$

Since β^+ subsequently undergoes an annihilation with an electron, the mass difference between Be⁷ and Li⁷ must at least equal the rest mass of 2 electrons, i.e. $2 \times .00055 = .00110\ u$. The actual mass difference between parent and daughter nuclides is $(7.01693 - 7.01601)$, or $.00092\ u$. We thus see that positron emission in this case is impossible. By elimination, we conclude that Be⁷ undergoes K capture.

Note that we have predicted only that Be⁷ *should* decay by K capture into Li⁷. We have said nothing about the rate of such a process. *Measurements* show the half-life of the process to be 54 days.

20.12. N^{13} decays by β^+ emission. The maximum kinetic energy of the β^+ is 1.19 MeV. What is the nuclidic mass of N^{13}?

The reaction is $\qquad _7N^{13} \longrightarrow {}_6C^{13} + \beta^+$

This is the type of process, mentioned in Problem 20.7, in which a simple difference of whole atom masses is not the desired quantity.

$$\begin{aligned}
\text{Mass difference} &= (M_n \text{ for } _7N^{13}) - (M_n \text{ for } C^{13}) - m_e \\
&= [(M \text{ for } _7N^{13}) - 7m_e] - [(M \text{ for } _6C^{13}) - 6m_e] - m_e \\
&= (M \text{ for } N^{13}) - (M \text{ for } C^{13}) - 2m_e \\
&= (M \text{ for } N^{13}) - 13.00335 - 2(.00055) \\
&= (M \text{ for } N^{13}) - 13.00445
\end{aligned}$$

This expression must equal the mass equivalent of the maximum kinetic energy of the β^+, $\dfrac{1.19 \text{ MeV}}{931 \text{ MeV}/u} = .00128\ u$. Then

$.00128 = (M \text{ for } N^{13}) - 13.00445 \qquad$ or $\qquad M \text{ for } N^{13} = 13.00445 + .00128 = 13.00573$

The additional $.00110\ u$ is not lost until the subsequent annihilation reaction occurs.

20.13. An isotopic species of lithium hydride, Li^6H^2, is a potential nuclear fuel on the basis of the following reaction.

$$_3Li^6 + {}_1H^2 \longrightarrow 2\ _2He^4$$

Calculate the expected power production, in kilowatts, associated with the consumption of 1.00 gram of Li^6H^2 per day. Assume 100% efficiency in the process.

The change of mass for the reaction is first computed.

$_3Li^6$	6.01513
$_1H^2$	2.01410
Total mass of reactants	8.02923
$2\ _2He^4 = 2 \times 4.00260$	8.00520
Loss in mass	0.02403

The energy of the reaction is $931.5 \text{ MeV}/u \times 0.02403\ u = 22.38 \text{ MeV}$. In kcal/mole, this is $22.38 \times 10^6 \text{ eV} \times 23.06 \text{ (kcal/mole)/eV} = 51.6 \times 10^7 \text{ kcal/mole}$. The energy per gram of LiH is then $\dfrac{51.6 \times 10^7 \text{ kcal/mole}}{8.03 \text{ g/mole}} = 6.43 \times 10^7 \text{ kcal/g}$. This can be expressed in kilowatt-hours by the following conversions.

$$6.43 \times 10^7 \text{ kcal} \times \frac{4.184 \times 10^3 \text{ joules}}{1 \text{ kcal}} \times \frac{1 \text{ watt-sec}}{1 \text{ joule}} \times \frac{1 \text{ hr}}{3.6 \times 10^3 \text{ sec}}$$

$$= 7.47 \times 10^7 \text{ watt-hr} = 7.47 \times 10^4 \text{ kw-hr}$$

The power, or rate of energy production, is $\dfrac{7.47 \times 10^4 \text{ kw-hr}}{24 \text{ hr}} = 3.11 \times 10^3 \text{ kw.}$

20.14. F^{18} is found to undergo 90% radioactive decay in 366 minutes. What is its half-life?

The fraction remaining after 366 minutes is one-tenth. This is between $1/8 \ [= (1/2)^3]$ and $1/16 \ [= (1/2)^4]$. Somewhere between 3 and 4 half-lives must have elapsed to reduce the sample to 10% of its initial value. A quick estimate shows that the half-life is between $366/3 \ [= 122 \min]$ and $366/4 \ [= 92 \min]$. The exact solution is obtained by using the logarithmic equation.

$$T = \frac{-0.301 \, t}{\log f} = \frac{-0.301 \times 366 \min}{\log .10} = + 0.301 \times 366 \min = 110 \min$$

20.15. A sample of river water was found to contain 8×10^{-18} tritium atoms, $_1H^3$, per atom of ordinary hydrogen. Tritium decomposes radioactively with a half-life of 12.3 years. (a) What will be the ratio of tritium to normal hydrogen atoms 49 years after the original sample was taken if the sample is stored in a place where additional tritium atoms cannot be formed? (b) How many individual tritium atoms would 10 grams of such a sample contain 40 years after the initial sampling?

(a) 49 years is almost exactly four half-lives, 4×12.3 years, or $4T$. It is simpler in this case to avoid the logarithmic formula. The fraction of the tritium atoms remaining would be $\frac{1}{2} \times \frac{1}{2} \times \frac{1}{2} \times \frac{1}{2} = \frac{1}{16}$. The normal hydrogen atoms are not radioactive and therefore their number does not change. Thus the final ratio of tritium to normal hydrogen will be $\frac{1}{16}(8 \times 10^{-18}) = 5 \times 10^{-19}$.

(b) 10 grams of water contains $\dfrac{10 \text{ g}}{18 \text{ g/mole}}$ or 10/18 moles of water or 10/9 gram-atoms of hydrogen. The total number of hydrogen atoms is $(10/9)(6 \times 10^{23}) = 6.7 \times 10^{23}$. Of these, the original number of tritium atoms is $(8 \times 10^{-18})(6.7 \times 10^{23}) = 5 \times 10^6$. Only a fraction of this number, f, remains after 40 years.

$$\log f = -\frac{0.301 \, t}{T} = -\frac{0.301 \times 40 \text{ years}}{12.3 \text{ years}} = -0.979 = -1 + 0.021$$

from which $f = 0.10$. The number of remaining tritium atoms is $0.10(5 \times 10^6) = 5 \times 10^5$.

20.16. A sample of uraninite, a uranium containing mineral, was found on analysis to contain 0.214 gram of lead for every gram of uranium. Assuming that the lead all resulted from the radioactive disintegration of the uranium since the geological formation of the uraninite, and that all isotopes of uranium other than U^{238} can be neglected, estimate the date when the mineral was formed in the earth's crust. The half-life of U^{238} is 4.5×10^9 years.

The radioactive decay of U^{238} leads, after 14 steps, to the stable lead isotope, Pb^{206}. The first of these steps, the α-decay of U^{238} with a 4.5×10^9 year half-life, is intrinsically over 10,000 times as slow as any of the subsequent steps. As a result, the time required for the first step accounts for essentially all the time required for the entire 14-step sequence.

Number of gram-atoms of Pb^{206} per gram of uranium $= \dfrac{0.214 \text{ g Pb}}{206 \text{ g/g-atom}} = 1.04 \times 10^{-3}$ g-at Pb.

Number of gram-atoms of U^{238} in same sample $= \dfrac{1.000 \text{ g U}}{238 \text{ g/g-atom}} = 4.20 \times 10^{-3}$ g-at U.

If each atom of lead in the mineral today is the daughter of a uranium atom that existed at the time of the formation of the mineral, then the original number of gram-atoms of uranium in the sample would have been $(1.04 + 4.20) \times 10^{-3} = 5.24 \times 10^{-3}$. In the half-life equation, $f = \dfrac{4.20 \times 10^{-3}}{5.24 \times 10^{-3}} = 0.802$, and t is the elapsed time from the formation of the mineral in the earth's crust to the present time.

$$t = -\frac{T \log f}{0.301} = -\frac{(4.5 \times 10^9 \text{ years})(\log 0.802)}{0.301} = -\frac{(4.5 \times 10^9)(-1 + 0.904)}{0.301}$$

$$= -\frac{(4.5 \times 10^9)(-0.096)}{0.301} = 1.4 \times 10^9 \text{ years}$$

20.17. A sample of $C^{14}O_2$ was to be mixed with ordinary CO_2 for a biological tracer experiment. In order that 10 cc (S.T.P.) of the diluted gas should have 10^4 disintegrations per minute, how many curies of radioactive carbon are needed to prepare 60 liters of the diluted gas?

$$\text{Total activity} = \frac{10^4 \text{ dis/min}}{10 \text{ cc}} \times \frac{60 \; l \times 10^3 \text{ cc}/l}{60 \text{ sec/min}}$$

$$= 10^6 \text{ dis/sec} \times \frac{1 \text{ curie}}{3.7 \times 10^{10} \text{ dis/sec}} = 2.7 \times 10^{-5} \text{ curie}$$

Supplementary Problems

20.18. Find the wavelength λ in the indicated units for light of the following frequencies: (a) 55 megacycles (λ in m), (b) 1000 cycles (λ in cm), (c) 7.5×10^{15} sec^{-1} (λ in Å).
Ans. (a) 5.5 m, (b) 2.998×10^7 cm, (c) 4.0×10^2 Å

20.19. The critical wavelength for producing the photoelectric effect in tungsten is 2.6×10^3 Å. (a) What is the energy of a quantum at this wavelength in ergs and in eV? (b) What wavelength would be necessary to produce photoelectrons from tungsten having twice the kinetic energy of those produced at 2.2×10^3 Å?   *Ans.* (a) 8×10^{-12} erg, 5 eV; (b) 1.9×10^3 Å

20.20. In a measurement of the quantum efficiency of photosynthesis in green plants, it was found that 8 quanta of red light at 6850 Å were needed to evolve one molecule of O_2. The average energy storage in the photosynthetic process is 112 kcal per mole of O_2 evolved. What is the energy conversion efficiency in this experiment.   *Ans.* 34%

20.21. O_2 undergoes photochemical dissociation into one normal oxygen atom and one oxygen atom 1.967 eV more energetic than normal. The dissociation of O_2 into two normal oxygen atoms is known to require 117.2 kcal per mole O_2. What is the maximum wavelength effective for the photochemical dissociation of O_2?   *Ans.* 1758 Å

20.22. The dye acriflavine, when dissolved in water, has its maximum light absorption at 4530 Å, and its maximum fluorescence emission at 5080 Å. The number of fluorescence quanta is, on the average, 53% of the number of quanta absorbed. Using the wavelengths of maximum absorption and emission, what percentage of absorbed energy is emitted as fluorescence?   *Ans.* 47%

20.23. Determine the number of (a) nuclear protons, (b) nuclear neutrons, (c) electrons, in each of the following atoms: (1) Ge^{70}, (2) Ge^{72}, (3) Be^9, (4) U^{235}.
Ans. (1): (a) 32, (b) 38, (c) 32   (3): (a) 4, (b) 5, (c) 4
(2): (a) 32, (b) 40, (c) 32   (4): (a) 92, (b) 143, (c) 92

20.24. Write the complete nuclear symbol for natural fluorine and natural arsenic. Each has only one stable isotope.   *Ans.* $_9F^{19}$, $_{33}As^{75}$

20.25. By natural radioactivity, U^{238} emits an α-particle. The heavy residual nucleus is called UX_1. UX_1 in turn emits a β-particle. The heavy residual nucleus from this radioactive process is called UX_2. Determine the atomic numbers and mass numbers of (a) UX_1 and (b) UX_2.
Ans. (a) 90, 234; (b) 91, 234

20.26. By radioactivity, $_{93}Np^{239}$ emits a β-particle. The residual heavy nucleus is also radioactive and gives rise to U^{235} by the radioactive process. What small particle is emitted simultaneously with the formation of U^{235}?   *Ans.* α-particle

20.27. Complete the following equations.
(a) $_{11}Na^{23} + {}_2He^4 \longrightarrow {}_{12}Mg^{26} + ?$ (c) $Ag^{106} \longrightarrow Cd^{106} + ?$
(b) $_{29}Cu^{64} \longrightarrow \beta^+ + ?$ (d) $_5B^{10} + {}_2He^4 \longrightarrow {}_7N^{13} + ?$
Ans. (a) $_1H^1$, (b) $_{28}Ni^{64}$, (c) β^-, (d) $_0n^1$

20.28. Complete the notations for the following nuclear processes.
(a) $Mg^{24}(d, \alpha)$?   (c) $Ar^{40}(\alpha, p)$?   (e) $Te^{130}(d, 2n)$?   (g) $Co^{59}(n, \alpha)$?
(b) $Mg^{26}(d, p)$?   (d) $C^{12}(d, n)$?   (f) $Mn^{55}(n, \gamma)$?
Ans. (a) Na^{22}, (b) Mg^{27}, (c) K^{43}, (d) N^{13}, (e) I^{130}, (f) Mn^{56}, (g) Mn^{56}

20.29. If an element in Group IA of the periodic table undergoes radioactive decay by emitting positrons, what is the valence expected for the resulting element?   *Ans.* Zero

20.30. An alkaline earth element is radioactive. It and its daughter elements decay by emitting 3 alpha particles in succession. In what Group should the resulting element be found?   *Ans.* Group IV

20.31. If an atom of U^{235}, after absorption of a slow neutron, undergoes fission to form an atom of Xe^{139} and an atom of Sr^{94}, what other particles are produced, and how many?   *Ans.* 3 neutrons

20.32. Which is more unstable of each of the following pairs, and in each case what type of process could the unstable nucleus undergo: (a) C^{16}, N^{16}; (b) F^{18}, Ne^{18}?
Ans. (a) C^{16}, β^-; (b) Ne^{18}, both β^+ and K electron capture are possibilities on the basis of data given here.

20.33. One of the stablest nuclei is Mn^{55}. Its nuclidic mass is 54.938. Determine its total binding energy and average binding energy per nucleon.   *Ans.* 482 MeV, 8.77 MeV per nucleon

20.34. How much energy is released during each of the following reactions? (a) $_1H^1 + {}_3Li^7 \longrightarrow 2\,{}_2He^4$;
(b) $_1H^3 + {}_1H^2 \longrightarrow {}_2He^4 + {}_0n^1$ *Ans.* (a) 17.4 MeV, (b) 17.6 MeV

20.35. C^{14} is believed to be made in the upper atmosphere by an (n, p) process on N^{14}. What is Q for this reaction?   *Ans.* 0.62 MeV

20.36. In the reaction $N^{14}(n, \gamma)N^{15}$ with slow neutrons, the γ is produced with an energy of 10.9 MeV. What is the nuclidic mass of N^{15}?   *Ans.* 15.0001

20.37. If a β^+ and a β^- annihilate each other and their rest masses are converted into two γ rays of equal energy, what is the energy in MeV and the wavelength of each γ?
Ans. 0.51 MeV, 0.024 Å

20.38. The combustion of a mole of ethylene in oxygen is exothermic to the extent of 337 kcal. What would be the loss in mass (expressed in u) accompanying the oxidation of one molecule of ethylene?
Ans. $1.57 \times 10^{-8}\ u$ (This value is so small that the change in mass, as in all chemical reactions, is ordinarily not taken into account.)

20.39. The sun's energy is believed to come from a series of nuclear reactions, the overall result of which is the transformation of 4 hydrogen atoms into one helium atom. How much energy is released in the formation of one helium atom? (Include the annihilation energy of the two positrons formed in the nuclear reactions with two electrons.)   *Ans.* 26.7 MeV

20.40. It is proposed to use the nuclear fusion reaction

$$2\ _1H^2 \longrightarrow\ _2He^4 + energy$$

to produce industrial electric power. If the output is to be 50,000 kilowatts and the energy of the above reaction is used with 30% efficiency, how many grams of deuterium fuel will be needed per day?   *Ans.* 25 grams

20.41. A pure radiochemical preparation was observed to disintegrate at the rate of 4280 counts per minute at 1:35 p.m. At 4:55 p.m. of the same day, the disintegration rate of the sample was only 1070 counts per minute. The disintegration rate is proportional to the number of radioactive atoms in the sample. What is the half-life of the material?   *Ans.* 100 min

20.42. An atomic battery for pocket watches has been developed which uses the beta particles from Pm^{147} as the primary energy source. The half-life of Pm^{147} is 2.65 years. How long would it take for the rate of beta emission in the battery to be reduced to 10% of its initial value?   *Ans.* 8.8 years

20.43. The half-life of C^{14} is 5720 years. (*a*) What fraction of its original C^{14} would a sample of $CaCO_3$ have after 11,440 years of storage in a locality where additional radioactivity could not be produced? (*b*) What fraction of the original C^{14} would still remain after 13,000 years?
Ans. (*a*) 0.25, (*b*) 0.21

20.44. All naturally occurring rubidium ores contain Sr^{87}, resulting from the beta decay of Rb^{87}. In naturally occurring rubidium, 278 of every 1000 rubidium atoms are Rb^{87}. A mineral containing 0.85% rubidium was analyzed and found to contain 0.0098% strontium. Assuming that all of the strontium originated by radioactive decay of Rb^{87}, estimate the age of the mineral. Rb^{87} has a half-life of 5.7×10^{10} years.   *Ans.* 3.4×10^9 years

20.45. Prior to the use of nuclear weapons, the specific activity of C^{14} in soluble ocean carbonates was found to be 16 disintegrations per minute per gram of carbon. The amount of carbon in these carbonates has been estimated as 5×10^{13} tons. How many megacuries of C^{14} did the ocean carbonates contain?   *Ans.* 3.3×10^2 megacuries

20.46. How much heat would be developed per hour from a 1 curie C^{14} source if all the energy of the β^- decay were imprisoned?   *Ans.* 0.8 cal

Exponents

A. The following is a partial list of powers of 10.

$10^0 = 1$

$10^1 = 10$

$10^2 = 10 \times 10 = 100$

$10^3 = 10 \times 10 \times 10 = 1000$

$10^4 = 10 \times 10 \times 10 \times 10 = 10,000$

$10^5 = 10 \times 10 \times 10 \times 10 \times 10 = 100,000$

$10^6 = 10 \times 10 \times 10 \times 10 \times 10 \times 10 = 1,000,000$

$10^{-1} = \dfrac{1}{10} = 0.1$

$10^{-2} = \dfrac{1}{10^2} = \dfrac{1}{100} = 0.01$

$10^{-3} = \dfrac{1}{10^3} = \dfrac{1}{1000} = 0.001$

$10^{-4} = \dfrac{1}{10^4} = \dfrac{1}{10,000} = 0.0001$

In the expression 10^5, the *base* is 10 and the *exponent* is 5.

B. In multiplication, exponents of like bases are added.

(1) $a^3 \times a^5 = a^{3+5} = a^8$

(2) $10^2 \times 10^3 = 10^{2+3} = 10^5$

(3) $10 \times 10 = 10^{1+1} = 10^2$

(4) $10^7 \times 10^{-3} = 10^{7-3} = 10^4$

(5) $(4 \times 10^4)(2 \times 10^{-6}) = 8 \times 10^{4-6} = 8 \times 10^{-2}$

(6) $(2 \times 10^5)(3 \times 10^{-2}) = 6 \times 10^{5-2} = 6 \times 10^3$

C. In division, exponents of like bases are subtracted.

(1) $\dfrac{a^5}{a^3} = a^{5-3} = a^2$

(2) $\dfrac{10^2}{10^5} = 10^{2-5} = 10^{-3}$

(3) $\dfrac{8 \times 10^2}{2 \times 10^{-6}} = \dfrac{8}{2} \times 10^{2+6} = 4 \times 10^8$

(4) $\dfrac{5.6 \times 10^{-2}}{1.6 \times 10^4} = \dfrac{5.6}{1.6} \times 10^{-2-4} = 3.5 \times 10^{-6}$

D. Any number may be expressed as an integral power of 10, or as the product of two numbers one of which is an integral power of 10 (e.g., $300 = 3 \times 10^2$).

(1) $22,400 = 2.24 \times 10^4$

(2) $7,200,000 = 7.2 \times 10^6$

(3) $454 = 4.54 \times 10^2$

(4) $0.454 = 4.54 \times 10^{-1}$

(5) $0.0454 = 4.54 \times 10^{-2}$

(6) $0.00006 = 6 \times 10^{-5}$

(7) $0.00306 = 3.06 \times 10^{-3}$

(8) $0.0000005 = 5 \times 10^{-7}$

Moving the decimal point one place to the right is equivalent to multiplying a number by 10; moving the decimal point two places to the right is equivalent to multiplying by 100, and so on. Whenever the decimal point is moved to the right by n places, compensation can be achieved by *dividing* at the same time by 10^n, and the overall value of the number remains unchanged. Thus $0.0325 = \dfrac{3.25}{10^2} = 3.25 \times 10^{-2}$.

Moving the decimal point one place to the left is equivalent to dividing by 10. Whenever the decimal point is moved to the left n places, compensation can be achieved by *multiplying* at the same time by 10^n, and the overall value of the number remains unchanged. For example, $7296 = 72.96 \times 10^2 = 7.296 \times 10^3$.

E. An expression with an exponent of zero is equal to 1.

(1) $a^0 = 1$ (2) $10^0 = 1$ (3) $(3 \times 10)^0 = 1$ (4) $7 \times 10^0 = 7$ (5) $8.2 \times 10^0 = 8.2$

F. A factor may be transferred from the numerator to the denominator of a fraction, or vice versa, by changing the sign of the exponent.

(1) $10^{-4} = \dfrac{1}{10^4}$ (2) $5 \times 10^{-3} = \dfrac{5}{10^3}$ (3) $\dfrac{7}{10^{-2}} = 7 \times 10^2$ (4) $-5a^{-2} = -\dfrac{5}{a^2}$

G. The meaning of the fractional exponent is illustrated by the following.

(1) $10^{2/3} = \sqrt[3]{10^2}$ (2) $10^{3/2} = \sqrt{10^3}$ (3) $10^{1/2} = \sqrt{10}$ (4) $4^{3/2} = \sqrt{4^3} = \sqrt{64} = 8$

H. (1) $(10^3)^2 = 10^{3 \times 2} = 10^6$ (2) $(10^{-2})^3 = 10^{-2 \times 3} = 10^{-6}$ (3) $(a^3)^{-2} = a^{-6}$

I. To extract the square root of a power of 10, divide the exponent by 2. If the exponent is an odd number it should be increased or decreased by 1, and the coefficient adjusted accordingly. To extract the cube root of a power of 10, adjust so that the exponent is divisible by 3; then divide the exponent by 3. The coefficients are treated independently.

(1) $\sqrt{90{,}000} = \sqrt{9 \times 10^4} = \sqrt{9} \times \sqrt{10^4} = 3 \times 10^2$ or 300

(2) $\sqrt{3.6 \times 10^3} = \sqrt{36 \times 10^2} = \sqrt{36} \times \sqrt{10^2} = 6 \times 10^1$ or 60

(3) $\sqrt{4.9 \times 10^{-5}} = \sqrt{49 \times 10^{-6}} = \sqrt{49} \times \sqrt{10^{-6}} = 7 \times 10^{-3}$ or 0.007

(4) $\sqrt[3]{8 \times 10^9} = \sqrt[3]{8} \times \sqrt[3]{10^9} = 2 \times 10^3$ or 2000

(5) $\sqrt[3]{1.25 \times 10^5} = \sqrt[3]{125 \times 10^3} = \sqrt[3]{125} \times \sqrt[3]{10^3} = 5 \times 10$ or 50

J. Multiplication and division of numbers expressed as powers of ten.

(1) $8000 \times 2500 = (8 \times 10^3)(2.5 \times 10^3) = 20 \times 10^6 = 2 \times 10^7$ or $20{,}000{,}000$

(2) $\dfrac{48{,}000{,}000}{1200} = \dfrac{48 \times 10^6}{12 \times 10^2} = 4 \times 10^{6-2} = 4 \times 10^4$ or $40{,}000$

(3) $\dfrac{0.0078}{120} = \dfrac{7.8 \times 10^{-3}}{1.2 \times 10^2} = 6.5 \times 10^{-5}$ or 0.000065

(4) $(4 \times 10^{-3})(5 \times 10^4)^2 = (4 \times 10^{-3})(5^2 \times 10^8) = 4 \times 5^2 \times 10^{-3+8} = 100 \times 10^5 = 1 \times 10^7$

(5) $\dfrac{(6{,}000{,}000)(0.00004)^4}{(800)^2 (0.0002)^3} = \dfrac{(6 \times 10^6)(4 \times 10^{-5})^4}{(8 \times 10^2)^2 (2 \times 10^{-4})^3} = \dfrac{6 \times 4^4}{8^2 \times 2^3} \times \dfrac{10^6 \times 10^{-20}}{10^4 \times 10^{-12}}$

$\qquad = \dfrac{6 \times 256}{64 \times 8} \times \dfrac{10^{6-20}}{10^{4-12}} = 3 \times \dfrac{10^{-14}}{10^{-8}} = 3 \times 10^{-6}$

(6) $(\sqrt{4.0 \times 10^{-6}})(\sqrt{8.1 \times 10^3})(\sqrt{0.0016}) = (\sqrt{4.0 \times 10^{-6}})(\sqrt{81 \times 10^2})(\sqrt{16 \times 10^{-4}})$

$\qquad = (2 \times 10^{-3})(9 \times 10^1)(4 \times 10^{-2})$

$\qquad = 72 \times 10^{-4} = 7.2 \times 10^{-3}$ or 0.0072

(7) $(\sqrt[3]{6.4 \times 10^{-2}})(\sqrt[3]{27{,}000})(\sqrt[3]{2.16 \times 10^{-4}}) = (\sqrt[3]{64 \times 10^{-3}})(\sqrt[3]{27 \times 10^3})(\sqrt[3]{216 \times 10^{-6}})$

$\qquad = (4 \times 10^{-1})(3 \times 10^1)(6 \times 10^{-2})$

$\qquad = 72 \times 10^{-2}$ or 0.72

Exercises

(a) Express the following in powers of 10.

(1) 320

(2) 32,600

(3) 1006

(4) 36,000,000

(5) 0.831

(6) 0.03

(7) 0.000002

(8) 0.000706

(9) $\sqrt{640,000}$

(10) $\sqrt{0.000081}$

(11) $\sqrt[3]{8,000,000}$

(12) $\sqrt[3]{0.000027}$

Ans. (1) 3.2×10^2

(2) 3.26×10^4

(3) 1.006×10^3

(4) 3.6×10^7

(5) 8.31×10^{-1}

(6) 3×10^{-2}

(7) 2×10^{-6}

(8) 7.06×10^{-4}

(9) 8.0×10^2

(10) 9.0×10^{-3}

(11) 2×10^2

(12) 3×10^{-2}

(b) Evaluate the following and express the results in powers of 10.

(1) 1500×260

(2) $220 \times 35,000$

(3) $40 \div 20,000$

(4) $82,800 \div 0.12$

(5) $\dfrac{1.728 \times 17.28}{0.0001728}$

(6) $\dfrac{(16,000)(0.0002)(1.2)}{(2000)(0.006)(0.00032)}$

(7) $\dfrac{0.004 \times 32,000 \times 0.6}{6400 \times 3000 \times 0.08}$

(8) $(\sqrt{14,400})(\sqrt{0.000025})$

(9) $(\sqrt[3]{2.7 \times 10^7})(\sqrt[3]{1.25 \times 10^{-4}})$

(10) $(1 \times 10^{-3})(2 \times 10^5)^2$

(11) $\dfrac{(3 \times 10^2)^3 (2 \times 10^{-5})^2}{3.6 \times 10^{-8}}$

(12) $8(2 \times 10^{-2})^{-3}$

Ans. (1) 3.9×10^5

(2) 7.7×10^6

(3) 2×10^{-3}

(4) 6.9×10^5

(5) 1.728×10^5

(6) 1×10^3

(7) 5×10^{-5}

(8) 6.0×10^{-1}

(9) 1.5×10^1

(10) 4×10^7

(11) 3×10^5

(12) 1×10^6

Significant Figures

INTRODUCTION

The following discussion is intended to give the elementary student some workable rules of procedure, but is not intended to replace the rigorous treatment of a text on the theory of measurements.

The numerical value of every observed measurement is an approximation. No physical measurement, such as mass, length, time, volume, velocity, is ever absolutely correct. The accuracy (reliability) of every measurement is limited by the reliability of the measuring instrument, which is never absolutely reliable.

Consider that the length of an object is recorded as 15.7 cm. By convention, this means that the length was measured to the *nearest* tenth of a centimeter and that its exact value lies between 15.65 and 15.75 cm. If this measurement were exact to the nearest hundredth of a centimeter, it would have been recorded as 15.70 cm. The value 15.7 cm represents *three significant figures* (1, 5, 7), while the value 15.70 represents *four significant figures* (1, 5, 7, 0). A significant figure is one which is known to be reasonably reliable.

Similarly, a recorded weight of 3.4062 g, observed with an analytical balance, means that the object was weighed to the nearest tenth of a milligram and represents five significant figures (3, 4, 0, 6, 2), the last figure (2) being reasonably correct and guaranteeing the certainty of the preceding four figures.

A 50-ml buret has markings 1/10 ml apart, and the hundredths of a ml are estimated. A recorded volume of 41.83 ml represents four significant figures. The last figure (3), being estimated, may be in error by one or two digits in either direction. The preceding three figures (4, 1, 8) are completely certain.

In elementary measurements in chemistry and physics, the last figure is estimated and is also considered as a significant figure.

ZEROS

A recorded volume of 28 ml represents two significant figures (2, 8). If this same volume were written as 0.028 liter it would still contain only two significant figures. Zeros appearing as the first figures of a number are not significant, since they merely locate the decimal point. However, the values 0.0280 l and 0.280 l represent three significant figures (2, 8, and the last zero); the value 1.028 l represents four significant figures (1, 0, 2, 8); and the value 1.0280 l represents five significant figures (1, 0, 2, 8, 0). Similarly, the value 19.00 for the atomic weight of fluorine contains four significant figures.

The statement that a body of ore weighs 9800 lb does not indicate definitely the accuracy of the weighing. The last two zeros may have been used merely to locate the decimal point. If it was weighed to the nearest hundred pounds, the weight contains only two significant figures and may be written exponentially as 9.8×10^3 lb. If weighed to the nearest ten pounds it may be written as 9.80×10^3 lb, which indicates that the value is accurate to three significant figures. Since the zero in this case is not needed to locate the decimal point, it must be a significant figure. If the object was weighed to the nearest pound, the weight could be written as 9.800×10^3 lb (four significant figures). Likewise, the statement that the velocity of light is 186,000 mi/sec is accurate to three significant figures, since this value is accurate only to the nearest thousand miles per second; to avoid confusion, it may be written as 1.86×10^5 mi/sec. (It is customary to place the decimal point after the first significant figure.)

ROUNDING OFF

A number is rounded off to the desired number of significant figures by dropping one or more digits to the right. When the first digit dropped is less than 5, the last digit retained should remain unchanged; when it is more than 5, 1 is added to the last digit retained. When it is exactly 5, 1 is added to the last digit retained if that digit is odd; otherwise it is dropped. Thus successive approximations to 3.14159 are 3.1416, 3.142, 3.14, 3.1, 3. The quantity 51.75 g may be rounded off to 51.8 g, 51.65 g to 51.6 g, 51.85 g to 51.8 g.

ADDITION AND SUBTRACTION

The answer should be rounded off after adding or subtracting, so as to retain digits only as far as the first column containing estimated figures. (Remember that the last significant figure is estimated.)

Examples.   Add the following quantities expressed in grams.

(1)	25.340	(2)	58.0	(3)	4.20	(4)	415.5
	5.465		0.0038		1.6523		3.64
	0.322		0.00001		0.015		0.238
	31.127 g (*Ans.*)		58.00381		5.8673		419.378
			= 58.0 g (*Ans.*)		= 5.87 g (*Ans.*)		= 419.4 g (*Ans.*)

An alternative procedure is to round off the individual numbers before performing the arithmetic operation, retaining only as many columns to the right of the decimal as would give a digit in every item to be added or subtracted. Examples (2), (3), and (4) above would be done as follows:

(2)	58.0	(3)	4.20	(4)	415.5
	0.0		1.65		3.6
	0.0		0.02		0.2
	58.0		5.87		419.3

Note that the answer to (4) differs by one in the last place from the previous answer. The last place, however, is known to have some uncertainty in it.

MULTIPLICATION AND DIVISION

The answer should be rounded off to contain only as many significant figures as are contained in the least exact factor. For example, when multiplying 7.485×8.61, or when dividing $0.1642 \div 1.52$, the answer should be given in three significant figures.

This rule is an approximation to a more exact statement that the fractional or percentage error of a product or quotient cannot be any less than the fractional or percentage error of any one factor. For this reason, numbers whose first significant figure is 1 (or occasionally 2) must contain an additional significant figure to have a given fractional error in comparison with a number beginning with 8 or 9.

Consider the division $\frac{9.84}{9.3} = 1.06$. By the approximate rule, the answer should be 1.1 (two significant figures). However, a difference of 1 in the last place of 9.3 (9.3 ± 0.1) results in an error of about 1%, while a difference of 1 in the last place of 1.1 (1.1 ± 0.1) yields an error of roughly 10%. Thus the answer 1.1 is of much lower percentage accuracy than 9.3. Hence in this case the answer should be 1.06, since a difference of 1 in the last place of the least exact factor used in the calculation (9.3) yields a percentage of error about the same (about 1%) as a difference of 1 in the last place of 1.06 (1.06 ± 0.01). Similarly, $0.92 \times 1.13 = 1.04$.

In nearly all academic and commercial calculations, a precision of only two to four significant figures is required. Therefore the student is advised to use an inexpensive 10-inch slide rule which is accurate to three significant figures, or the table of logarithms in the Appendix, which is accurate to four significant figures. The efficient use of a slide rule or log table will save very much time in calculations without sacrificing accuracy.

Exercises

1.   How many significant figures are given in the following quantities?

(a)   454 g   (e)   0.0353 ft   (i) 1.118×10^{-3} g

(b)   2.2 lb   (f)   1.0080 g   (j)   1030 g/cm^2

(c)   2.205 lb   (g)   14.0 ml   (k)   125,000 lb

(d)   0.3937 in.   (h) 9.3×10^7 mi

Ans.   (a) 3,   (b) 2,   (c) 4,   (d) 4,   (e) 3,   (f) 5,   (g) 3,   (h) 2,   (i) 4,   (j) 3 or 4,   (k) 3, 4, 5, or 6

2. Add: (a) 703   g   (b) 18.425 cm   (c) 0.0035 l (d) 4.0 lb
              7   g   7.21 cm   0.097 l 0.632 lb
             0.66 g   5.0 cm   0.225 l 0.148 lb

Ans. (a) 711 g, (b) 30.6 cm, (c) 0.326 l, (d) 4.8 lb

3. Subtract: (a) 7.26 lb   (b) 562.4 ft   (c) 34 kg
              0.2 lb   16.8 ft   0.2 kg

Ans. (a) 7.1 lb, (b) 545.6 ft, (c) 34 kg

4. Multiply: (a) 2.21 × 0.3   (c) 2.02 × 4.113   (e) 12.4 × 84
            (b) 72.4 × 0.084   (d) 107.87 × 0.610 (f) 7.24 × 8.6

Ans. (a) 0.7, (b) 6.1, (c) 8.31, (d) 65.8, (e) 1.04×10^3, (f) 62

5. Divide: (a) $\dfrac{97.52}{2.54}$ (b) $\dfrac{14.28}{0.714}$ (c) $\dfrac{0.032}{0.004}$ (d) $\dfrac{9.8}{9.3}$

Ans. (a) 38.4, (b) 20.0, (c) 8, (d) 1.05

Logarithms

DEFINITION OF TERMS

The logarithm of a positive number is the exponent, or power, of a given base that is required to produce that number. For example, since $1000 = 10^3$, $100 = 10^2$, $10 = 10^1$, $1 = 10^0$, then the logarithms of 1000, 100, 10, 1, to the base 10 are respectively 3, 2, 1, 0.

The system of logarithms whose base is 10 (called the common or Briggsian system) may be used in all numerical computations.

It is obvious that $10^{1.5377}$ will give some number greater than 10 (which is 10^1) but smaller than 100 (10^2). Actually, $10^{1.5377} = 34.49$; hence $\log 34.49 = 1.5377$. The digit before the decimal point is the *characteristic* of the log, and the decimal fraction part is the *mantissa* of the log. In the above example, the characteristic is 1 and the mantissa is .5377.

The mantissa of the log of a number is found in tables, printed without the decimal point. Each mantissa in the tables is understood to have a decimal point preceding it, and the mantissa is always considered positive.

THE CHARACTERISTIC

The characteristic is determined by inspection from the number itself according to the following rules.

(1) For a number greater than 1, the characteristic is positive and is one *less* than the number of digits before the decimal point. For example:

Number	5297	348	900	34.8	60	4.764	3
Characteristic	3	2	2	1	1	0	0

(2) For a positive number less than 1, the characteristic is negative and is one *more* than the number of zeros immediately following the decimal point. The negative sign of the characteristic is written in either of these two ways: (*a*) above the characteristic, as $\bar{1}$, $\bar{2}$, and so on; (*b*) as 9. -10, 8. -10, and so on. Thus the characteristic of the logarithm of 0.3485 is $\bar{1}$, or 9. -10; of the logarithm of 0.0513 is $\bar{2}$, or 8. -10.

(3) Negative numbers do not have logarithms.

TO FIND THE LOGARITHM OF A NUMBER BY USE OF TABLES OF LOGARITHMS IN THE APPENDIX

Suppose it is required to find the complete log of the number 728. In the table of logarithms in the Appendix glance down the N column to 72, then horizontally to the right to column 8 and note the entry 8621 which is the required mantissa. Since the characteristic is 2, $\log 728 = 2.8621$. (This means that $728 = 10^{2.8621}$).

The mantissa for $\log 72.8$, for $\log 7.28$, for $\log 0.728$, for $\log 0.0728$, etc., is .8621, but the characteristics differ. Thus:

$$\log 728 = 2.8621 \qquad \log 0.728 = \bar{1}.8621 \text{ or } 9.8621 - 10$$
$$\log 72.8 = 1.8621 \qquad \log 0.0728 = \bar{2}.8621 \text{ or } 8.8621 - 10$$
$$\log 7.28 = 0.8621 \qquad \log 0.00728 = \bar{3}.8621 \text{ or } 7.8621 - 10$$

To find $\log 46.38$: Glance down the N column to 46, then horizontally to column 3 and note the mantissa 6656. Moving farther to the right along the same line, the figure 7 is found under column 8 of Proportional Parts. The required mantissa is $.6656 + .0007 = .6663$. Since the characteristic is 1, $\log 46.38 = 1.6663$.

The mantissa for log 4638, for log 463.8, for log 46.38, etc., is .6663, but the characteristics differ. Thus:

log 4638 = 3.6663	log 0.4638 = $\bar{1}$.6663 or 9.6663 − 10
log 463.8 = 2.6663	log 0.04638 = $\bar{2}$.6663 or 8.6663 − 10
log 46.38 = 1.6663	log 0.004638 = $\bar{3}$.6663 or 7.6663 − 10
log 4.638 = 0.6663	log 0.0004638 = $\bar{4}$.6663 or 6.6663 − 10

Exercises. Find the logarithms of the following numbers.

(1) 454	(6) 0.621	*Ans.* (1) 2.6571	(6) $\bar{1}$.7931 or 9.7931 − 10
(2) 5280	(7) 0.9463	(2) 3.7226	(7) $\bar{1}$.9760 or 9.9760 − 10
(3) 96,500	(8) 0.0353	(3) 4.9845	(8) $\bar{2}$.5478 or 8.5478 − 10
(4) 30.48	(9) 0.0022	(4) 1.4840	(9) $\bar{3}$.3424 or 7.3424 − 10
(5) 1.057	(10) 0.0002645	(5) 0.0241	(10) $\bar{4}$.4224 or 6.4224 − 10

Sometimes the log of a number must be used in an algebraic equation, such as $y = 7.5 \log x$, or in graphs. If x is greater than 1, $\log x$ is positive and there is no special problem. If x is less than 1, however, $\log x$ is negative. This negative log, according to the above rules, is written as the sum of a positive mantissa and a negative characteristic. For algebraic manipulations it is preferable to treat $\log x$ as a single number with a definite sign, either positive or negative. For such a purpose, $\bar{2}$.7486 would be written as −1.2514, obtained by adding −2 and +.7486 algebraically.

Exercises. Write the logarithms of the following numbers as quantities suitable for insertion in an algebraic equation.

(1) 0.275   (2) 0.000394   (3) 0.0149   *Ans.*   (1) −0.5607   (2) −3.4045   (3) −1.8268

ANTILOGARITHMS

The antilogarithm is the number corresponding to a given logarithm. "The antilog of 3" means "the number whose log is 3"; that number is obviously 1000.

Suppose it is required to find the antilog of 2.6747, i.e. the number whose log is 2.6747. The characteristic is 2 and the mantissa is .6747. Using the table of Antilogarithms in the Appendix, locate 67 in the first column, then move horizontally to column 4 and note the digits 4721. Moving farther to the right along the same line, the entry 8 is found under column 7 of Proportional Parts. Adding 8 to 4721 gives 4729. Since the characteristic is 2, there are three digits to the left of the decimal point. Hence 472.9 is the required number.

It should be understood that the antilog of 1.6747 is 47.29; the antilog of 0.6747 is 4.729; the antilog of 9.6747 − 10 is 0.4729; etc. On the other hand, the antilog of −1.6747 must be rewritten as antilog of $\bar{2}$.3253, or 8.3253 − 10, before the tables may be used, because only positive mantissas are found in the tables.

Exercises. Find the numbers corresponding to the following logarithms.

(1) 3.1568	(7) 0.0008	*Ans.* (1) 1435	(7) 1.002
(2) 1.6934	(8) 9.7507 − 10 or $\bar{1}$.7507	(2) 49.37	(8) 0.5632
(3) 5.6934	(9) 8.0034 − 10 or $\bar{2}$.0034	(3) 493,700	(9) 0.01008
(4) 2.5000	(10) 7.2006 − 10 or $\bar{3}$.2006	(4) 316.2	(10) 0.001587
(5) 2.0436	(11) −0.2436	(5) 110.6	(11) 0.5707
(6) 0.9142	(12) −3.7629	(6) 8.208	(12) 0.0001726

BASIC PRINCIPLES OF LOGARITHMS

Since logarithms are exponents, all properties of exponents are also properties of logarithms.

A. The logarithm of the product of two numbers is the sum of their logarithms.

$$\log ab = \log a + \log b \qquad \log (5280 \times 48) = \log 5280 + \log 48$$

B. The logarithm of the quotient of two numbers is the logarithm of the numerator minus the logarithm of the denominator.

$$\log \frac{a}{b} \;=\; \log a - \log b \qquad\qquad \log \frac{536}{24.5} \;=\; \log 536 - \log 24.5$$

C. The log of the nth power of a number is n times the log of the number.

$$\log a^n \;=\; n \log a \qquad\qquad \log (4.28)^3 \;=\; 3 \log 4.28$$

D. The log of the nth root of a number is the log of the number divided by n.

$$\log \sqrt[n]{a} \;=\; \frac{1}{n} \log a \qquad \log \sqrt{32} \;=\; \frac{1}{2} \log 32 \qquad \log \sqrt[3]{792} \;=\; \frac{1}{3} \log 792$$

This is a special case of **C**, since the nth root of a number is the number raised to the $(1/n)$th power.

LOGARITHMS OF NUMBERS EXPRESSED IN THE DECIMAL NOTATION

The log of 4.50×10^7 $\;=\; \log 4.50 + \log 10^7 \;=\; 7 + \log 4.50$
$$= 7 + 0.6532 \;=\; 7.6532.$$

The log of 4.50×10^{-4} $\;=\; \log 4.50 + \log 10^{-4} \;=\; -4 + \log 4.50$
$$= -4 + 0.6532 \;=\; \overline{4}.6532 \;=\; 6.6532 - 10.$$

In general, if a number is expressed as a product of two factors, one being a number between 1 and 10, and the other being an integral power of ten, the logarithm has as its mantissa the logarithm of the first factor and as its characteristic the exponent of ten in the second factor.

Exercises.   Find the logarithms of the following numbers.

(1) 3.75×10^2	*(3)* 6.6×10^{-27}	*(5)* 60.3×10^{-8}
(2) 6.02×10^{23}	*(4)* 0.75×10^4	*(6)* 2.09×10^{-15}

Ans.
(1) 2.5740	*(3)* $\overline{27}.8195 = 3.8195 - 30$	*(5)* $\overline{7}.7803 = 3.7803 - 10$
(2) 23.7796	*(4)* 3.8751	*(6)* $\overline{15}.3201 = 5.3201 - 20$

Conversely, an antilogarithm can be expressed directly in the decimal notation. The power of 10 is the characteristic; the coefficient of the power of 10 is the antilog of the mantissa, with the decimal point in the antilog following the first digit.

antilog 3.8420 $= (\text{antilog } .8420) \times (\text{antilog } 3) = 6.95 \times 10^3$
antilog $\overline{3}.8420$ $= (\text{antilog } .8420) \times [\text{antilog } (-3)] = 6.95 \times 10^{-3}$
antilog -3.8420 $= \text{antilog } \overline{4}.1580 = (\text{antilog } .1580) \times [\text{antilog } (-4)] = 1.439 \times 10^{-4}$

Exercises.   Write the antilogarithms of the following in the decimal notation.

(1) 10.4769	*(3)* 5.0403	*(5)* 7.6216
(2) $\overline{19}.2046$	*(4)* $4.1402 - 20$	*(6)* -8.2763

Ans.
(1) 2.998×10^{10}	*(3)* 1.097×10^5	*(5)* 4.184×10^7
(2) 1.602×10^{-19}	*(4)* 1.381×10^{-16}	*(6)* 5.292×10^{-9}

ILLUSTRATIONS OF THE USE OF LOGARITHMS

1. Find the value of $487 \times 2.45 \times 0.0387$.

Let $x = 487 \times 2.45 \times 0.0387$.

$\log x = \log 487 + \log 2.45 + \log 0.0387$
$\quad\;\;= 1.6644$

$x = \text{antilog } 1.6644 = 46.17 \text{ or } 46.2$
(to three significant figures)

log 487	$=$ 2.6875
log 2.45	$=$ 0.3892
log 0.0387	$=$ 8.5877 − 10 (add)
log x	$=$ 11.6644 − 10
	or 1.6644

2. Find $x = \dfrac{136.3}{65.38}$.

$$\log x = \log 136.3 - \log 65.38 = 0.3191$$

$$x = \text{antilog } 0.3191 = 2.085$$

$$\log 136.3 = 2.1345$$
$$\log 65.38 = \underline{1.8154} \quad \text{(subtract)}$$
$$\log x = 0.3191$$

3. Find $x = \dfrac{1}{22.4}$.

$$\log x = \log 1 - \log 22.4 = 8.6498 - 10$$

$$x = \text{antilog } 8.6498 - 10 = 0.04465 \text{ or } 0.0446$$
$$\text{(three significant figures)}$$

$$\log 1 = 0 = 10.0000 - 10$$
$$\log 22.4 = \underline{1.3502} \quad \text{(subtract)}$$
$$\log x = 8.6498 - 10$$

Adding or subtracting $10.0000 - 10$, $20.0000 - 20$, etc., from any logarithm does not change its value.

4. Find $x = \dfrac{17.5 \times 1.92}{0.283 \times 0.0314}$.

$$\log x = (\log 17.5 + \log 1.92) - (\log 0.283 + \log 0.0314)$$

$$\log 17.5 = 1.2430$$
$$\log 1.92 = \underline{0.2833} \quad \text{(add)}$$
$$1.5263 \text{ or } 11.5263 - 10$$

$$\log 0.283 = 9.4518 - 10$$
$$\log 0.0314 = \underline{8.4969 - 10} \quad \text{(add)}$$
$$17.9487 - 20 \text{ or } 7.9487 - 10$$

$$\log 17.5 + \log 1.92 = 11.5263 - 10$$
$$\log 0.283 + \log 0.0314 = \underline{7.9487 - 10} \quad \text{(subtract)}$$
$$\log x = 3.5776$$
$$x = \text{antilog } 3.5776 = 3781 \text{ or } 3.78 \times 10^3$$

5. Find $x = (6.138)^3$.

$$\log x = 3(\log 6.138) = 3(0.7881) = 2.3643$$
$$x = \text{antilog } 2.3643 = 231.4$$

6. Find $x = \sqrt{7514}$ or $(7514)^{1/2}$.

$$\log x = \tfrac{1}{2}(\log 7514) = \tfrac{1}{2}(3.8758) = 1.9379$$
$$x = \text{antilog } 1.9379 = 86.68$$

7. Find $x = \sqrt[3]{0.0592}$ or $(0.0592)^{1/3}$.

$$\log x = \tfrac{1}{3} \log 0.0592 = \tfrac{1}{3}(8.7723 - 10) = \tfrac{1}{3}(28.7723 - 30) = 9.5908 - 10$$
$$x = \text{antilog } 9.5908 - 10 = 0.3898 \text{ or } 0.390 \text{ (three significant figures)}$$

8. Find $x = \sqrt{(152)^3}$.

$$\log x = \tfrac{1}{2}(3 \log 152) = \tfrac{1}{2}(3 \times 2.1818) = \tfrac{1}{2}(6.5454) = 3.2727$$
$$x = \text{antilog } 3.2727 = 1874 \text{ or } 1.87 \times 10^3 \text{ (three significant figures)}$$

9. Find $(6.8 \times 10^{-4})^3$ or $(6.8)^3 \times 10^{-12}$.

$$\log (6.8)^3 = 3(\log 6.8) = 3(0.8325) = 2.4975$$
$$(6.8)^3 = \text{antilog } 2.4975 = 314.4 \text{ or } 3.1 \times 10^2 \text{ (two significant figures)}$$

Then $(6.8 \times 10^{-4})^3 = 3.1 \times 10^2 \times 10^{-12} = 3.1 \times 10^{-10}$

10. Find $\sqrt{8.31 \times 10^{-11}}$ or $\sqrt{83.1 \times 10^{-12}}$ or $\sqrt{83.1 \times 10^{-6}}$.

$\log \sqrt{83.1} = \frac{1}{2}(\log 83.1) = \frac{1}{2}(1.9196) = 0.9598$

$\sqrt{83.1} = $ antilog $0.9598 = 9.116$ or 9.12

Then $\sqrt{8.31 \times 10^{-11}} = 9.12 \times 10^{-6}$

11. Find $x = 97^{1.665}$.

$\log x = 1.665(\log 97)$

$= 1.665(1.9868) = 3.309$

$x = $ antilog $3.309 = 2.04 \times 10^3$

$\log 1.665 = .2214$

$\log 1.987 = \underline{.2983}$

$\log (1.665 \times 1.987) = .5197$

$1.665 \times 1.987 = 3.309$

Exercises. Evaluate each of the following, using logarithms. Follow the rules for significant figures.

(1) 28.32×0.08254

(2) $573 \times 6.96 \times 0.00481$

(3) $\dfrac{79.28}{63.57}$

(4) $\dfrac{65.38}{225.2}$

(5) $\dfrac{1}{239}$

(6) $\dfrac{0.572 \times 31.8}{96.2}$

(7) $47.5 \times \dfrac{779}{760} \times \dfrac{273}{300}$

(8) $(8.642)^2$

(9) $(0.08642)^2$

(10) $(11.72)^3$

(11) $(0.0523)^3$

(12) $\sqrt{9463}$

(13) $\sqrt{946.3}$

(14) $\sqrt{0.00661}$

(15) $\sqrt[3]{1.79}$

(16) $\sqrt[4]{0.182}$

(17) $\sqrt{643} \times (1.91)^3$

(18) $(8.73 \times 10^{-2})(7.49 \times 10^6)$

(19) $(3.8 \times 10^{-5})^2 (1.9 \times 10^{-5})$

(20) $\dfrac{8.5 \times 10^{-45}}{1.6 \times 10^{-22}}$

(21) $\sqrt{2.54 \times 10^6}$

(22) $\sqrt{9.44 \times 10^5}$

(23) $\sqrt{7.2 \times 10^{-13}}$

(24) $\sqrt[3]{7.3 \times 10^{-14}}$

(25) $\sqrt{\dfrac{(1.1 \times 10^{-23})(6.8 \times 10^{-2})}{1.4 \times 10^{-24}}}$

(26) $2.04 \log 97.2$

(27) $37 \log 0.0298$

(28) $6.30 \log 2.95 \times 10^3$

(29) $8.09 \log 5.68 \times 10^{-16}$

(30) $(9.35)^{3.86}$

(31) $7.82 \times 0.183^{.248}$

Ans.

(1) 2.337	(9) 0.007468	(17) 177	(25) 0.73
(2) 19.18	(10) 1611	(18) 6.54×10^5	(26) 4.05
(3) 1.247	(11) 0.0001431	(19) 2.7×10^{-14}	(27) −56
(4) 0.2903	(12) 97.27	(20) 5.3×10^{-23}	(28) 21.9
(5) 0.00418	(13) 30.76	(21) 1.59×10^3	(29) −123.3
(6) 0.1890	(14) 0.0813	(22) 9.72×10^2	(30) 5.60×10^3
(7) 44.3	(15) 1.21	(23) 8.5×10^{-7}	(31) 5.1
(8) 74.68	(16) 0.653	(24) 4.2×10^{-5}	

FOUR-PLACE LOGARITHMS

N	0	1	2	3	4	5	6	7	8	9	Proportional Parts								
											1	2	3	4	5	6	7	8	9
10	0000	0043	0086	0128	0170	0212	0253	0294	0334	0374	4	8	12	17	21	25	29	33	37
11	0414	0453	0492	0531	0569	0607	0645	0682	0719	0755	4	8	11	15	19	23	26	30	34
12	0792	0828	0864	0899	0934	0969	1004	1038	1072	1106	3	7	10	14	17	21	24	28	31
13	1139	1173	1206	1239	1271	1303	1335	1367	1399	1430	3	6	10	13	16	19	23	26	29
14	1461	1492	1523	1553	1584	1614	1644	1673	1703	1732	3	6	9	12	15	18	21	24	27
15	1761	1790	1818	1847	1875	1903	1931	1959	1987	2014	3	6	8	11	14	17	20	22	25
16	2041	2068	2095	2122	2148	2175	2201	2227	2253	2279	3	5	8	11	13	16	18	21	24
17	2304	2330	2355	2380	2405	2430	2455	2480	2504	2529	2	5	7	10	12	15	17	20	22
18	2553	2577	2601	2625	2648	2672	2695	2718	2742	2765	2	5	7	9	12	14	16	19	21
19	2788	2810	2833	2856	2878	2900	2923	2945	2967	2989	2	4	7	9	11	13	16	18	20
20	3010	3032	3054	3075	3096	3118	3139	3160	3181	3201	2	4	6	8	11	13	15	17	19
21	3222	3243	3263	3284	3304	3324	3345	3365	3385	3404	2	4	6	8	10	12	14	16	18
22	3424	3444	3464	3483	3502	3522	3541	3560	3579	3598	2	4	6	8	10	12	14	15	17
23	3617	3636	3655	3674	3692	3711	3729	3747	3766	3784	2	4	6	7	9	11	13	15	17
24	3802	3820	3838	3856	3874	3892	3909	3927	3945	3962	2	4	5	7	9	11	12	14	16
25	3979	3997	4014	4031	4048	4065	4082	4099	4116	4133	2	3	5	7	9	10	12	14	15
26	4150	4166	4183	4200	4216	4232	4249	4265	4281	4298	2	3	5	7	8	10	11	13	15
27	4314	4330	4346	4362	4378	4393	4409	4425	4440	4456	2	3	5	6	8	9	11	13	14
28	4472	4487	4502	4518	4533	4548	4564	4579	4594	4609	2	3	5	6	8	9	11	12	14
29	4624	4639	4654	4669	4683	4698	4713	4728	4742	4757	1	3	4	6	7	9	10	12	13
30	4771	4786	4800	4814	4829	4843	4857	4871	4886	4900	1	3	4	6	7	9	10	11	13
31	4914	4928	4942	4955	4969	4983	4997	5011	5024	5038	1	3	4	6	7	8	10	11	12
32	5051	5065	5079	5092	5105	5119	5132	5145	5159	5172	1	3	4	5	7	8	9	11	12
33	5185	5198	5211	5224	5237	5250	5263	5276	5289	5302	1	3	4	5	6	8	9	10	12
34	5315	5328	5340	5353	5366	5378	5391	5403	5416	5428	1	3	4	5	6	8	9	10	11
35	5441	5453	5465	5478	5490	5502	5514	5527	5539	5551	1	2	4	5	6	7	9	10	11
36	5563	5575	5587	5599	5611	5623	5635	5647	5658	5670	1	2	4	5	6	7	8	10	11
37	5682	5694	5705	5717	5729	5740	5752	5763	5775	5786	1	2	3	5	6	7	8	9	10
38	5798	5809	5821	5832	5843	5855	5866	5877	5888	5899	1	2	3	5	6	7	8	9	10
39	5911	5922	5933	5944	5955	5966	5977	5988	5999	6010	1	2	3	4	5	7	8	9	10
40	6021	6031	6042	6053	6064	6075	6085	6096	6107	6117	1	2	3	4	5	6	8	9	10
41	6128	6138	6149	6160	6170	6180	6191	6201	6212	6222	1	2	3	4	5	6	7	8	9
42	6232	6243	6253	6263	6274	6284	6294	6304	6314	6325	1	2	3	4	5	6	7	8	9
43	6335	6345	6355	6365	6375	6385	6395	6405	6415	6425	1	2	3	4	5	6	7	8	9
44	6435	6444	6454	6464	6474	6484	6493	6503	6513	6522	1	2	3	4	5	6	7	8	9
45	6532	6542	6551	6561	6571	6580	6590	6599	6609	6618	1	2	3	4	5	6	7	8	9
46	6628	6637	6646	6656	6665	6675	6684	6693	6702	6712	1	2	3	4	5	6	7	7	8
47	6721	6730	6739	6749	6758	6767	6776	6785	6794	6803	1	2	3	4	5	5	6	7	8
48	6812	6821	6830	6839	6848	6857	6866	6875	6884	6893	1	2	3	4	4	5	6	7	8
49	6902	6911	6920	6928	6937	6946	6955	6964	6972	6981	1	2	3	4	4	5	6	7	8
50	6990	6998	7007	7016	7024	7033	7042	7050	7059	7067	1	2	3	3	4	5	6	7	8
51	7076	7084	7093	7101	7110	7118	7126	7135	7143	7152	1	2	3	3	4	5	6	7	8
52	7160	7168	7177	7185	7193	7202	7210	7218	7226	7235	1	2	2	4	5	6	7	7	
53	7243	7251	7259	7267	7275	7284	7292	7300	7308	7316	1	2	2	3	4	5	6	6	7
54	7324	7332	7340	7348	7356	7364	7372	7380	7388	7396	1	2	2	3	4	5	6	6	7
N	0	1	2	3	4	5	6	7	8	9	1	2	3	4	5	6	7	8	9

FOUR-PLACE LOGARITHMS

N	0	1	2	3	4	5	6	7	8	9	Proportional Parts								
											1	2	3	4	5	6	7	8	9
55	7404	7412	7419	7427	7435	7443	7451	7459	7466	7474	1	2	2	3	4	5	5	6	7
56	7482	7490	7497	7505	7513	7520	7528	7536	7543	7551	1	2	2	3	4	5	5	6	7
57	7559	7566	7574	7582	7589	7597	7604	7612	7619	7627	1	2	2	3	4	5	5	6	7
58	7634	7642	7649	7657	7664	7672	7679	7686	7694	7701	1	1	2	3	4	4	5	6	7
59	7709	7716	7723	7731	7738	7745	7752	7760	7767	7774	1	1	2	3	4	4	5	6	7
60	7782	7789	7796	7803	7810	7818	7825	7832	7839	7846	1	1	2	3	4	4	5	6	6
61	7853	7860	7868	7875	7882	7889	7896	7903	7910	7917	1	1	2	3	4	4	5	6	6
62	7924	7931	7938	7945	7952	7959	7966	7973	7980	7987	1	1	2	3	3	4	5	6	6
63	7993	8000	8007	8014	8021	8028	8035	8041	8048	8055	1	1	2	3	3	4	5	5	6
64	8062	8069	8075	8082	8089	8096	8102	8109	8116	8122	1	1	2	3	3	4	5	5	6
65	8129	8136	8142	8149	8156	8162	8169	8176	8182	8189	1	1	2	3	3	4	5	5	6
66	8195	8202	8209	8215	8222	8228	8235	8241	8248	8254	1	1	2	3	3	4	5	5	6
67	8261	8267	8274	8280	8287	8293	8299	8306	8312	8319	1	1	2	3	3	4	5	5	6
68	8325	8331	8338	8344	8351	8357	8363	8370	8376	8382	1	1	2	3	3	4	4	5	6
69	8388	8395	8401	8407	8414	8420	8426	8432	8439	8445	1	1	2	2	3	4	4	5	6
70	8451	8457	8463	8470	8476	8482	8488	8494	8500	8506	1	1	2	2	3	4	4	5	6
71	8513	8519	8525	8531	8537	8543	8549	8555	8561	8567	1	1	2	2	3	4	4	5	5
72	8573	8579	8585	8591	8597	8603	8609	8615	8621	8627	1	1	2	2	3	4	4	5	5
73	8633	8639	8645	8651	8657	8663	8669	8675	8681	8686	1	1	2	2	3	4	4	5	5
74	8692	8698	8704	8710	8716	8722	8727	8733	8739	8745	1	1	2	2	3	4	4	5	5
75	8751	8756	8762	8768	8774	8779	8785	8791	8797	8802	1	1	2	2	3	3	4	5	5
76	8808	8814	8820	8825	8831	8837	8842	8848	8854	8859	1	1	2	2	3	3	4	5	5
77	8865	8871	8876	8882	8887	8893	8899	8904	8910	8915	1	1	2	2	3	3	4	4	5
78	8921	8927	8932	8938	8943	8949	8954	8960	8965	8971	1	1	2	2	3	3	4	4	5
79	8976	8982	8987	8993	8998	9004	9009	9015	9020	9025	1	1	2	2	3	3	4	4	5
80	9031	9036	9042	9047	9053	9058	9063	9069	9074	9079	1	1	2	2	3	3	4	4	5
81	9085	9090	9096	9101	9106	9112	9117	9122	9128	9133	1	1	2	2	3	3	4	4	5
82	9138	9143	9149	9154	9159	9165	9170	9175	9180	9186	1	1	2	2	3	3	4	4	5
83	9191	9196	9201	9206	9212	9217	9222	9227	9232	9238	1	1	2	2	3	3	4	4	5
84	9243	9248	9253	9258	9263	9269	9274	9279	9284	9289	1	1	2	2	3	3	4	4	5
85	9294	9299	9304	9309	9315	9320	9325	9330	9335	9340	1	1	2	2	3	3	4	4	5
86	9345	9350	9355	9360	9365	9370	9375	9380	9385	9390	1	1	2	2	3	3	4	4	5
87	9395	9400	9405	9410	9415	9420	9425	9430	9435	9440	0	1	1	2	2	3	3	4	4
88	9445	9450	9455	9460	9465	9469	9474	9479	9484	9489	0	1	1	2	2	3	3	4	4
89	9494	9499	9504	9509	9513	9518	9523	9528	9533	9538	0	1	1	2	2	3	3	4	4
90	9542	9547	9552	9557	9562	9566	9571	9576	9581	9586	0	1	1	2	2	3	3	4	4
91	9590	9595	9600	9605	9609	9614	9619	9624	9628	9633	0	1	1	2	2	3	3	4	4
92	9638	9643	9647	9652	9657	9661	9666	9671	9675	9680	0	1	1	2	2	3	3	4	4
93	9685	9689	9694	9699	9703	9708	9713	9717	9722	9727	0	1	1	2	2	3	3	4	4
94	9731	9736	9741	9745	9750	9754	9759	9763	9768	9773	0	1	1	2	2	3	3	4	4
95	9777	9782	9786	9791	9795	9800	9805	9809	9814	9818	0	1	1	2	2	3	3	4	4
96	9823	9827	9832	9836	9841	9845	9850	9854	9859	9863	0	1	1	2	2	3	3	4	4
97	9868	9872	9877	9881	9886	9890	9894	9899	9903	9908	0	1	1	2	2	3	3	4	4
98	9912	9917	9921	9926	9930	9934	9939	9943	9948	9952	0	1	1	2	2	3	3	4	4
99	9956	9961	9965	9969	9974	9978	9983	9987	9991	9996	0	1	1	2	2	3	3	3	4
N	0	1	2	3	4	5	6	7	8	9	1	2	3	4	5	6	7	8	9

ANTILOGARITHMS

	0	1	2	3	4	5	6	7	8	9	1	2	3	4	5	6	7	8	9
.00	1000	1002	1005	1007	1009	1012	1014	1016	1019	1021	0	0	1	1	1	1	2	2	2
.01	1023	1026	1028	1030	1033	1035	1038	1040	1042	1045	0	0	1	1	1	1	2	2	2
.02	1047	1050	1052	1054	1057	1059	1062	1064	1067	1069	0	0	1	1	1	1	2	2	2
.03	1072	1074	1076	1079	1081	1084	1086	1089	1091	1094	0	0	1	1	1	1	2	2	2
.04	1096	1099	1102	1104	1107	1109	1112	1114	1117	1119	0	1	1	1	1	2	2	2	2
.05	1122	1125	1127	1130	1132	1135	1138	1140	1143	1146	0	1	1	1	1	2	2	2	2
.06	1148	1151	1153	1156	1159	1161	1164	1167	1169	1172	0	1	1	1	1	2	2	2	2
.07	1175	1178	1180	1183	1186	1189	1191	1194	1197	1199	0	1	1	1	1	2	2	2	2
.08	1202	1205	1208	1211	1213	1216	1219	1222	1225	1227	0	1	1	1	1	2	2	2	3
.09	1230	1233	1236	1239	1242	1245	1247	1250	1253	1256	0	1	1	1	1	2	2	2	3
.10	1259	1262	1265	1268	1271	1274	1276	1279	1282	1285	0	1	1	1	1	2	2	2	3
.11	1288	1291	1294	1297	1300	1303	1306	1309	1312	1315	0	1	1	1	2	2	2	2	3
.12	1318	1321	1324	1327	1330	1334	1337	1340	1343	1346	0	1	1	1	2	2	2	3	3
.13	1349	1352	1355	1358	1361	1365	1368	1371	1374	1377	0	1	1	1	2	2	2	3	3
.14	1380	1384	1387	1390	1393	1396	1400	1403	1406	1409	0	1	1	1	2	2	2	3	3
.15	1413	1416	1419	1422	1426	1429	1432	1435	1439	1442	0	1	1	1	2	2	2	3	3
.16	1445	1449	1452	1455	1459	1462	1466	1469	1472	1476	0	1	1	1	2	2	2	3	3
.17	1479	1483	1486	1489	1493	1496	1500	1503	1507	1510	0	1	1	1	2	2	2	3	3
.18	1514	1517	1521	1524	1528	1531	1535	1538	1542	1545	0	1	1	1	2	2	2	3	3
.19	1549	1552	1556	1560	1563	1567	1570	1574	1578	1581	0	1	1	1	2	2	3	3	3
.20	1585	1589	1592	1596	1600	1603	1607	1611	1614	1618	0	1	1	1	2	2	3	3	3
.21	1622	1626	1629	1633	1637	1641	1644	1648	1652	1656	0	1	1	2	2	2	3	3	3
.22	1660	1663	1667	1671	1675	1679	1683	1687	1690	1694	0	1	1	2	2	2	3	3	3
.23	1698	1702	1706	1710	1714	1718	1722	1726	1730	1734	0	1	1	2	2	2	3	3	4
.24	1738	1742	1746	1750	1754	1758	1762	1766	1770	1774	0	1	1	2	2	2	3	3	4
.25	1778	1782	1786	1791	1795	1799	1803	1807	1811	1816	0	1	1	2	2	2	3	3	4
.26	1820	1824	1828	1832	1837	1841	1845	1849	1854	1858	0	1	1	2	2	3	3	3	4
.27	1862	1866	1871	1875	1879	1884	1888	1892	1897	1901	0	1	1	2	2	3	3	3	4
.28	1905	1910	1914	1919	1923	1928	1932	1936	1941	1945	0	1	1	2	2	3	3	4	4
.29	1950	1954	1959	1963	1968	1972	1977	1982	1986	1991	0	1	1	2	2	3	3	4	4
.30	1995	2000	2004	2009	2014	2018	2023	2028	2032	2037	0	1	1	2	2	3	3	4	4
.31	2042	2046	2051	2056	2061	2065	2070	2075	2080	2084	0	1	1	2	2	3	3	4	4
.32	2089	2094	2099	2104	2109	2113	2118	2123	2128	2133	0	1	1	2	2	3	3	4	4
.33	2138	2143	2148	2153	2158	2163	2168	2173	2178	2183	0	1	1	2	2	3	3	4	4
.34	2188	2193	2198	2203	2208	2213	2218	2223	2228	2234	1	1	2	2	3	3	4	4	5
.35	2239	2244	2249	2254	2259	2265	2270	2275	2280	2286	1	1	2	2	3	3	4	4	5
.36	2291	2296	2301	2307	2312	2317	2323	2328	2333	2339	1	1	2	2	3	3	4	4	5
.37	2344	2350	2355	2360	2366	2371	2377	2382	2388	2393	1	1	2	2	3	3	4	4	5
.38	2399	2404	2410	2415	2421	2427	2432	2438	2443	2449	1	1	2	2	3	3	4	4	5
.39	2455	2460	2466	2472	2477	2483	2489	2495	2500	2506	1	1	2	2	3	3	4	5	5
.40	2512	2518	2523	2529	2535	2541	2547	2553	2559	2564	1	1	2	2	3	4	4	5	5
.41	2570	2576	2582	2588	2594	2600	2606	2612	2618	2624	1	1	2	2	3	4	4	5	5
.42	2630	2636	2642	2649	2655	2661	2667	2673	2679	2685	1	1	2	2	3	4	4	5	6
.43	2692	2698	2704	2710	2716	2723	2729	2735	2742	2748	1	1	2	3	3	4	4	5	6
.44	2754	2761	2767	2773	2780	2786	2793	2799	2805	2812	1	1	2	3	3	4	4	5	6
.45	2818	2825	2831	2838	2844	2851	2858	2864	2871	2877	1	1	2	3	3	4	5	5	6
.46	2884	2891	2897	2904	2911	2917	2924	2931	2938	2944	1	1	2	3	3	4	5	5	6
.47	2951	2958	2965	2972	2979	2985	2992	2999	3006	3013	1	1	2	3	3	4	5	5	6
.48	3020	3027	3034	3041	3048	3055	3062	3069	3076	3083	1	1	2	3	4	4	5	6	6
.49	3090	3097	3105	3112	3119	3126	3133	3141	3148	3155	1	1	2	3	4	4	5	6	6
	0	**1**	**2**	**3**	**4**	**5**	**6**	**7**	**8**	**9**	1	2	3	4	5	6	7	8	9

ANTILOGARITHMS

	0	1	2	3	4	5	6	7	8	9	Proportional Parts								
											1	2	3	4	5	6	7	8	9
.50	3162	3170	3177	3184	3192	3199	3206	3214	3221	3228	1	1	2	3	4	4	5	6	7
.51	3236	3243	3251	3258	3266	3273	3281	3289	3296	3304	1	2	2	3	4	5	5	6	7
.52	3311	3319	3327	3334	3342	3350	3357	3365	3373	3381	1	2	2	3	4	5	5	6	7
.53	3388	3396	3404	3412	3420	3428	3436	3443	3451	3459	1	2	2	3	4	5	6	6	7
.54	3467	3475	3483	3491	3499	3508	3516	3524	3532	3540	1	2	2	3	4	5	6	6	7
.55	3548	3556	3565	3573	3581	3589	3597	3606	3614	3622	1	2	2	3	4	5	6	7	7
.56	3631	3639	3648	3656	3664	3673	3681	3690	3698	3707	1	2	3	3	4	5	6	7	8
.57	3715	3724	3733	3741	3750	3758	3767	3776	3784	3793	1	2	3	3	4	5	6	7	8
.58	3802	3811	3819	3828	3837	3846	3855	3864	3873	3882	1	2	3	4	4	5	6	7	8
.59	3890	3899	3908	3917	3926	3936	3945	3954	3963	3972	1	2	3	4	5	5	6	7	8
.60	3981	3990	3999	4009	4018	4027	4036	4046	4055	4064	1	2	3	4	5	6	6	7	8
.61	4074	4083	4093	4102	4111	4121	4130	4140	4150	4159	1	2	3	4	5	6	7	8	9
.62	4169	4178	4188	4198	4207	4217	4227	4236	4246	4256	1	2	3	4	5	6	7	8	9
.63	4266	4276	4285	4295	4305	4315	4325	4335	4345	4355	1	2	3	4	5	6	7	8	9
.64	4365	4375	4385	4395	4406	4416	4426	4436	4446	4457	1	2	3	4	5	6	7	8	9
.65	4467	4477	4487	4498	4508	4519	4529	4539	4550	4560	1	2	3	4	5	6	7	8	9
.66	4571	4581	4592	4603	4613	4624	4634	4645	4656	4667	1	2	3	4	5	6	7	9	10
.67	4677	4688	4699	4710	4721	4732	4742	4753	4764	4775	1	2	3	4	5	7	8	9	10
.68	4786	4797	4808	4819	4831	4842	4853	4864	4875	4887	1	2	3	4	6	7	8	9	10
.69	4898	4909	4920	4932	4943	4955	4966	4977	4989	5000	1	2	3	5	6	7	8	9	10
.70	5012	5023	5035	5047	5058	5070	5082	5093	5105	5117	1	2	4	5	6	7	8	9	11
.71	5129	5140	5152	5164	5176	5188	5200	5212	5224	5236	1	2	4	5	6	7	8	10	11
.72	5248	5260	5272	5284	5297	5309	5321	5333	5346	5358	1	2	4	5	6	7	9	10	11
.73	5370	5383	5395	5408	5420	5433	5445	5458	5470	5483	1	3	4	5	6	8	9	10	11
.74	5495	5508	5521	5534	5546	5559	5572	5585	5598	5610	1	3	4	5	6	8	9	10	12
.75	5623	5636	5649	5662	5675	5689	5702	5715	5728	5741	1	3	4	5	7	8	9	10	12
.76	5754	5768	5781	5794	5808	5821	5834	5848	5861	5875	1	3	4	5	7	8	9	11	12
.77	5888	5902	5916	5929	5943	5957	5970	5984	5998	6012	1	3	4	5	7	8	10	11	12
.78	6026	6039	6053	6067	6081	6095	6109	6124	6138	6152	1	3	4	6	7	8	10	11	13
.79	6166	6180	6194	6209	6223	6237	6252	6266	6281	6295	1	3	4	6	7	9	10	11	13
.80	6310	6324	6339	6353	6368	6383	6397	6412	6427	6442	1	3	4	6	7	9	10	12	13
.81	6457	6471	6486	6501	6516	6531	6546	6561	6577	6592	2	3	5	6	8	9	11	12	14
.82	6607	6622	6637	6653	6668	6683	6699	6714	6730	6745	2	3	5	6	8	9	11	12	14
.83	6761	6776	6792	6808	6823	6839	6855	6871	6887	6902	2	3	5	6	8	9	11	13	14
.84	6918	6934	6950	6966	6982	6998	7015	7031	7047	7063	2	3	5	6	8	10	11	13	15
.85	7079	7096	7112	7129	7145	7161	7178	7194	7211	7228	2	3	5	7	8	10	12	13	15
.86	7244	7261	7278	7295	7311	7328	7345	7362	7379	7396	2	3	5	7	8	10	12	13	15
.87	7413	7430	7447	7464	7482	7499	7516	7534	7551	7568	2	3	5	7	9	10	12	14	16
.88	7586	7603	7621	7638	7656	7674	7691	7709	7727	7745	2	4	5	7	9	11	12	14	16
.89	7762	7780	7798	7816	7834	7852	7870	7889	7907	7925	2	4	5	7	9	11	13	14	16
.90	7943	7962	7980	7998	8017	8035	8054	8072	8091	8110	2	4	6	7	9	11	13	15	17
.91	8128	8147	8166	8185	8204	8222	8241	8260	8279	8299	2	4	6	8	9	11	13	15	17
.92	8318	8337	8356	8375	8395	8414	8433	8453	8472	8492	2	4	6	8	10	12	14	15	17
.93	8511	8531	8551	8570	8590	8610	8630	8650	8670	8690	2	4	6	8	10	12	14	16	18
.94	8710	8730	8750	8770	8790	8810	8831	8851	8872	8892	2	4	6	8	10	12	14	16	18
.95	8913	8933	8954	8974	8995	9016	9036	9057	9078	9099	2	4	6	8	10	12	15	17	19
.96	9120	9141	9162	9183	9204	9226	9247	9268	9290	9311	2	4	6	8	11	13	15	17	19
.97	9333	9354	9376	9397	9419	9441	9462	9484	9506	9528	2	4	7	9	11	13	15	17	20
.98	9550	9572	9594	9616	9638	9661	9683	9705	9727	9750	2	4	7	9	11	13	16	18	20
.99	9772	9795	9817	9840	9863	9886	9908	9931	9954	9977	2	5	7	9	11	14	16	18	20
	0	1	2	3	4	5	6	7	8	9	1	2	3	4	5	6	7	8	9

INDEX

International Atomic Weights

(Based on Carbon-12)

	Symbol	Atomic Number	Atomic Weight		Symbol	Atomic Number	Atomic Weight
Actinium	Ac	89	...	Mercury	Hg	80	200.59
Aluminum	Al	13	26.9815	Molybdenum	Mo	42	95.94
Americium	Am	95	...	Neodymium	Nd	60	144.24
Antimony	Sb	51	121.75	Neon	Ne	10	20.183
Argon	Ar	18	39.948	Neptunium	Np	93	...
Arsenic	As	33	74.9216	Nickel	Ni	28	58.71
Astatine	At	85	...	Niobium	Nb	41	92.906
Barium	Ba	56	137.34	Nitrogen	N	7	14.0067
Berkelium	Bk	97	...	Nobelium	No	102	...
Beryllium	Be	4	9.0122	Osmium	Os	76	190.2
Bismuth	Bi	83	208.980	Oxygen	O	8	15.9994
Boron	B	5	10.811	Palladium	Pd	46	106.4
Bromine	Br	35	79.909	Phosphorus	P	15	30.9738
Cadmium	Cd	48	112.40	Platinum	Pt	78	195.09
Calcium	Ca	20	40.08	Plutonium	Pu	94	...
Californium	Cf	98	...	Polonium	Po	84	...
Carbon	C	6	12.01115	Potassium	K	19	39.102
Cerium	Ce	58	140.12	Praseodymium	Pr	59	140.907
Cesium	Cs	55	132.905	Promethium	Pm	61	...
Chlorine	Cl	17	35.453	Protactinium	Pa	91	...
Chromium	Cr	24	51.996	Radium	Ra	88	...
Cobalt	Co	27	58.9332	Radon	Rn	86	...
Copper	Cu	29	63.54	Rhenium	Re	75	186.2
Curium	Cm	96	...	Rhodium	Rh	45	102.905
Dysprosium	Dy	66	162.50	Rubidium	Rb	37	85.47
Einsteinium	Es	99	...	Ruthenium	Ru	44	101.07
Erbium	Er	68	167.26	Samarium	Sm	62	150.35
Europium	Eu	63	151.96	Scandium	Sc	21	44.956
Fermium	Fm	100	...	Selenium	Se	34	78.96
Fluorine	F	9	18.9984	Silicon	Si	14	28.086
Francium	Fr	87	...	Silver	Ag	47	107.870
Gadolinium	Gd	64	157.25	Sodium	Na	11	22.9898
Gallium	Ga	31	69.72	Strontium	Sr	38	87.62
Germanium	Ge	32	72.59	Sulfur	S	16	32.064
Gold	Au	79	196.967	Tantalum	Ta	73	180.948
Hafnium	Hf	72	178.49	Technetium	Tc	43	...
Helium	He	2	4.0026	Tellurium	Te	52	127.60
Holmium	Ho	67	164.930	Terbium	Tb	65	158.924
Hydrogen	H	1	1.00797	Thallium	Tl	81	204.37
Indium	In	49	114.82	Thorium	Th	90	232.038
Iodine	I	53	126.9044	Thulium	Tm	69	168.934
Iridium	Ir	77	192.2	Tin	Sn	50	118.69
Iron	Fe	26	55.847	Titanium	Ti	22	47.90
Krypton	Kr	36	83.80	Tungsten	W	74	183.85
Lanthanum	La	57	138.91	Uranium	U	92	238.03
Lead	Pb	82	207.19	Vanadium	V	23	50.942
Lithium	Li	3	6.939	Xenon	Xe	54	131.30
Lutetium	Lu	71	174.97	Ytterbium	Yb	70	173.04
Magnesium	Mg	12	24.312	Yttrium	Y	39	88.905
Manganese	Mn	25	54.9380	Zinc	Zn	30	65.37
Mendelevium	Md	101	...	Zirconium	Zr	40	91.22

Catalog

If you are interested in a list of SCHAUM'S
OUTLINE SERIES in Science, Mathematics,
Engineering and other subjects, send your name
and address, requesting your free catalog, to:

SCHAUM'S OUTLINE SERIES, Dept. C
McGRAW-HILL BOOK COMPANY
1221 Avenue of Americas
New York, N.Y. 10020